Russia in Space
The failed frontier?

Springer
London
Berlin
Heidelberg
New York
Barcelona
Hong Kong
Milan
Paris
Santa Clara
Singapore
Tokyo

Brian Harvey

Russia in Space

The failed frontier?

Springer

Published in association with
Praxis Publishing
Chichester, UK

Brian Harvey, M.A., H.D.E., F.B.I.S.
2 Rathdown Crescent
Terenure
Dublin 6W
Ireland

SPRINGER-PRAXIS BOOKS IN ASTRONOMY AND SPACE SCIENCES
SUBJECT *ADVISORY EDITOR*: John Mason B.Sc., Ph.D.

ISBN 1-85233-203-4 Springer-Verlag Berlin Heidelberg New York

British Library Cataloguing in Publication Data
   Harvey, Brian, 1953–
     Russia in space: the failed frontier? – (Springer-Praxis
     books in astronomy and space sciences)
     1. Astronautics – Russia (Federation) – History
     2. Astronautics – Soviet Union – History
     I. Title
     629.4'0947
     ISBN 1852332034

Library of Congress Cataloging-in-Publication Data
   Harvey, Brian, 1953–
     Russia in space: the failed frontier?/Brian Harvey.
       p.cm. – (Springer-Praxis books in astronomy and space sciences)
     Includes bibliographical references and index.
     ISBN 1-85233-203-4 (alk. paper)
     1. Astronautics – Russia (Federation) – History.
     2. Astronautics – Soviet Union – History.
     I. Title. II. Series.
     TL789.8.R8 H37 2001
     629.4'0947–dc21   00-049273

Cover design: Jim Wilkie
Typesetting: Originator publishing services, Gt. Yarmouth, Norfolk, UK

Printed on acid-free paper supplied by Precision Publishing Papers Ltd, UK

Contents

List of illustrations

List of tables

Introduction

1986–7 was the heyday of the Soviet space programme. American's astronauts were grounded, following the tragic accident which took the lives of the seven Challenger astronauts in January 1986. Less than a month after the space shuttle exploded in the skies over Cape Canaveral, the Soviet Union launched the Mir space station, the first space station with six docking ports. It was the era of *glasnost* and so confident were the Russians in their rockets that they showed live on television the launch of their first crew of the Mir space station, Soyuz T-15, comprising Leonid Kizim and Vladimir Solovyov. The two men chased Mir in orbit, docked and moved into the new orbital station. They spent two months testing its systems, debugging minor problems and watched the Earth pass below them from their personalized cabins.

They left Mir on 6 May and boarded their Soyuz T-15 transport craft. However, instead of returning to the Earth, they crossed thousands of kilometres of empty space and began pursuit of the ageing Soviet space station, Salyut 7, arriving the following day. Here they carried out a spacewalk and collected old scientific equipment for transfer to Mir. After four weeks, they undocked, crossed space again and rejoined the new Mir complex. After several more weeks there, they returned to Earth in July 1986 after 125 days orbiting the Earth in two different orbital space stations.

This virtuoso performance was only one of several Soviet space achievements that year. Across the solar system, two 5-tonne Vega probes had reached their target. Launched in 1984, the Vegas had headed for Venus. Here they had dropped the first balloons into the atmosphere of Venus and landed probes on the surface. Altering course, the mother crafts sped away from the hothouse of Venus and set an interception course for Comet Halley, which was visiting the solar system for the first time in 76 years. Vega 1 and Vega 2 sped into the tail of the comet, taking red and orange pictures of the coloured hues of the snowball-shaped comet's core. Even as they did so, engineers on the ground in Moscow were preparing new probes to land on Mars's moon Phobos.

Back on Earth, half a million people were employed in the Soviet space industry. They lived and worked in the design and manufacturing plants in Moscow and

Leningrad, in the two sprawling launch centres of Plesetsk and Baikonour, at the tracking stations scattered across the Soviet Union and some lived on board ship on the communications ships (comships) that sailed around the world and, whatever the weather, tracked spaceships speeding to their cosmic destinations.

During 1986–7, construction engineers toiled over two giant new complexes at Baikonour cosmodrome. These were to be launching points for the Soviet Union's latest rocket, the giant, hydrogen-fuelled, Energiya. It was not called 'Energiya', or 'energy' for nothing, for Energiya would put into orbit the Soviet space shuttle Buran or unmanned cargoes or space stations weighing up to 140 tonnes. Energiya would give the Soviet Union undisputed space supremacy for 50 years and send the first cosmonauts to Mars. A new space complex, Mir 2, was already on the drawing board, the prelude to the first space city, Kosmograd.

Not that the Soviet space programme was just about spectacular feats of space navigation like Vega or crossing from one space station to another, like Kizim and Solovyov, or about powerful prestigious launchers. The unmanned space programme saw, on average, a launching every week from Baikonour and Plesetsk. One year, the Soviet Union launched 102 spacecraft, more than the rest of the world put together. These ranged from small scientific spacecraft to navigation satellites, from photoreconnaissance to ocean surveillance platforms, from weather satellites to instrumented probes to study the stars.

Fifteen years later, however, all was so different. The Soviet space programme in its glory days was a faded memory. The programme had already begun to contract in the final years of the Soviet Union, from 1990 to 1991. Money, resources and funding had become tight and programmes had been delayed. Although the stunning Energiya performed flawlessly twice and impressed the world, it found itself without customers. The Buran space shuttle made a perfect first flight, was suspended and the gliding space plane never flew again. Mir, designed to fly for five years, was stretched to keep going for 14. Mir 2 was scrapped, the concept improbably merged with the programme of the old rival, the United States. Following the collapse of the Soviet Union in 1991, money almost dried up. Budgets were cut and cut, halved and halved again and promised funds did not arrive as often as they did. Soviet space memorabilia from the early days of triumph were sold off at auctions in a desperate attempt to raise money to keep the programme afloat. The scientific programme almost disappeared. The military suffered as much as the civilian space programme and by the late 1990s there were periods of months when the Russian Federation had no operational spy satellites in orbit. Communications satellites exceeded twice their normal lifetime and when they broke down they were simply not replaced. Russian design bureaux had to transform themselves into commercial companies, touting business from their former competitors in Europe and the United States. By the turn of the century, Russia was launching less than 30 satellites a year. Within the country, spaceflight now longer excited public attention. For people trying to survive the harsh realities of post-communist economics, space travel was a distraction at best and to others an absurdity. Cosmonauts could at least do their shopping without being recognized, because so few people cared who they were. All this was a far cry from the mid-1980s when the USSR was supreme in space exploration. The dreams of Tsiolkovsky were now faded indeed. Was the collapse of the Soviet space programme the final, delayed footnote for the fall of the Soviet Union?

President John F. Kennedy had proclaimed space travel as the ultimate frontier. His country had conquered the first landmark on that frontier when, less than 10 years later, American astronauts had planted the flag Old Glory in the dusty surface of the Sea of Tranquillity. By the new century, the United States was supreme in space again. The loss of the Challenger had proved to be a tragic event, but only a temporary setback. In contrast to its Russian rival, permanently consigned to the hangar, the American space shuttle was flying several times a year. American unmanned probes had flown to the furthest ends of the solar system and had even landed a small pathfinding rover on the red surface of the planet Mars. Its unmanned rockets were lofting commercial and military payloads every other week. The Hubble space telescope had taken images from the furthest end of the time–space continuum.

For the Soviet Union, space had proved to be the final, failed frontier—a delusion as great as that of the communist ideal on Earth.

Or had it?

Author's acknowledgements

I wish to thank all those who contributed to this book by the provision of advice, information and other assistance. I would particularly like to thank:

Clive Horwood, Praxis Publishing, Chichester; Rex Hall, London; and Phil Clark, Molniya Space Consultancy, Hastings.

I also wish to thank those who provided photographs for this and the previous editions, in particular:

Novosti Press Agency
NPO Molniya, Moscow
Roman Turkenich, NPO-PM, Krasnoyarsk
Nigel Evans
John Mason
Boris Poletaev, Designer General, KB Arsenal, St Petersburg
SS Nazarko, OKB Polyot, Omsk
Centre Nationale d'Études Spatiales, France
Galina Zatcossova, Design Bureau of Transport Machinery (KBTM), Moscow
Deutsches Zentrum für Luft und Raumfahrt, Cologne, Germany
European Space Agency (Cluster)
Deborah Rickman, Kistler Aerojet
Orly König Lopez, International Launch Services
Paula Korn, Sea Launch

This book should be read in conjunction with other books in the Springer/Praxis series in astronomy and space sciences. Recommended are David M. Harland: *The Mir space station—a precursor to space colonization* (Wiley/Praxis, 1997); David J. Shayler: *Disasters and accidents in manned spaceflight* (Springer/Praxis, 2000); and (forthcoming) Rex Hall and David M. Shayler: *Off the planet—the flights of the first cosmonauts* and *Soyuz—a universal spacecraft.*

1

The Soviet space programme 1921–1991

The Soviet space programme began in 1921 when a scientist quite unknown to the outside world, Nikolai Tikhomirov, set up what was called the Gas Dynamics Laboratory (GDL). The title 'gas dynamics' was rather a misnomer, as Tikhomirov was really interested in discovering if, by mixing fuels under pressure, a rocket engine could be built to fly as successfully in the vacuum of space as on our own planet.

Tikhomirov persuaded the new Soviet government to let him undertake some experiments in the old, now disused, Peter & Paul fortress in Leningrad. Here, he could carry out rocket experiments in private in a disused building without risk to the citizenry at large. Not long after he arrived there, he was presented with the graduation paper of a young engineering student in Leningrad University, Valentin Petrovich Glushko. This paper outlined how electrically propelled engines could

Valentin Glushko

project rockets across the vastness of space. This so impressed Tikhomirov that he took on the young Glushko.

Here, in the Petropavlovskaya fortress, they built the first liquid fuel engine in the Soviet Union. This was the ORM-1, one of the earliest liquid-propellant rocket engines. The first version had a thrust of only 20 kg, but later ones were more powerful, at 300 kg. By 1933 Glushko had developed a technique of regenerative cooling which ensured that liquid fuel engines could fire for a long duration without overheating. In 1936 he developed the first throttle-type liquid fuel engine. Glushko developed new turbine pumps and engines which gimballed. They experimented with toxic, storable fuels, such as nitric acid, and the following year with fuels which ignited spontaneously on contact with one another. Thousands of static tests were made: the engines had variable thrusts and achieved rates of efficiency that the west did not match until the 1950s. The burn marks from these early tests are still visible. The roar from the fortress as the static tests went on startled passing citizens and also probably mystified them. GDL itself expanded from its initial staff of ten in 1928 to 200 people by 1933.

This was the beginning of a space programme that was to stun the world. The Soviet Union in the 1920s was alive with young scientists and experimental groups. Amateur societies were founded dedicated to the conquest of space. An exhibition was held in Moscow in 1927 with designs, models, drawings and even a spacesuit on display. A Latvian engineer, Friedrich Tsander, designed a space shuttle, put forward the idea of hydrogen-fuelled rockets and drew up a schedule for a manned flight to Mars. A former White army officer, Alexander Shargei, who took on the alias of Yuri Kondratyuk to avoid the curiosity of the Bolsheviks, described multistaged rockets, orbiting space stations, the best way to land on the Moon, thermal protection systems and using the gravitational force of one planet to fly on to the next. An encyclopaedia of space travel, in nine volumes, was published in Moscow in 1929.

These young scientists based their work on theories dating back to the Tsar's Russia. In 1883, Konstantin Tsiolokovsky, a shy, deaf teacher in the town of Kaluga, had begun to write, in the course of the school summer holidays, a series of books and papers outlining how staged rockets could be built, how small satellites could circle the Earth, how space stations could one day be built, and the best ways to cope with the problems of weightlessness. He even considered how atomic power would be necessary for longer space journeys.

FIRST ROCKETS

This intellectual ferment was matched by practical action. Sergei Korolev, a young technical school graduate from Ukraine, gathered together a like-minded group of colleagues called the GIRD, or the Group for the Study of Reactive Propulsion, and in 1933 they built a 2-metre-long rocket which they fired just above the top of the trees of an adjoining forest. The government took an interest and asked them to design and build rocket-powered gliders for the air force. Three Red Army fliers took

Bust of Sergei Korolev, the chief designer

a balloon, called the USSR 1, up to what was then a world record of 19 km to where they could see the sky turn deep violet.

The Soviet Union's rocketeers suffered badly when Stalin's purges broke out in 1938. Because many were now working for army generals who fell under suspicion, several were shot; many (e.g. Korolev) were sent to the camps and others (e.g. Glushko) were put under house arrest. When war broke out in 1941, they were brought back to work in aircraft design offices. In 1944, as Red Army troops moved westward, they discovered that the Germans had managed, in the town of Peenemünde, to build and fly what was a real, modern, powerful rocket. Glushko and Korolev spent 1946 in Germany picking over the remains of the A-4 rocket, known and feared as the V-2 that had devastated London. Two hundred German scientists were shipped east to see just how much they knew.

In April 1946 Stalin ordered the beginning of the Soviet post-war rocket programme and Sergei Korolev was appointed chief designer, with responsibility for the first rocket design bureau. German A-4s were brought back to the Soviet Union and launched from the first cosmodrome, Kapustin Yar, near Volgograd. The first German rockets were fired from the Volgograd station in autumn 1947. The Russians built their own version of the A-4, called the R-1 (R for rocket) and then a more powerful R-2. The first rocket of wholly Soviet design was built in 1953, the R-5: it was used to fire animals on up-and-down trajectories.

In the early 1950s discussion began within the Academy of Sciences of the idea of an Earth satellite. Korolev argued the case and, with Glushko, won over the academicians in 1955. Korolev began building a much more powerful launcher, the R-7, based on clustering together five rockets with 20 of Glushko's engines at take-off. To launch the satellite, a new site was selected in the Kazakhstan desert and construction of a new cosmodrome began there in 1955 at Tyuratam, though it was called Baikonour.

The building of a Soviet Earth satellite owed much to a unique alliance of two men—new Soviet leader Nikita Khrushchev and chief designer Sergei Korolev. Khrushchev wanted rockets in order to build a rocket strike force able to hit the United States with nuclear bombs. Korolev wanted permission to build rockets, albeit for another purpose—the conquest of space. Khrushchev brought together four of Korolev's R-7 rockets and installed them at a new base in Plesetsk, in northern Russia near the arctic circle, in 1959.

By October 1957, Korolev was ready to launch his satellite. His R-7 lofted what he called the Preliminary Satellite into orbit on 4 October 1957. When it came over the cosmodrome an orbit later, the scientists gathered excitedly to listen for the famous 'beep! beep! beep!' of its transmitter. They were not disappointed. Korolev assembled them in the hanger and told them that the conquest of space had begun.

Initially, the new satellite, now called Sputnik, aroused little comment, but as Korolev's engineers took the train to Moscow they were mobbed by excited crowds who could talk about nothing but the satellite. The United States was utterly shocked by the Soviet Union launching the first Earth satellite, although they should not have been, because it had been announced many times in advance in the Soviet press. Their politicians fumed that their adversaries were now ahead of them. A delighted Nikita Khrushchev recalled Korolev from holiday and ordered him to launch another Sputnik straight away, with an animal on board. Korolev built another Sputnik, attached a dog cabin from one of the early atmospheric flights, and put the dog Laika into orbit only three weeks later.

From that point, the race was on, but the United States did not get its Explorer satellite into orbit until the following year. The R-7 put Sputnik 3 in orbit in May 1958, weighing 1.5 tonnes, at a time when the United States had just launched a satellite, Vanguard, weighing not much more than 1.5 kg. Khrushchev, overjoyed at the impact of the Soviet Union's space progress on public opinion at home and world opinion abroad, gave Korolev a blank cheque. In 1958, Korolev devised plans for sending a small spacecraft to the Moon, and plans were also drawn up for spacecraft to fly to Mars and Venus. The construction of a spycraft, the Zenit, which was able to take pictures of American military installations, had already begun.

The first Moonshot took place in January 1959, but missed. At the next attempt, in September 1959, the USSR hit the middle to the Moon with Luna 2. Three weeks later, Luna 3 circled around the far side of the Moon and transmitted pictures of its hidden face. The largest crater was named Tsiolkovsky after the man who had inspired them all.

'I saw, with my own eyes, the spherical shape of the Earth'

By this time, Korolev's thoughts had turned to putting the first man into space. He adapted the design of the Zenit into a 4-tonne spacecraft, called Vostok, which was

Yuri Gagarin's Vostok 1

capable of carrying a cosmonaut into orbit. The new spaceship was a ball-shaped object with a service module, able to orbit the Earth for a week. The first Vostoks were launched in 1960, one carrying dogs and other animals into orbit for a day. The first group of cosmonauts was recruited—20 young jet fighter pilots.

By 12 April 1961, after five unmanned Vostok missions, all was ready for the first flight of a man into space. The honour fell to a 27-year-old Air Force major, Yuri Gagarin. On that day, he made the first ever orbit of the Earth, parachuting down near Saratov just 108 minutes after he went up. In orbit, he relayed stunning impressions of the blues of our planet, sunrise and sunset, and how he could make out the spherical shape of the Earth. Gagarin flew to Moscow two days later, to the greatest celebration in the capital since the end of the Second World War. Millions crowded the streets, Khrushchev challenged the capitalist nations to catch up—if they could—and Gagarin went on to make a world tour. It was the high summer of communist achievement.

Andrian Nikolayev (left) with Yuri Gagarin

Gagarin's flight led to an intensification of the space race. The newly elected American president, Kennedy, made no effort to downplay the Soviet achievement and asked for advice as to how best to beat the USSR for world leadership. Ask the janitor if he knows the answer, he once exclaimed in desperation. His vice-president, Lyndon Johnson, one of the sharpest critics of America's inadequate response to Sputnik, proposed that the United States should send a man to the Moon. A month later, Kennedy duly announced the project to go to the Moon but, although approved, his proposal received a muted response.

The Russians took no notice of the Kennedy speech. They probably felt that they had no need to. In 1961, Korolev had dispatched his first probe to Venus, equipped to float on its oceans. Later that year, his second cosmonaut, Gherman Titov, spent a full day in space, his spaceship being seen over Europe and the United States as a bright star crossing silently through the August night sky.

By 1962, Korolev had put together plans for a new spaceship, the Soyuz complex, whereby his spaceships would join together in Earth orbit and then fly around the Moon. In August that year, the USSR launched its third spaceman, Andrian Nikolayev, and only a day later, Vostok 4 with Pavel Popovich. They circled the Earth together for three days, initially only a few kilometres apart. Meantime, the central committee, possibly catching wind of the efforts of American women to join NASA, decided that if America put the first woman in space, that would be an insult to Soviet womanhood. In June 1963, two days after the launch of Vostok 5, the Soviet Union launched the first spacewoman, 26-year-old loom operator and amateur parachutist Valentina Terreskhova. Her three days in space were more than what America's six astronauts had achieved together by this time.

The early 1960s were a period of flat-out effort by the Soviet space programme. In 1962, the first spycraft, Zenit, was launched, to snap pictures of the United States, following which they were recovered in the main landing zone. These missions grew ever more frequent. The first spaceship to the red planet, Mars 1, was sent long before the Americans were able to undertake a comparable mission. In 1963, Korolev built the first probes to soft land on the Moon, although the first missions were far from successful. Whenever opportunities came to send spacecraft to Mars and Venus, Korolev routinely organized two, three or even four launches. At a time when the Americans were thinking of sending probes to fly past Mars and Venus and take pictures, several of Korolev's probes were ambitious missions, designed to parachute down and search for signs of life. His efforts were frustrated by repeated failures of the upper stage of his Molniya launcher, and for every successful mission there were normally two or even three that failed but were not announced.

SOVIET SUPREMACY COMPLETE

There was little sign of an effective American challenge to what became utter Soviet supremacy in space. On virtually every single space objective in the early 1960s, the Russians beat the Americans—to the Moon, Mars, Venus, the first man in space, the first woman, and the first rendezvous. Later, on reflection, these became known as 'the golden years of Soviet rocketry', and it's not difficult to see why.

In autumn 1962, President Kennedy toured American space facilities. Already, the infrastructure necessary to support the Man-on-the-Moon project was beginning to take shape, like the huge new facilities at Cape Canaveral for the Saturn V Moon rocket. At last, the Moon effort began to get a public response. President Kennedy made a habit of visiting student campuses, talking about noble causes such as the peace corps, to which they listened politely; but whenever he began to speak about going to the Moon, they got to their feet and he was repeatedly interrupted by applause.

What neither he nor the students knew was that the Russians still did not have a plan for going to the Moon. When Nikita Khrushchev told journalists in 1963 that the Soviet Union was not racing to the Moon, no one believed him and thought it was a trick. He was being truthful. It was not until August 1964 that a resolution of the government and the central committee committed the Soviet Union to sending a man to the Moon. Korolev was made responsible for the construction of a giant Moon rocket, the N-1, which would fire no less than 30 engines at take-off and put a single cosmonaut on the Moon. A design bureau led by Vladimir Chelomei would develop another powerful rocket, the Proton, to send a man looping around the Moon and return him to Earth six days later.

Soviet space supremacy was extended over the next three years. In autumn 1964, the United States was preparing its first two-man spacecraft, Gemini. Korolev converted the Vostok into a three-man spaceship called Voskhod and, in October 1964, launched a pilot, engineer and doctor into orbit in a shirtsleeve environment. The next spring, just five days before Gemini was actually launched, Korolev's team launched Voskhod 2. On its second orbit, Alexei Leonov left the safety of his cabin and made the first walk in space for 10 minutes, televised across the entire world. In

January 1966, the first soft-landing on the Moon was achieved: little Luna 9 touched down in the Ocean of Storms, relaying back over the next four days pictures of its rocky and cratered surface. March 1966 saw a double triumph. On 1 March, Venera 3 hit Venus and became the first spaceship to reach the surface of another planet. On 31 March, Luna 10 reached the Moon, and as it swung into lunar orbit it relayed the bars of the *Internationale* to the assembled delegates of the communist party congress then gathered in Moscow. This was greeted by thunderous applause.

DEATH OF THE CHIEF DESIGNER

By this time, the genius behind these stupendous achievements was dead. Sergei Korolev entered hospital for routine surgery in January 1966. It was an operation for the removal of colon tumours, and no less a person than the Minister for Health, Dr Boris Petrovsky, carried out the operation. Mid-way through he discovered a more serious tumour, but continued the operation. A large blood vessel burst; haemorrhaging began; and Sergei Korolev's heart—weakened as it had been from the toil of the labour camps—collapsed. Attempts to ventilate him were made more difficult by his jaw having been broken during the gulag years. Frantic efforts were made to revive him but on 14 January he was pronounced dead.

He was buried in the wall of the Kremlin on 16 January 1966. In the eulogies which followed, no efforts were spared to tell of his boundless energy, iron will, limitless imagination and engineering genius. This could have been mistaken for nostalgia but it was not. With Korolev's death, the Soviet space programme was never the same again. The driving force went out of it and with it Korolev's unique ability to command, inspire, bargain, lead, design and attend to detail. After 1966 the programme had many excellent designers, planners, politicians, administrators and prophets, but never again in just one person. Korolev was succeeded by his deputy, Vasili Mishin, who had worked with him since 1945.

During 1966, the American investment in the Moon race in the early 1960s at last began to yield concrete results. The Americans concluded their Gemini series of spacecraft. During 1965–6, Gemini demonstrated space rendezvous, docking, space-walking and long-duration missions of up to two weeks. During ten missions, in the course of which not a single cosmonaut left the ground, the techniques essential for a manned landing on the Moon were demonstrated time and time again. In 1966, the Americans made five Surveyor soft-landings on the Moon and put five orbiters into lunar orbit.

Over 1967–9, the Moon race entered its final stages. In early 1967, both pro-grammes of the space superpowers were set back by terrible tragedies. In January, the United States lost Gus Grissom, Ed White and Roger Chaffee in a flash fire in a ground test in their Apollo 1 on the launch pad during a rehearsal. In April, the Soviet Union tried to recover the lost ground which the Americans had won back by Gemini. On 23 April 1967, the USSR launched Soyuz 1 on what was to be a 72-hour spectacular in the course of which two Soyuz would link together, spacewalk, exchange crews and do all the Russian manoeuvres necessary for a Moon flight.

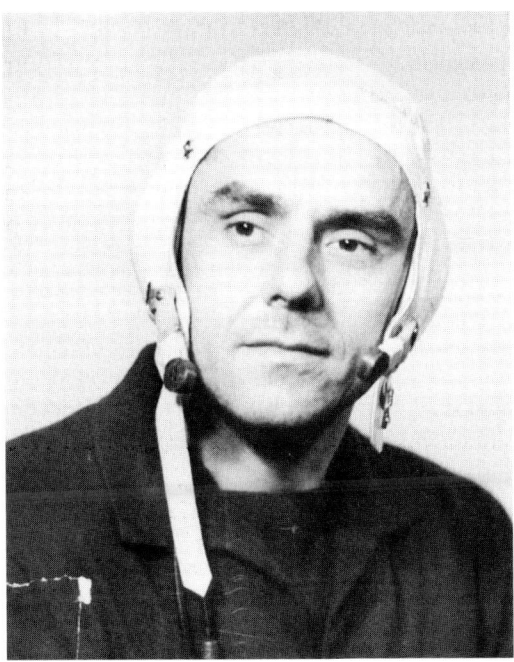

Vladimir Komarov

The new Soyuz spacecraft was a radical advance with a service module, acorn-shaped descent cabin and orbital module, but had failed numerous pre-launch tests. The mission went ahead despite warnings and foreboding. Once in orbit, Vladimir Komarov had great difficulty controlling his spacecraft. One of the solar panels failed to open, depriving him of power. The launch of Soyuz 2 was cancelled. Against all the odds, he managed, on the third attempt, to fire his retrorockets for re-entry. As the spacecraft came in to land, near Orenburg south of the Urals, the parachute snagged, and Soyuz 1 crashed to Earth at great speed, exploding on landing.

Investigation found that the parachute failure was entirely separate from the many problems that had dogged Soyuz in orbit. The parachute design was faulty, as was its manufacturing, sealing and testing process, and the investigation came to the chilling conclusion that had Soyuz 2 been launched, that crew would also have perished on its return. Were corners being cut by both sides as the Moon race entered its final lap?

FINAL LAP TO THE MOON

Both the United States and the Soviet Union fought frantically to regain the lead. The Soviet Union had long planned to go around the Moon first, on a figure-of-eight trajectory using the new Proton booster of Vladimir Chelomei. In September 1967, all was ready to send a stripped-down unmanned Soyuz called Zond around the Moon. The Proton crashed in 60 seconds, and a second attempt also failed that November. After many more frustrations, the Soviet Union eventually flew a Zond around the Moon in September 1968, one of the last of its many firsts, the spacecraft splashing down for recovery in the Indian Ocean.

The following months saw hectic neck-and-neck competition. America's first Apollo flew on 11 October. Soyuz at last made its first successful safe flight on 26 October. The United States had already decided it would send the next Apollo, Apollo 8, around the Moon at Christmas, but a launching window opened in Russia on the 7 December. In November, the Soviet Union sent Zond 6 to fly around the Moon to pave the way for a cosmonaut, and all went well until the return. Zond 6 re-entered over the Indian ocean, but used its heat shield to skip across the atmosphere and soared back into space before coming down over the Soviet Union. On its second re-entry, things began to go badly wrong. First, a gasket blew, depressurizing the cabin—which would have been fatal if unsuited cosmonauts had been on board. Second, the parachutes deployed prematurely and the cabin crashed with such violence that a human crew would have died outright. However, its film of the Moon was salvaged and exhibited to the world as proof of a completely successful mission, and only years later was a picture of the badly battered Moon cabin released.

With the problems encountered by Zond 6, there could be no Soviet flight around the Moon ahead of Apollo. Although a launch window opened on 7–8 December, the Russians chose not to go until Zond was fixed. Cosmonauts pleaded to make the risky mission, but the safety-conscious Russians refused. It was thus Apollo 8 that first flew men to the Moon—and went down in history.

The Moon race did not end at this point. A new joint resolution of the party and the Council of Ministers was passed on 1 January 1969. It authorized the manned Moon programme to continue, commanded the urgent construction of an unmanned soil sample return mission to beat an American manned landing on the Moon, and ordered all speed ahead with the development of a manned space station.

In January 1969, with the American landing on the Moon drawing ever closer, the Soviet Union stepped up its efforts. On 16 January, Soyuz 4 and 5 met in Earth orbit and exchanged crews, just as Soyuz 1 should have done. This lifted Soviet spirits, but they were quickly set back on 20 January when the next Zond mission aborted before it even reached Earth orbit. Worse, when on 21 February the first N-1 Moon rocket took off, a pipe burst at 66 seconds into launch and the whole monster blew up. The first Luna mission to scoop Moon rocks ahead of the Americans crashed in April; the second, in June, failed to get above Earth orbit. These attempts coincided with the worst period in the development of the powerful Proton rocket—though it later proved to be a reliable launcher.

LAST, DESPERATE MEASURES

The game seemed to be up. In early July, the Russians wheeled out to the pad their third Moonscooper, which, if it succeeded, would bring back lunar soil just 12 hours

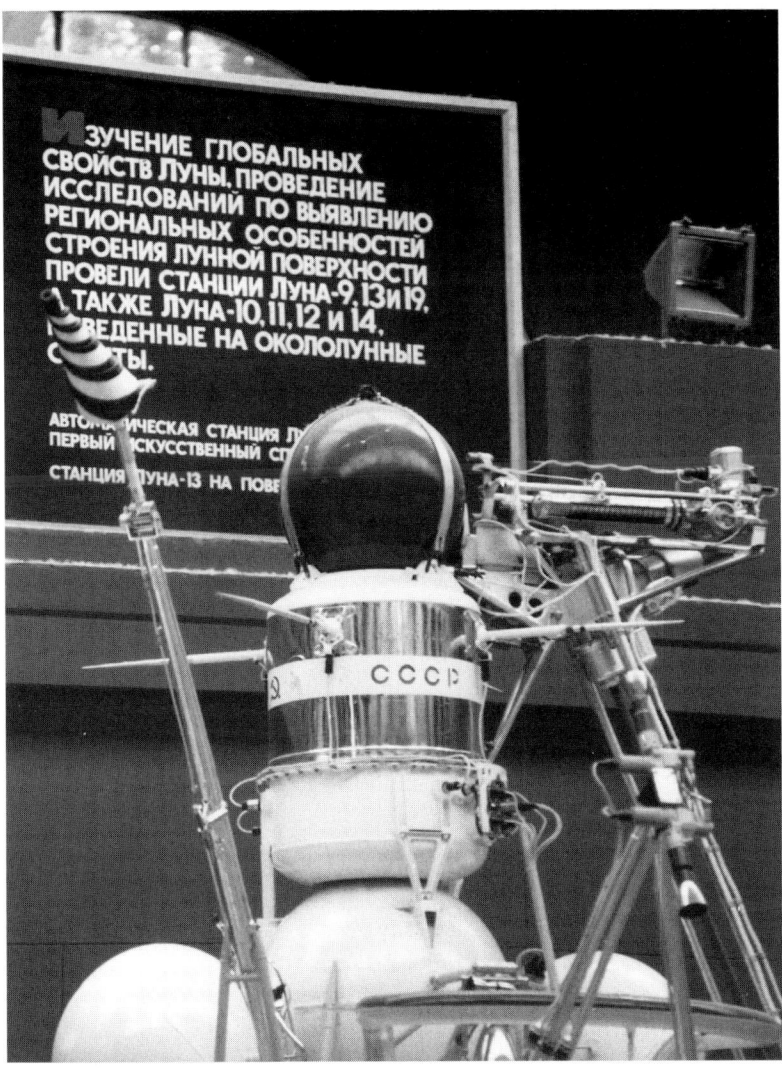

ИЗУЧЕНИЕ ГЛОБАЛЬНЫХ
СВОЙСТВ ЛУНЫ, ПРОВЕДЕНИЕ
ИССЛЕДОВАНИЙ ПО ВЫЯВЛЕНИЮ
РЕГИОНАЛЬНЫХ ОСОБЕННОСТЕЙ
СТРОЕНИЯ ЛУННОЙ ПОВЕРХНОСТИ
ПРОВЕЛИ СТАНЦИИ ЛУНА-9, 13 и 19,
А ТАКЖЕ ЛУНА-10, 11, 12 и 14,
ВЕДЕННЫЕ НА ОКОЛОЛУННЫЕ
ТЫ.

АВТОМАТИЧЕСКАЯ СТАНЦИЯ ЛУ
ПЕРВЫЙ ИСКУССТВЕННЫЙ СП
СТАНЦИЯ ЛУНА-13 НА ПОВЕ

The return craft of the Soviet Moonscooper, with the drill on the right

ahead of the American astronauts of Apollo 11. They brought to the pad their
second N-1 rocket. Since an engineering model was already at an adjoining pad,
prying American spy satellites must have picked up the two giant N-1s and the
waiting Luna—and caused apoplexy in Washington.

As it was, the second N-1 did even worse than its predecessor. An engine pump
exploded as it cleared the pad and the N-1 fell back, exploding and destroying the
surrounding area—all the spy photographs could pick out was blackened, charred

debris and scorch marks. Ironically, against all the odds, the next Luna got away smoothly. Luna 15 entered Moon orbit a day after the Apollo 11 crew left Earth. It manoeuvred extensively over the next four days as it adjusted its orbit for landing. Just as Neil Armstrong and Buzz Aldrin prepared to leave the Moon, it came in for landing in the Sea of Crisis. Somehow, the automatic equipment misjudged the approach and it smashed into the Moon at 480 kilmetres per hour.

The Soviet Union was beaten decisively in the Moon race. That evening, Moscow television showed the Russian people a lengthy broadcast of the American Moon walk. For once, Soviet scientists did not resort to the party line that their 'machines were better', but ungrudgingly, though probably wistfully, commended the supreme American triumph.

It all looked so inevitable in the end. But what if Zond 6 had succeeded and cosmonauts had gone around the Moon first? What if the N-1 had worked? Or if Luna 15 had managed to get back Moon rocks first, as it nearly did? Or if the American venture had faltered? If the Russian flag had been the first to be planted on the dusty surface of the Moon, the effect on world opinion would have been immense. Subsequent global political history might have turned out differently and the communist experiment might have gone from strength to strength. As it was, the ebb tide had begun to set in.

The Soviet people were not given the opportunities to puzzle these might-have-beens, but were instead told that the Soviet Union had never been in the Moon race anyway. The story of the N-1 was kept quiet for almost 20 years.

Just as Apollo 8 did not mark the end of Soviet attempts to win the around-the-Moon race, Apollo 11 did not bring to an end Soviet lunar ambitions. Far from it. The Russians still hoped to send cosmonauts around the Moon. In August 1969, less than a month after Apollo 11 returned to Earth, Zond 7 flew a perfect figure-of-eight around the Moon and landed effortlessly in the summer wheatfields of Kazakhstan. It was decided to fly cosmonauts around the Moon in April 1970 to mark the Lenin centenary, but the Kremlin baulked. Better never than late.

MOONBASE?

As for the Moon landing, the Soviet Union tested out all the hardware for its lunar landing. Cosmos 379, 398 and 434 put the lunar landing module through its paces, and Cosmos 382 the lunar orbiter. Everything went perfectly. The last remaining item was the N-1 launcher. Again, the N-1 was brought down to the pad. On 27 June 1971, the third N-1 started to roll 51 seconds into its mission, crashed and gouged out a crater 20 kilometres downrange. On its fourth mission in 1972 the N-1 managed to fly for 107 seconds before a fire broke out and the debris began to fall.

Undismayed, the Soviet Union prepared new N-1s for launch in 1974. By this time, the Americans had left the Moon. The absence of pressure meant that the Russians could reassemble their equipment without deadlines and pressure. Chief designer Mishin presented a plan whereby, with two N-1s, the Soviet Union could establish a proper Moon base by the end of the 1970s. The Americans might be first to the Moon, but the Soviet Union would be first to build a base and live there.

However, it was not to be. In May 1974, following so many disappointments, Mishin's enemies moved in and had him removed from his post as chief designer. In his place they appointed the great engine designer Valentin Glushko, who was to hold the position until his death. Valentin Glushko tried to persuade the party and government to support a lunar colony and a huge new rocket called the Vulkan to build it, but they had had enough. Instead, learning that the Americans were building a space shuttle, they ordered him to do the same. If the Americans needed one, the Russians also needed one.

The space race had come full circle! From being the undisputed masters of the cosmos in the mid-1960s, the Kremlin now felt that if the Americans were doing it, the Russians should follow. By the 1970s, the Soviet Union had developed an inferiority complex to American achievement. Losing the Moon race cost the Russians their self-confidence. In fact, the loss of the Moon race was not a technological failure at all. The Russians managed to master Zond, Soyuz, and even the temperamental Proton launcher, and, from what we now know, were probably quite close to achieving success with the N-1. They tested all the lunar equipment in Earth orbit and it functioned perfectly. The Russians lost because they didn't realize there was a race until it was too late (three years too late) and they failed to mobilize their resources in time. Writing about these events years later, Mishin blamed under-investment, lack of financial control, the dispersal of effort between design bureaux and poor management of the 26 government departments and 500 enterprises involved. The Soviet investment was only $4.5bn compared to America's $24bn. They underestimated the technical difficulties and should have done proper ground testing. In mythology, the Russians lost the Moon race because they didn't have the equipment to do the job—but the real reasons were political, financial and organizational.

The Moon race had two sequels. First, the unmanned programme ordered in January 1969 was brought to a conclusion over the years 1970-6. In September 1970, a year after Apollo 11, Luna 16 soft-landed in the Sea of Fertility. An arm was extended to the surface of the Moon where it drilled into the lunar surface, scooped up rock and lifted it up into a small capsule. A day later, the ascent stage fired, sending the small cabin on a 3-day return to the Earth, where beacons guided the recovery forces to its location. The Soviet Union retrieved more samples in 1972 from the crater Apollonius (Luna 20) and in 1976 a core sample from the Sea of Crisis (Luna 24). Also in the unmanned programme, two spacecraft were put in lunar orbit to map landing zones for later unmanned and manned missions (Luna 19 and 22). The missions that made the most impact were the Lunokhods (Luna 17 and 21). Soon after landing, a 760-kg rover, looking like a bathtub on wheels, rolled down a ramp onto the lunar surface. On the top was a lid which was swung back to receive the Sun's rays for electricity. On the front were cameras, lasers and, at various points on the vehicle, equipment for studying Moon soil, radiation and the heavens above. The Lunokhods were steered by a ground crew of five in mission control who used commands to make the rover drive, swivel, cross craters

The Lunokhod

and explore the terrain. Lunokhod 2 spent five months roving around the Sea of Rains and travelled 37 km before its power gave out.

SPACE STATION SALYUT

These automatic missions were considerable achievements in their own right, even if they paled in comparison to Apollo. The second sequel to Apollo was the space station. Here, the Soviet Union build up an expertise in space exploration which it has maintained ever since. In the immediate aftermath of Apollo, the official objective of the Soviet space programme was the building of space stations and all past and present missions were presented as being in pursuit of this objective. In January 1969, the rival Korolev and Chelomei bureaux had been ordered to sink their differences and together build a new space station.

Eventually launched in April 1971, it was named Salyut, in order to salute Yuri Gagarin's first flight around the Earth 10 years earlier. Although the first crew (Soyuz 10) was frustrated in its efforts to board the station, the second crew,

Soyuz 11: first crew on Salyut 1

Soyuz 11, docked with and entered the station in June 1971. Georgi Dobrovolski, Vladislav Volkov and Viktor Patsayev spent 24 days on board Salyut 1 in 1971. Daily they sent back pictures of life aboard the world's first orbiting space station, and viewers saw them floating in the spacious cabin, observing the Earth through a huge camera, growing plants, watching the stars and making medical tests. They manoeuvred the station into a higher orbit, wore special gravity suits, tested for radiation, observed stars and watched tadpoles swim in jars in zero-gravity. As they fired their retrorockets to return to what would have been a heroes' welcome, they caught a last glimpse of Salyut sailing high above, the Sun reflecting off its solar panels.

The Mil recovery helicopters in the landing area were disturbed by the lack of radio contact during the descent, but the parachutes opened and the Soyuz 11 cabin made a normal touchdown. As they rushed forward to get the cosmonauts out, rescuers found them staring straight ahead, dead in their seats. Investigation found that high above the atmosphere a valve opened, the air had leaked out and they had died in seconds.

The next years were a period of unrelieved gloom for the Soviet space programme. It took a year to redesign the Soyuz spacecraft. The next space station crashed in July 1972. In the following year, with the Americans due to orbit their Skylab space station in mid-May, the Soviet Union tried to get two space stations into orbit simultaneously. The first, Salyut 2, entered orbit in April 1973 but suffered a violent explosion on 12 April and had to be abandoned. The next, less than a month later, discharged all its manoeuvring fuel on its first orbit. Thus in May 1973, no less than two Soviet orbital stations crashed in flames. By vivid contrast, the American Skylab was a tremendous success, with three sets of astronauts spending 28, 59 and 84 days on board, conducting an enormous range of experiments into Earth resources, astrophysics and medicine. The Soviet Union had been humiliated not just in the Moon race but in the space station race too.

However, it was at this stage that a poorly appreciated aspect of the Soviet and Russian psyche began to come into play. Far from giving up, the organizers of the space programme patiently rebuilt their programme. An improved, safer Soyuz flew again in autumn 1973. In the following year, with Salyut 3, the USSR at last flew a full mission on an orbital station and brought the men home safely. With Salyut 4 in 1975, they matched the achievements of the American Skylab and put crews on board for 29 and 63 days.

Although the manned programme of the Soviet Union was the one which attracted the most media and public attention at home and abroad, it was only the tip of what had become a huge enterprise by the 1970s. The first dedicated scientific programme was announced in 1964 when the first of four Elektron spacecraft was launched to study the Earth's radiation belts. In 1965, the first of four large Proton spacecraft (up to 17 tonnes) was launched to study cosmic rays. The first applications programme began in April 1965, when the first Molniya spacecraft was launched to relay television programmes from Moscow to the far-flung regions of Siberia and further afield. A programme of weather satellites, called Meteor, was initiated in 1969. To facilitate the expansion of the space programme, the Plesetsk missile base in northern Russia was turned into a full-scale cosmodrome. Volgograd station continued to be used for smaller satellites on scientific missions.

The space programme had now become so large that it could no longer be handled by one design bureau. The Korolev OKB-1 bureau was rivalled for contracts by the OKB-52 of Vladimir Chelomei and the OKB-586 of Mikhail Yangel in Dnepropetrovsk, Ukraine. Communications satellites were spun off to a company in Krasnoyarsk, the NPO-PM of Mikhail Reschetnev. Upper stages became the speciality of the bureau of Semion Kosberg. Spy satellites, a growing area of work, went to the central design bureau in Kyubyshev.

VENUS CONQUERED

The Soviet Union conquered Venus, which, it soon discovered, was not covered in steamy oceans. In its attempt to land probes on the surface, the Lavotchkin Design Bureau, which had responsibility for Venus probes, had designed space probes to descend quickly through the atmosphere under parachute. The theory was that a rapid descent would get a probe down through the acid-filled atmosphere to the hot and pressure-dense surface before it succumbed to the hostile conditions. In October 1967, Venera 7 had successfully entered the Venusian atmosphere but the probe had been crushed 20 km above its surface where the temperature reached 272 °C and the pressure was 22 atmospheres. Using parachutes half the size, Venera 5 and 6 entered the Venusian atmosphere in May 1969, but had been destroyed at 15 km and 16 km respectively.

Soviet scientists believed that Venera 7 had suffered a similar fate in December

Venera soft lander

1970, but a technician going through signals six weeks later managed to find a faint signal that had been transmitted from the surface after landing. The signals were consistent, showing the temperature constant at 475 °C and the pressure constant at 90 atmospheres, so the probe could only be steady, on the surface and at rest. The probe had sat there, cooking gradually in a temperature able to melt lead or zinc before it succumbed.

Venera 9 arrived in October 1975. The lander fell to 50 km, when the parachute was discarded. This was the radically new design step. The lander would now fall unaided but for a disc brake—fast enough to get it to the surface quickly, yet hopefully not too fast to damage it. After 75 minutes the capsule was on the surface, its impact cushioned by shock absorbers. Then the real work began. Caps dropped off the camera covers and showed that there were rocks everywhere—sharp, round and curved on a dark black surface. They were the first pictures from the surface of another planet. Venera 10 touched some days later some 2,200 km away—another triumph.

When Venera 13 landed in March 1982, a mechanical ladder immediately extended to the surface and began to analyse the rocks using screw drills. The system had to work against time, for there was the possibility that the whole probe might collapse under the pressure any time. Within minutes scientists had a read-out on the composition of the soil: 45% silica oxide, 4% potassium oxide, 7% calcium oxide. The general composition was basalt. Remarkable though these findings were, they had none of the dramatic impact of the pictures taken by the cameras of Venera 13. They sent back eight separate panoramas, scanned in red, green and blue. The cameras revealed a rolling stony plateau and the curved horizon in the distance. Scattered on it were stones, pebbles and flat rocks. Looking skywards, the cameras found a bright orange sky, not a blue one.

Venera 14, landing on another part of the planet at the same time, sent down a mechanical drilling arm which reached out for samples, scooped them into a hermetically sealed chamber and subjected them to X-ray and fluorescent analysis. The Venera 13 and 14 missions were triumphs by any standards. The scientific data returned were unambiguous and the pictures were a feast to geologists' eyes. Later, in 1985, Vegas 1 and 2 repeated their triumphs—and their mother craft then flew on to Comet Halley.

LONGER AND LONGER FLIGHTS

By this time, the manned space programme had made a full recovery. September 1977 saw the launching of what was called the second generation of orbiting space stations. Salyut 6 was an innovation because it had two docking ports: one could receive a Soyuz crew, while the other was designed to receive an unmanned Soyuz turned into a freighter, able to bring up supplies of air, water, food, fuel, scientific equipment and personal items. This enabled the length of flight to be extended far beyond anything that had previously been possible.

The first resident crew aboard Salyut 6 comprised Yuri Romanenko and Georgi Grechko, who spent 96 days on board over the winter of 1977/8. Their flight was significant for three reasons. First, they broke the existing American endurance record of 84 days and since then the Soviet Union, and then Russia, have always held the title for long flights, never looking back. Second, they received the first Progress unmanned freighter, ushering in a system whereby these unmanned cargo flights were sent up to replenish orbiting space stations several months apart. Third, they welcomed on board the first non-American, non-Russian to fly in space: a Czech. In 1976, the Soviet Union had invited the socialist countries to send suitably qualified people to train for flights to orbital stations. In the course of 1976–82, representatives of all the socialist bloc countries visited Salyut, not just those in eastern and central Europe but also Cuba and Vietnam. From 1982, the system was extended to other countries. From then until the end of the Soviet period, there were visits to Soviet orbiting stations by representatives of India, France, Syria

Georgi Grechko

and Afghanistan. In addition, the Tokyo Broadcasting Service paid for a visit by its foreign news editor and there was a Soviet–British project called Juno, whereby Helen Sharman spent a week in space.

The Salyut 6 crews gradually pushed back the endurance record: 139 days by Vladimir Kovalyonok and Alexander Ivanchenkov; 175 days by Vladimir Lyakhov and Valeri Ryumin; and 185 days by Leonid Popov and Valeri Ryumin. Starting with Kovalyonok and Ivanchenkov, spacewalking became a regular feature of space station operations, the crews exiting to place and retrieve experimental packages on the outside, to carry out repairs and, later, to add solar panels sent up from Earth. Salyut 7 was launched on the anniversary of Salyut 1—19 April 1982. Similar in dimensions and design, it was home to crews spending ever longer on board. Anatoli Berezovoi and Valentin Lebedev, the first Salyut 7 crew, spent 211 days on Salyut 7, to be followed by the trio of Leonid Kizim, Vladimir Solovyov and medical doctor Oleg Atkov, who pushed the longest mission to 237 days. Salyut 7 also received large, experimental modules, Cosmos 1443 and 1686, with additional room, equipment and cargoes, demonstrating the possibilities of assembling modular space stations in Earth orbit.

The operation of these two long-stay orbital space stations was not without mishap. One visiting crew, Soyuz 33, had to make a scary emergency descent to Earth when the manoeuvring engine failed. There were leaks and other problems which required repair. Soyuz T-8's radar failed, requiring a premature return. Dramatically, Soyuz T-10 went on fire on the pad just seconds before lift-off. Flames licked around the base of the rocket and it was obvious that it was about to explode. In the nick of time, the escape tower fired at the top, whooshing Vladimir Titov and Gennadiy Strkekalov up to 1,000 metres in seconds. The Soyuz cabin fell out of its shroud, parachutes opened and gave them a few seconds of air before touchdown. A shot of vodka helped their recovery from what must have been a terrifying ordeal. In early 1985, the Salyut 7 space station lost its orientation: the electrical systems died

Salyut 7 orbital space station

and all the environmental systems froze. Two cosmonauts flew a Soyuz up to Salyut 7 on a rescue mission, docked with the drifting station, entered it and brought all its systems back to life.

Typically, Salyut would be unoccupied for months between these long-stay missions. Crews would return home with the often voluminous results of their research into space, medicine, astronomy, the atmosphere, materials processing, the Earth's resources and the weather—and new missions would be prepared. In September 1985, for the first time, a new crew came on board just as the old one was concluding its mission. As Vladimir Dzhanibekov and Georgi Grechko returned to Earth on Soyuz T-13, Vladimir Vasyutin and Alexander Volkov were already on board with existing resident Viktor Savinyikh, demonstrating the ability of Salyut to provide for the permanent handover of crews.

Thus by the mid-1980s, the Soviet Union had built up its programme slowly and patiently to the point that the permanent occupation of orbiting space stations had become a possibility. The long-duration missions had pushed back the knowledge of how humans adapted to weightlessness to a degree unimaginable a few years earlier. The Soviet Union had accumulated considerable hands-on experience not just in operating stations, but in overcoming one operational problem after another, with the inherent difficulties and dangers.

International collaboration: German cosmonauts in training

HIGH TIDE

This period, the mid-1980s, was the high water mark of the Soviet space programme. The most public face of the space programme was the manned programme and the automatic space probes sent to the Moon, Mars, Venus and Comet Halley. There were 50 men and women in the cosmonaut squad. About 400,000 people worked for the Soviet space programme—in the assembly lines in Moscow, in rocket factories in Dnepropetrovsk and Kyubyshev, in scientific institutes scattered around the country, in production plants and in the world-wide land- and sea-based tracking network.

By the mid-1980s, the Soviet Union was launching up to two satellites a week. Most were given the Cosmos label, meaning that they were military satellites. Official bulletins gave their orbital data, but said little or nothing about their purpose. In fact, Western analysts were able to break down their various categories. They comprised new generations of photoreconnaissance satellites (Yantar, Kometa), electronic intelligence satellites (Tselina), hunter–killer satellites able to intercept and destroy missile targets and nuclear-powered naval reconnaissance radar platforms able to track American ships at sea. The military had the benefit of space-borne navigation satellites (Tsikada and Parus). Although the manned and deep space missions attracted the most publicity and public interest in the Soviet Union and abroad, the unmanned programme probably consumed greater resources. During the time of Leonid Brezhnev, there was a huge expansion of the Soviet air force, missile forces and navy and the military space programme benefited equally from this largesse. Whether the Soviet Union could really afford this was another question.

Turning to applications and science, Molniya communications satellites were supplemented by a new generation of communications satellites operating from 24-hour orbit—Raduga (military), Gorizont (civilian) and Ekran (direct broadcast). Meteor weather satellites were orbited at regular intervals. Three astronomical observatories were launched: Astron, Granat and Gamma. The Vostok cabin was adapted for use as a biological cabin carrying monkeys and other animals (Bion) and to carry out materials-processing in zero gravity (Foton). Scientific missions with small satellites were flown under the Cosmos designation and in a joint programme with the socialist countries (Intercosmos). The *National Geographic Magazine*, which had hitherto covered only American space activities, acknowledged the huge scale of the Russian effort when it devoted the main feature of its October 1986 issue to 'The Soviets in space'.[1]*

By 1986, the Soviet space programme was ready for two new leaps forward: The introduction of the third generation of space stations, and the introduction of the Soviet space shuttle.

INTRODUCING MIR

The new station, originally to be called Salyut 8, had a crucial modification: instead of one docking port at each end (like Salyuts 6 and 7) it had six—one at the rear end, and five at the front, enabling it to take modules, manned and unmanned supply spacecraft. As well as being permanently manned, a possibility demonstrated by Salyut 7, this would make it the first-ever modular space station, one which could be assembled in sections. Close to the ports was an arm called Lappa which could manipulate modules from one port to another. The station itself had no scientific equipment, only control mechanisms and living quarters. Each cosmonaut had an individualized cabin with bunk, couch and table.[2]

The new space station coincided with a change in the leadership of the Soviet Union. Mikhail Gorbachev had become General Secretary of the Communist Party in 1985 and he introduced the policies of *glasnost* (openness) and *perestroika* (economic and social transformation). To symbolize the fresh start, he ordered

* Notes and references are listed on pages 317—322.

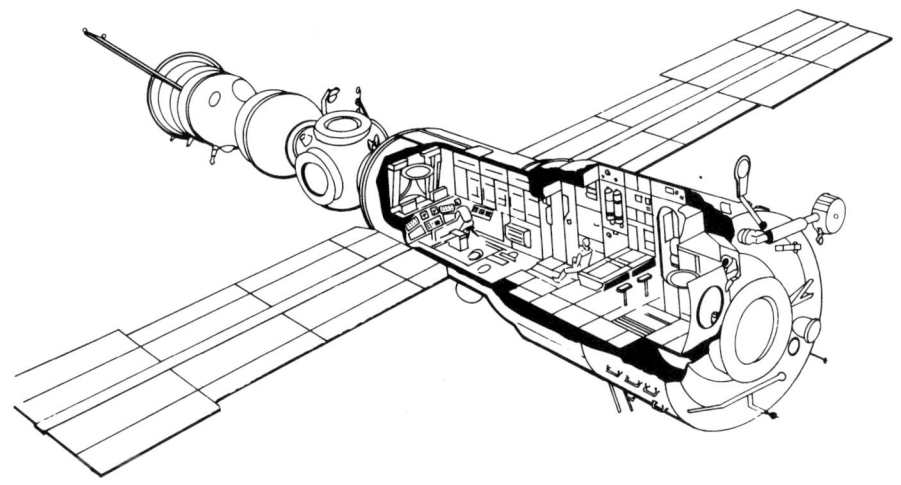

Mir base block

that the station be given a new name, Mir, a word which in Russian means 'peace', but also has connotations of community and harmony.

Mir was originally designed to fly in late 1986, but the mission was brought forward by Mikhail Gorbachev to mark the congress of the Communist Party of the Soviet Union. Mir had been in design for 10 years and the project had been approved by government as far back as 1976. Originally, small 7-tonne Soyuz-class modules were to have docked with Mir, but the Mir project was scaled up, with the addition of four 20-tonne modules: Kvant 2, Krystall, Spektr and Priroda. This approach of using large modules the size of a space station was to make the station grow much larger than its designers ever intended.

The launching of Mir took place on 20 February 1986. That spring, cosmonauts Leonid Kizim and Vladimir Solovyov boarded Mir, de-bugging the systems and getting the station ready for permanent occupation. In February 1987, the station duly received its first permanent crew, Yuri Romanenko and Alexander Laveikin. An astrophysical module, Kvant, was docked with Mir in April and the process of orbital construction had begun. Even as Romanenko and Laveikin worked on Mir, the new station was eclipsed by dramatic developments at Baikonour.

LENIN'S TRIUMPH

On 14–17 May 1987, general secretary Mikhail Gorbachev visited Baikonour. There, he was given a tour of Soviet space potential, in the course of which he inspected launch pads, assembly workshops, military facilities and the administrative centre. Gorbachev trumpeted:

Energiya, May 1987

Lenin's dream of making our State a great industrial power has come true. Every-thing here is Soviet-made, everything is of high quality and of modern technological standard. There is no need for us to go cap in hand to foreign lands.

Pride of place for the visit was the new Soviet super-rocket which Valentin Glushko had spent 10 years building. Ever since he had been given the order to build a space shuttle in 1976, he had toiled both on a shuttle design and on a powerful rocket able to put it into orbit. The super-rocket was called Energiya, or 'energy' in English. It was an appropriate title. Energiya used a liquid hydrogen-fuelled central stage, assisted by four Zenit rockets on the side, using the most powerful rocket engines ever built, the new RD-170s. The new Energiya took 12 seconds to build up to full thrust, but once on its way it was able to put a staggering 140 tonnes into Earth orbit.

Night turned to day as the countdown of Energiya climaxed on 15 May. Weighing 2,000 tonnes, 60 metres high, with eight engines and a thrust of 170 million horse-power, Energiya illuminated gantries, observers and towers for miles around as it headed skywards. Although the orbital insertion manoeuvre did not go entirely to plan, this first successful flight laid the ghost of the N-1 to rest: the Soviet Union could, and just had, built the most powerful rocket ever known.

The objective of Energiya was to put into orbit the Soviet space shuttle, now ten years in design. The Soviet Union had a long experience from the 1950s of designing and flying shuttle models and spaceplanes, though few projects had been brought to conclusion before. This time was different. It was called Buran, the Russian word for 'snowstorm'. A huge runway constructed at Baikonour for its return was clearly visible to prying American satellites. The old N-1 facilities were converted as shuttle facilities. The N-1 pads were rebuilt, along with elaborate launch towers, escape chutes and tunnels. Twenty cosmonauts trained to fly it. Six one-eighth scale models were flown on sub-orbital tests from 1983 to 1988 out of Kapustin Yar at speeds of up to mach 16 as high as 120 km. Six full-scale analogues of Buran were built for tests on Earth: one was fitted with four engines (two jet engines, two ramjets) and used to test the shuttle's aerodynamic characteristics and landing profile.

Buran looked very like the American space shuttle—its dimensions were start-lingly similar—but it was quite different. Buran did not use its own engines during the ascent to orbit, which meant that it could carry heavier cargoes. It could also fly automatically and for its first test flight it would be unmanned.

BURAN: FLIGHT OF THE SOVIET SPACE SHUTTLE

On 14 November 1988, gigantic billowing clouds erupted all around the Energiya tower. After an agonizingly long period of waiting, Buran was seen to rise slowly above the clouds of steam, dust and flame and head far into the night sky. Aircraft high above picked up Buran as it pierced the cloud deck, climbing on a pillar of blinding yellow flame. Buran took eight minutes to reach orbit and then used its own engines to take it from 160 km to 260 km.

Over Fiji, ground controllers activated the on-board television cameras to pan around the empty cabin and out of the windows to spot the blue Pacific down below. At the end of its second orbit, retrorockets fired. Buran came in over the Mediter-ranean at 25 times the speed of sound. Ground controllers and computers took

Buran

Buran touchdown

Buran through a series of flying turns as the space shuttle swung towards the cosmodrome where it had started off. Baikonour radar picked up Buran at an altitude of 40 km and a range of 400 km. Television picked up the shuttle as it fell gently from the early morning sky. Buran curved, rolled and aligned itself automatically with the 4.5-km-long runway. Landing speed was nearly 340 km/h. A jet fighter sped past Buran's tail as the wheels came down. A braking parachute shot out to slow the space shuttle's speed. Sixteen TV cameras watched Buran come to a halt, its snow white frame standing out against the wintry russet browns of wintertime Tyuratam. Buran came to a standstill less than 2 km from its touchdown point.

The chief designer of the Energiya–Buran system, Valentin Glushko, lived just long enough to see Energiya and Buran take to the skies: he died in Moscow on 10 January 1989, aged 81. He had been designing and building rocket engines since he was 13. The climax of his life was to leave the Soviet Union with a space shuttle which equalled, if not surpassed, the American shuttle and a rocket system that was the envy of the world.

MIR'S NEW RECORDS

Meanwhile, Mir was breaking fresh records. In December 1987, Yuri Romanenko had returned to the Soviet Union after a record-breaking 326 days circling the Earth. This was but a prelude to an even more ambitious target: a full year in space. As he returned, Vladimir Titov and Musa Manarov arrived on Mir. As well as their busy programme of work on the orbital station, they fought the debilitating effects of weightlessness on the body by exercising 5 km a day on the treadmill and 10 km a day on the bicycle. Late in their mission Valeri Poliakov, a doctor, came on board to check on their condition to ensure that they could safely return to Earth, which they did, coming down in winter snow in December 1988, 366 days after they had left the planet.

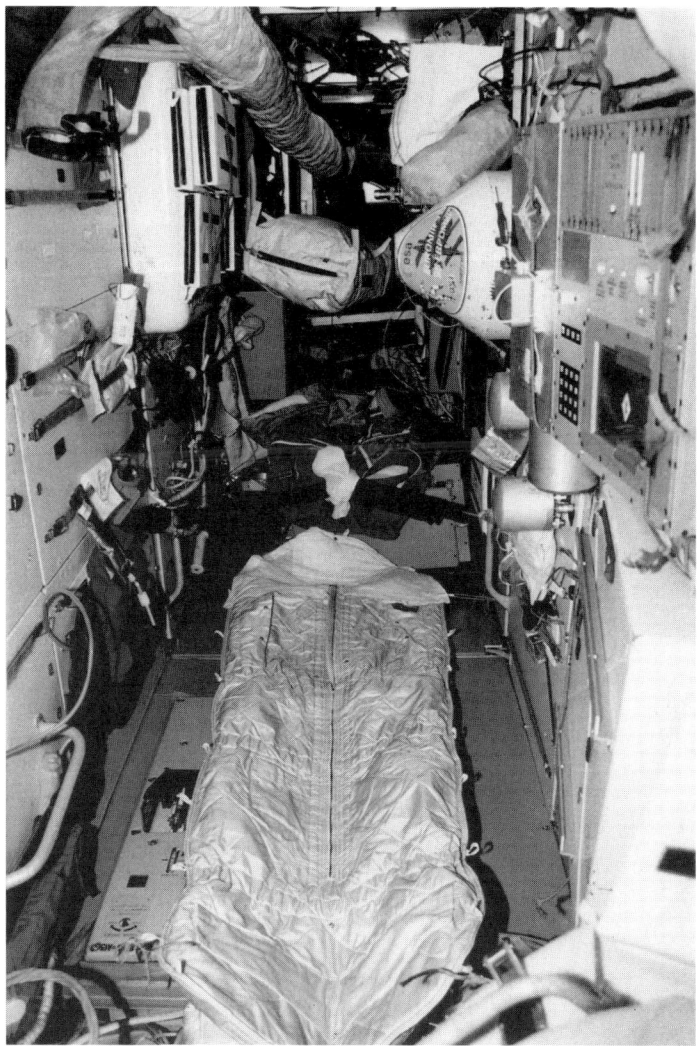

Interior of Krystall

The second phase of Mir operations began in autumn 1989. The station had been unoccupied for several months, waiting for new modules to be completed. The first of these, Kvant 2, arrived in November 1989, bringing up the first manned man-oeuvring unit, a jet backpack which cosmonauts could use to fly some distance from the station. The second large module, Krystall, arrived in June 1990, bringing up a veritable factory of materials processing furnaces going under the names of Kratar, Optizon and Zona, as well as two new docking ports, one of which was designed to receive the Buran space shuttle.

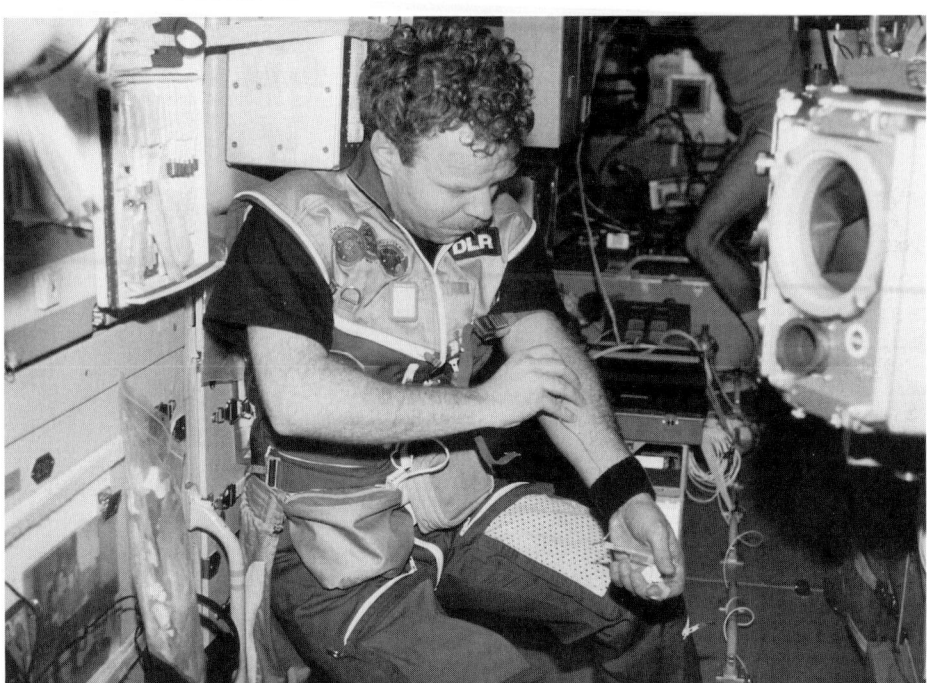

Blood test on Mir—an essential aspect of monitoring the effects of long-term flight

By 1991, Mir had settled down in a pattern of regular six-monthly resident missions, punctuated by the regular arrivals of Progress supply craft. Important work was carried out on Mir in the course of 1991. Although Mir still awaited its two last large modules, some final construction work was required on the outside of the station. That January, Viktor Afanasayev and Musa Manarov clambered along the hull to erect a 12-metre-long telescopic crane which could move large objects around the outside of the station. That July, Anatoli Artsebarski and Sergei Krikalev erected a 14-metre girder called sofora, designed to take a manoeuvring block and engine so as to effectively control the growing complex. Before they climbed back in to Mir, they had one final ceremonial function to carry out. On 27 July, to celebrate their achievement in completing the external construction of the Mir space station, they unfurled the hammer and sickle of the Soviet Union and attached it to the highest point of the girder.

It could never have occurred to them that the Soviet Union would last only a few more days.

2

The manned frontier
Space station Mir

The collapse of communism brought a host of unwelcome financial, organizational and political problems in its wake. For Russia, the challenge was to see if it could find ways of continuing to run its manned space station Mir, maintain the presence in space built up during the Soviet period and find a successor programme for the now ageing space station. First, Mir.

LAST SOVIET CITIZEN

During the coup against Gorbachev, mission control was politically even-handed and relayed up to the Mir space station news broadcasts from Soviet central TV (which supported the *putsch*) and Russian radio (which supported Yeltsin), probably confusing the cosmonauts as to what was really going on. The news that both Ukraine and Kazakhstan had taken advantage of the fighting to declare their independence was a footnote to the more dramatic events taking place in Moscow. Whatever their opinions, the cosmonauts then on Mir—Anatoli Artsebarski and Sergei Krikalev—wisely kept their counsel and made no comment one

Last Soviet citizen—Sergei Krikalev

way or the other. The operating manual specified that the cosmonauts must return in the event of war, but did not specify what was to happen in the event of a coup: the manual had never contemplated such an unlikely eventuality, so they stayed put until it blew over. Life continued on the Mir complex much as before. The cosmonauts rose at 8 a.m. as on a day on Earth. They divided their time between carrying out experiments, exercising for at least an hour every day and maintaining the space station. They took time off at weekends and this time was spent in television hook-ups with home, watching videos, reading novels in the Mir library and, the favourite occupation of all cosmonauts, watching the Earth roll by below.

Ultimately the political changes of August 1991 were to have profound conse-quences for the space programme. These were not yet apparent, but an inkling of what was to come was evident when Soyuz TM-13 was launched on 2 October, the take-off observed by Nasultan Nasurbayev, President of the now newly-independent Republic of Kazakhstan. Prior to the break-up of the Soviet Union, Kazakhstan, then a state within Russia, had been invited to send a cosmonaut to Mir. Toktar Aubakirov, the man chosen, was an aircraft engineer by profession, an experienced military pilot and a test pilot of MiGs and fast Sukhoi jets. His back-up, Talgat Musabayev, was Kazakh aerobatic champion and had flown everything from Antonov crop-dusters to Tupolev commercial jet aircraft.

Soyuz TM-13 had been originally planned as a joint mission with Austria, and Soyuz TM-14 had been intended to take a Kazakh cosmonaut some months later. However, financial shortages led to a merger of these two missions. Soyuz TM-13 duly flew with a Russian commander, Alexander Volkov; a Kazakh, Toktar Aubakirov; and an Austrian, Franz Viehböck. The Kazakh, in effect, took Krika-lev's return seat, so Krikalev was ordered to stay on the station indefinitely. Artse-barski returned on 10 October with the Kazakh and the Austrian after a week's experiments together. Sergei Krikalev remained on board with Alexander Volkov as the permanent crew of Mir.

The routine of maintenance and scientific experiments continued on Mir as the politically Earth-shaking events took place down below. Krikalev and Volkov welcomed in the new year on board Mir on 1 January 1992, aware that on Earth the hammer and sickle had just been pulled down from the Kremlin and replaced by the red-blue-and-white bars of the new Russian flag. This was the third year in a row that a crew had been on board Mir for the new year and, as normal, the crew sent new year greetings back to Earth by their television link, sitting at their control panel beside a small plastic fir tree. Their first major task in the new year was to load up a Raduga recoverable capsule on the Progress M-10 freighter to Mir. These small capsules were cone-shaped and able to take 150 kg of research results, film and reports back to Earth. The little Radugas were heated to a searing 2,200 °C during re-entry before a small parachute opened to cushion the final descent and a beacon popped out to help recovery workers find its location. Volkov and Krikalev sent back their first Raduga of the new year on 20 January.

Sergei Krikalev's extended stay on Mir was the clearest evidence that the growing financial shortages overtaking Russia were having a direct impact on the manned space programme. In the old Soviet command economy, credits had simply been

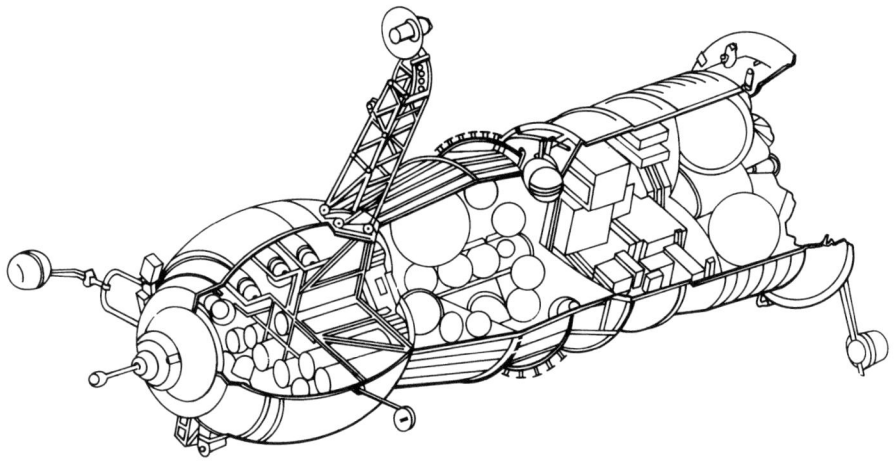

Progress freighter

voted for approved projects and the type of financial controls familiar to Western companies had never operated. During a celebrated incident in 1975 in the course of the Apollo–Soyuz Test Project, an American journalist once asked Intercosmos council chairman Boris Petrov how much the USSR was putting into the project. He rambled on for half an hour. In the end he gave up, saying he didn't know. 'What's the use?', he said. 'I don't count the money and there is still plenty of everything we need.' Despite his protestations, the Soviet Union had begun to be cost conscious in its final months, and more and more alert to the need to attract foreign currency earnings. Foreign countries looking for launching facilities had to pay more and more for a service hitherto given free in a spirit of socialist generosity. Tokyo Broadcasting System's Toyohiro Akiyama's flight to Mir had cost his company over $12 million. In the late 1980s, the Soviet Union had invited the European Space Agency countries, some individually and all collectively, to participate in flights on Mir. This price varied from $12 million to around $40 million, depending on the length and complexity of the mission. Originally these payments would have gone into what might be termed the 'Soviet Union general account' but as companies and corporations in the new Russia were required to be self-financing, it became payable to the Energiya company which was now effectively the owner of the Mir space station and its associated operations.

Sergei Krikalev still had his communist party membership card with him and, the last surviving member of a now-banned party, he became known as 'the last Soviet citizen'. The end of his long sojourn was in sight when Soyuz TM-14 was rolled out to the pad on 15 March, decorated with the flags of Russia, Kazakhstan and Germany. Launching took place two days later, carrying Alexander Viktorenko, Alexander Kaleri and Klaus-Dietrich Flade for a week of medical, biological and engineering experiments devised by the German Space Agency. On 25 March, Alexander Volkov, Sergei Krikalev and Klaus-Dietrich Flade dropped away from

Mir to land in calm, windless spring snow in Kazakhstan, to be met at once by recovery teams. Krikalev had been in orbit 312 days, the longest ever unintended stay in space.

The third joint Russian–French mission, project *Antares*, began with the launch of Soyuz TM-15 from Baikonour on 27 July 1992. On board were veteran commander Anatoli Soloviev, flight engineer Sergei Avdeev and Michel Tognini of the French Space Agency CNES (Centre Nationale des Etudes Spatiales). The Frenchman had been back-up to the previous French cosmonaut and in the course of his long period of training had married his physical education instructress, Elena Chechina. It was a two-week mission which concentrated on medical and biological research. In September, in the course of three spacewalks, Soloviev and Avdeev fitted a 700-kg manoeuvring and propulsion block called VDU on the sofora girder on the outside of Mir. The VDU had been brought up by Progress M-14, from which it had to be extracted and then moved into open space. They attached cables around the sofora to enable them to connect the electrical commands that would control the VDU. The manoeuvring block would reduce the use of propellant by the station by over a third.

LIGHTING UP THE NIGHT SKIES OF EARTH

The most politically symbolic moment of the mission took place on 7 September 1992 when the spacewalking Soloviev and Avdeev took down the Soviet flag from the sofora girder, replacing it with the Russian flag. It was the last Soviet flag flying anywhere in the solar system, though pennants and red stars from an earlier political epoch had by this stage been widely distributed on the Moon, Mars and Venus, presumably to be found by some future space archaeologist. Soloviev and Avdeev successfully hatched out quail chicks in the course of their mission. After 189 days in space, they were replaced by Gennadiy Manakov and Alexander Poleschuk, launched on Soyuz TM-16 on 24 January 1993.

There were two unusual features in the launch of Soyuz TM-16. It had been the norm for the Soyuz rocket to fly the national flag, but no one could agree as to whether the rocket, departing from Kazakhstan, should fly a Russian or a Kazakh flag, or both, so it flew neither. Second, Soyuz TM-16 docked not with the main Mir docking port, but with the docking unit on the Krystall module on its side, which had not yet been used. Krystall had an androgynous docking unit, similar to the one used in the Apollo–Soyuz Test Project in 1975. Rather than use a probe-and-drogue method, it carried a petal system more suitable for adaptation to other spaceships. Soyuz TM-16 carried the docking system to be used by the Buran space shuttle and docked with the port where Buran was due to come in. The test cleared the way for a flight by Buran to Mir, if the funding could be found for it. There were no problems with the docking on the Buran port, and, with the arrival of Soyuz TM-16, a record total of seven spacecraft were now docked with the orbiting space station.

These missions attracted little public attention, partly because of their routine nature and partly because of the economic crisis now gripping Russia. An

Nght-time Soyuz launch

exception was the mission of Progress M-15, the cargo craft which arrived at Mir on
29 October 1992. Progress M-15 undocked early on 4 February 1993, and an experiment which has been planned for years got under way. Filmed by Gennadiy
Manakov and Alexander Poleschuk, Progress released a 5-kg, 10-metre solar sail
called *Znamia* ('banner' or 'flag' in Russian) which popped out of the hatch of
Progress and unfurled like an umbrella. Progress began to spin against the blue
oceans of Earth. As it entered the night-time skies over Earth, *Znamia* was
pointed downward and began to reflect the Sun's light on Earth, testing the ability
of objects in the sky to light up the ground. The idea could be traced back to
Austrian space theorist Herman Oberth in the 1920s and subsequent theorists who

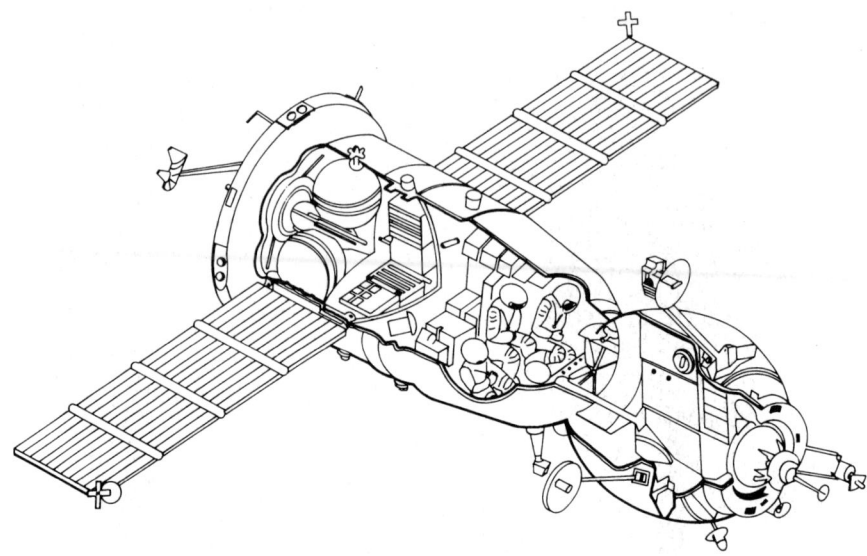

Soyuz TM spaceship

dreamed of using artificial sunlight to light cities in the polar midwinter. From an altitude of 250 km, *Znamia* cut a cone of light 5 km wide in a swathe from Toulouse through Geneva, Munich, Prague, Lodz, Brest and Gomel, where it could be seen as a bright diamond rapidly traversing the sky. The experiment was declared a complete success. The project had been funded by a Russian car company and a natural gas firm. Western experts had long doubted if it could be done.

THE PROBLEM OF BURAN

Even as Manakov and Poleschuk were illuminating the night skies of Europe and western Russia, the story of the Buran space shuttle was reaching its climax. Buran had made its first, highly successful maiden flight in November 1988. Strangely for such a huge programme, there was no blueprint of future missions for the new space shuttle. Rumours and announcements abounded during 1989–93 of forthcoming Buran and Energiya missions. There was an expectation that Buran would make a second, longer unmanned flight maybe a year later, before then making a first manned flight with a two-man crew. Plans were drawn up for Buran to fly up a series of Kvant modules called 37KB to Mir from 1993. Clearly, it would have a lead role in the bringing up of modules, supplies and large crews to Mir 2 later in the new decade.[3] In the meantime, however, there was no immediate role or need for the shuttle at a time of contracting finances. Although the Krystall module on the Mir space station carried a docking module designed to receive Buran, there was little Buran could do with Mir that other Soviet rockets and spacecraft did not already do.

Buran launch pad—idle, used only once

Buran still had some way to go before being man-rated: the electronics required upgrading, and life-support equipment had to be installed in the cabin.

In order to pave the way for a manned Buran mission and to gain essential in-flight experience, Igor Volk had already flown into orbit on Soyuz T-12 in 1984 and Anatoli Levchenko on Soyuz TM-4 in 1987. When Levchenko died less than a year after his space mission, it was deemed essential that his replacement, Rimantas Stankivicius, should have orbital experience before he flew Buran in orbit. Accordingly, he was assigned to Soyuz TM-8 in 1989 along with Mir cosmonauts Alexander Viktorenko and Alexander Balandin. At this stage, the first manned flight of Buran was set for the first quarter of 1992. However, when that mission was delayed due to the difficulties with getting the Krystall and Kvant modules aloft, he was dropped. The entire cosmonaut group wrote a letter to the Central Committee of the Communist Party of the Soviet Union asking for help, but to no avail.

Igor Volk and Rimantas Stankivicius would most likely have made the first Buran flight, whenever that took place, probably sometime beween 1992 and 1994. A plan was then put together for a mission called Soyuz 101 in which Buran would be launched unmanned for its second orbital mission. Once in orbit, cosmonauts

Anatoli Berezovoi, Alexander Ivanchenkov and Rimantas Stankivicius would take a Soyuz to rendezvous with Buran, fly it for several days and return to Soyuz, allowing Buran to return to its runway on automatic pilot. In a variation on this plan, Buran would also dock automatically with Mir. This mission was cancelled, though the actual Soyuz spacecraft in question later flew as Soyuz TM-16 in early 1993. Its docking with the Buran docking port on the Krystall module was the first time it had been so tested.[4]

Efforts continued to find a useful mission for Buran, one drawing showing Buran bringing up a small science module designed in the early 1980s. In April 1990, the Soviet space promotional agency Glavcosmos announced that the original Buran shuttle would be mothballed: it was simply too expensive to install avionics, life-support systems and fuel cells for manned operation. The first manned shuttle would be the second one built, called Baikal (after the famous Siberian lake). In October 1990 there were the first indications from Glavcosmos that the entire Energiya–Buran programme would be suspended, although in early 1993 there was still discussion of the possibility of a shuttle mission to visit Mir and dock on the Krystall docking port.

BURAN GROUNDED

All this came to nothing. In June 1993, the Council of Designers formally decided to shelve the Energiya–Buran programme. The funds available were only about 1% of what was required to prepare a mission and it was futile, they added, to carry on the pretence of preparing a flight that had no real chance of leaving the ground. The news was formally given on Radio 1 Moscow on 30 June. The following month, 52 directors of military complexes wrote a letter of protest at the closure of the shuttle programme. The cost of the programme had been R20bn, about €4.4bn. Plans for modules for Buran to deliver to orbit, variously called Tekhnologika, TelluraEKO and Ekologia, also fizzled. Russian Space Agency director Yuri Koptev formally ordered Buran to be mothballed that November.

Of the shuttle hardware, three flight models were fully built (Buran itself (1KI), its back-up (2KI) and Baikal (MT-004)). Eight others were in a various state of assembly or had been used for other purposes (ground testing, training). Baikal and 2KI remain in the fuelling hangar at Baikonour, accompanied by six complete Energiya sets and could, at least theoretically, be made man-ready should the need arise. Buran was bought by an Australian company headed by oceanographer astronaut Paul Scully Power and was brought to Sydney for display in the summer of 2000. The others found homes in museums and storage sheds. One was bought by a joint stock company headed by cosmonaut Gherman Titov and relocated in Moscow's Gorky Park where 48 visitors at a time pay R60,000 (about €20) for a two-hour multi-media space voyage in which they eat space food, travel through space and even fend off a meteor shower. Of the 20 cosmonauts who trained to fly Buran, only two made it into space and they flew on the venerable Soyuz cabin. Igor Volk never flew the shuttle to orbit and his colleague Rimantas Stankivicius died before he got the chance. The rest were disappointed.[5]

The assembly halls where Energiya and Buran engineers fussed over their charges were later converted to make the Russian–American International Space Station, and the Energiya–Buran launch pads are now being reclaimed by nature and it is no longer safe to walk nearby. Though the programme is cited as a monument to politics and waste, it is not the only rotting shuttle launch base. On the California coast, wild flowers, desert brush and bobcats are gradually overrunning the site where the US Air Force had once planned to send American military space shuttles into polar orbit, and built a huge facility to make it possible. A recent booklet on the Vandenberg Air Force Base made no mention of the facility's existence.

SHORTAGES, DELAYS

The Mir programme continued as normally as possible, with its six-month-long resident missions and visiting cosmonauts. The fourth Russian–French mission, lasting three weeks, took place in July 1993, when Jean-Pierre Hagnieré flew with Vasili Tsibliev and Alexander Serebrov on 1 July to replace Gennadiy Manakov and Alexander Poleschuk. Shortly before take-off they received a sharp reminder of the difficulties facing the new Russia: there was a power blackout at the launch pad. An hour after take-off, the electricity supply in all of the adjoining city of Leninsk failed.

Once in orbit, their mission had a busy start. In August, the station was peppered by tiny micrometeorite impacts from the Perseids meteor shower. In September, the cosmonauts spacewalked to erect a new truss on the outside of Mir. Called *rapana*, it was a 26-kg, 5-metre-long truss designed to carry antennæ, reflectors and solar gas turbines. The crew made five spacewalks and carried out materials processing experiments, astrophysical observations and observed the Earth.

YEAR-AND-A-QUARTER MISSION

The mission of Vasili Tsibliev and Alexander Serebrov lasted longer than anticipated, as the economic crisis hit home. A shortage of rocket motors prevented the launch of Soyuz TM-18, originally slated for November 1993. It was therefore delayed three months while they were finished. The new Soyuz did not get away until 8 January 1994 when Soyuz TM-18 carried up veteran cosmonaut Viktor Afanasayev, newcomer flight engineer Yuri Usachov, and veteran space doctor Valeri Poliakov. Five years after the record, year-long marathon by Vladimir Titov and Musa Manarov, Russian space doctors now felt it was possible to break the duration record again with a mission of 1½ years. Despite the parlous and declining state of the Russian space programme, it was now set for its most significant ever medical experiment. Chosen for the mission was 52-year-old Dr

Medical check in orbit

Valeri Poliakov, who had already spent 240 days on Mir between 1988 and 1989. He had spent 25 years of his life preparing cosmonauts for their missions and this was his chance to test out for himself the physiological and psychological preparations which his own institute, the Institute for Medical and Biological Problems, had made for previous space travellers.[6]

The delay over rocket engines which affected Soyuz TM-18 was repeated in a similar manner when preparations to launch Soyuz TM-19 began during the summer. This time the upcoming mission was postponed—though not for long—

due to a shortage of nosecone fairings. Soyuz TM-19 left Baikonour with Yuri Malenchenko, a 32-year-old air force pilot, and a second Kazakh cosmonaut, Talgat Musabayev, neither of whom had flown before. An all-novice crew was against the normal rules, but Musabayev's mission was part of a barter deal with Kazakhstan for the use of Baikonour cosmodrome (the Kazakhs were charged $150m). Afanasayev and Usachov returned eight days later, leaving Malenchenko on board with Musabayev and the long-flying Poliakov. Next on the manifest was the first of two missions scheduled in collaboration with the European Space Agency (ESA). Hitherto, the European missions with France, Austria and Germany had been with the individual, national space agencies and this was the first which involved the 19-member organization. ESA paid €45m for the two missions.

MALENCHENKO SAVES THE MISSION

A further reminder of the fragile state of the Russian space programme came at the end of August 1994 when Progress M-24 arrived alongside on 27 August with fresh supplies of food, water, air, fuel and experiments for the forthcoming first mission with the European Space Agency (ESA). Suddenly, the docking was aborted at a distance of 150 metres when the rendezvous system detected a misalignment. Mission controllers made a second attempt on 30 August, but the automatic control system on Progress again cancelled the manoeuvre and the two craft began to drift apart once more. Progress had fuel for only one more attempt. If it failed, Mir would have run out of food and the station would have to be abandoned in mid-September, with little chance that there would be resources to reoccupy it. A third, desperate attempt was made on 2 September. Station commander Yuri Malenchenko took over when Progress closed to 150 metres and guided the spacecraft in manually using a television monitor. All was well, and Progress maintained its 100% record of 67 successful consecutive dockings.

The first of the European Space Agency missions, *Euromir*, arrived on 6 October, when Soyuz TM-20 reached the station carrying mission commander Alexander Viktorenko, flight engineer Elena Kondakova and European cosmonaut Ulf Merbold, a German who had already flown on the American space shuttle. The three new cosmonauts could be seen streaking across dark European October night skies as a small fast pinprick of light chasing the much brighter, graceful Mir station. It lasted a full month. Merbold returned with docking hero Yuri Malenchenko and Kazakh Talgat Musabayev on 4 November. Before coming down, the three men tested the rendezvous and docking equipment by undocking from Mir, backing away to 190 metres and redocking, but they had a difficult return to Earth. Malenchenko recalled later that he had not buckled up during the re-entry—he thought he would do that when the parachutes emerged—and he was being kept busy enough looking after his colleagues Musabayev and Merbold. The gravity (g) forces built up to 3g, which was nothing for a jet fighter pilot, but after several months of weightlessness was almost impossible to bear and he could not move his arm to attach his seat belt. When the parachute emerged, the cabin

Progress launch: ignition

Progress launch: take-off

Progress launch: afterwards

Collaborative missions between USSR/Russia and European Space Agency or individual European Union national agencies

Date	Spacecraft	Cosmonaut	Country	Duration (days)	Mission title
Jun 1982	Soyuz T-6	Jean-Loup Chrétien	France	8	
Nov 1988	Soyuz TM-4	Jean-Loup Chrétien	France	30	Argatz
Oct 1991	Soyuz TM-13	Franz Viehböck	Austria	8	Austromir
Mar 1992	Soyuz TM-14	K-D Flade	Germany	8	Mir 92
Jul 1992	Soyuz TM-15	Michel Tognini	France	8	Antares
Jul 1993	Soyuz TM-17	J-P Hagniaré	France	10	Altair
Oct 1994	Soyuz TM-20	Ulf Merbold	ESA	8	*Euromir 1*
Sep 1995	Soyuz TM-22	Thomas Reiter	ESA	135	*Euromir 2*
Jul 1996	Soyuz TM-24	Claudie Deshays	France	8	Cassiopea
Feb 1997	Soyuz TM-25	Reinhold Ewald	Germany	21	Mir 97
Feb 1999	Soyuz TM-28	J-P Hagniaré	France	188	Perseus

In addition, Briton Helen Sharman flew to Mir in May 1991 (Project Juno).

swung so violently he was nearly thrown out of his seat and could barely see his shaking control panel in front of him. The wind caught the capsule as it came in to land and it bounced three times: 'like a blow to your head with a piece of firewood' he recalled afterwards. Musabayev distinguished himself during the mission to the extent that he was eventually made a full member of the Russian cosmonaut squad.

AS LONG AS A FLIGHT TO MARS: THE ACHIEVEMENT OF VALERI POLIAKOV

Elena Kondakova had now joined Poliakov in her own contribution to long-distance space flight. Just as Poliakov was trying to break the male long-duration record, she was trying to break the female long-duration record. Kondakova was only the third Russian woman in space. She was married to one of the leading mission controllers and Salyut space station veteran Valeri Ryumin, and there were whispers that he had used his position to get her the flight, and also when he was later booked on a shuttle flight. Indeed, her mission drew the most stinging rebuke from her predecessor woman cosmonaut, Svetlana Savitskaya, who described such a long flight for a woman as 'madness. What do you expect? She's married to Valeri Ryumin, who himself has made two half-year space journeys. He's the person who makes it rain and shine in Energiya. Once militantly opposed to women in space, he now wants his wife to join the cosmic legend.'[7]

The three-person Mir crew saw in the new year carrying out a range of biological, technological and Earth observation experiments, but also spending much of their time keeping the ageing space station going. Soyuz TM-21 was launched to join them on 14 March 1995, carrying novice mission commander Vladimir Dezhurov, veteran Gennadiy Strekhalov, and, for the very first time, an American astronaut, Dr

The view from Mir's window

Norman Thagard, under the new American–Russia programme for cooperation in space. Thagard's pioneering mission was to pave the way for a rotation of American visitors to Mir the following year. All six worked together on the station for six days. The cabin of Soyuz TM-20 re-entered the Earth's atmosphere on the morning of 22 March, coming down in snow 55 km northeast of Arkalyk, Kazakhstan. Kondakova had set a new women's duration record of 169 days, exactly twice as long as any American man had spent in space.

However, all eyes were on Poliakov who had circled the Earth for 438 days, seeing spring turn to summer, then to winter and a new spring begin again. He had orbited the Earth over 7,000 times, passing through 14,000 sunrises and sunsets. With no further long-duration missions planned for some time, the record was likely to stand well into the twenty-first century. Only doctors were permitted to approach him on his return, and journalists were held back. Openly flouting strong medical advice, he insisted on walking immediately, as if to prove it could be done. The three cosmonauts were flown back to Moscow later that day for a month's recuperation and debriefing. Poliakov, wrapped in a warm Russian winter anorak when he returned to Moscow, shrugged off worries about his condition. Doctors estimated that the bones of Valeri Poliakov would be completely recalcified within 15 months.

Endurance records set on board Soviet and Russian space stations, 1971–2000

Georgi Dobrovolski			
Vladislav Volkov			
Viktor Patsayev	Soyuz 11	June 1971	24 days
Alexei Gubarev			
Georgi Grechko	Soyuz 17	January 1975	30 days
Pyotr Klimuk			
Vitally Sevastianov	Soyuz 18B	May 1975	63 days
Yuri Romanenko			
Georgi Grechko	Soyuz 26	December 1977	96 days
Vladimir Kovalyonok			
Alexander Ivanchenkov	Soyuz 29	June 1978	139 days
Vladimir Lyakhov			
Valeri Ryumin	Soyuz 32	February 1979	175 days
Leonid Popov			
Valeri Ryumin	Soyuz 35	April 1980	185 days
Anatoli Berezovoi			
Valentin Lebedev	Soyuz T-5	May 1982	211 days
Leonid Kizim			
Vladimir Solovyov			
Oleg Atkov	Soyuz T-10	February 1984	237 days
Yuri Romanenko	Soyuz TM-2	February 1987	326 days
Vladimir Titov			
Musa Manarov	Soyuz TM-4	December 1987	366 days
Valeri Poliakov	Soyuz TM-18	January 1994	438 days

SPEKTR AND PRIRODA

As doctors on Earth fretted over the condition of Valeri Poliakov, the new Mir crew prepared to receive the fourth of the large specialized modules due to fly to Mir. Under the original Mir schedule, Spektr should have flown to Mir in 1989. Falling financial allocations had first delayed and then grounded Spektr and its companion module, Priroda. They would probably never have flown but for the new joint programme of American–Russian flights. Spektr would have carried remote-sensing and tracking equipment for the Russian defence ministry, called Oktava, part of the Soviet Union's star wars programme. In the event, this was removed and the module was fitted out with 700 kg of American biological equipment, 880 kg of laptops, an ergonometer and centrifuges for NASA, with living quarters for the arriving American astronauts and an additional two solar panels to power them. Priroda was similarly transmogrified.

Before Spektr arrived, however, there was work to do. On 19 April, Dezhurov and Strekhalov spring-launched a tiny reflecting German microsatellite through the Mir airlock. Prosaically called GFZ, the microsatellite was 21.5 cm in diameter, 20 kg in weight and carried 60 reflectors for geodetic experiments. Developed at a cost of

DM1m by the GeoForschungsZentrum of Potsdam, Berlin (hence its name), it was designed to test its reflectors on ground stations in Potsdam, Berlin and Cuba.

On 12 May 1995, Vladimir Dezhurov and Gennadiy Strekhalov carried out the first of three spacewalks to move 14-metre-long solar panels from Krystall to the Kvant 1 module. There was a danger that the panels would cause an obstruction to the planned docking between Mir and the American space shuttle later in the summer: they would continue to provide power from their new location further back on the space station. Norman Thagard remained inside, and for the first time a Russian space station became the exclusive responsibility of an American. Further spacewalks were carried out on 17, 22, 29 May and 2 June, the panels being moved with the aid of the station's manipulator arm. Thagard, a shuttle veteran, was struck by the different pace of life on Mir compared to his own previous shuttle experiences. On short shuttle flights, time was at a premium; on long-duration Mir flights, there was a normal working day with time off in the evenings, and 'weekends really were weekends'.

Spektr was eventually launched on 20 May on a seven-day rendezvous pattern, though it had fuel for 90 days of independent flight. It entered a 337 by 221 km, 89.8 minute, 51.7 degree orbit slightly ahead and above Mir and gradually descended to match the station's orbit. Despite the nerve-wracking attempts to dock the first three large modules with Mir, the new module Spektr docked flawlessly with the Mir manned orbital space station first time on 1 June. Spektr weighed 23.5 tonnes, including 1.7 tonnes of fuel, and had enough resources for three docking attempts. Spektr had an internal volume of $63 \, \text{m}^3$ and, in a marked addition to the station's electrical supplies, four solar panels with an area of $132 \, \text{m}^2$. It carried a small 2-metre manipulator arm, pelikan.

TEN YEARS IN ORBIT

On 20 February 1996, the space station Mir marked its tenth anniversary in orbit. Celebrating the event on board were three cosmonauts, Yuri Gidzenko and Sergei Avdeev of Russia and Thomas Reiter of Germany, who had arrived on Mir the previous September. Reiter was flying the second *Euromir* mission for the European Space Agency, obtaining long-duration experience for Europe's astronauts and at the same time providing crucial revenue for the Russian space programme. All three cosmonauts should have returned to Earth much sooner, but in a problem which was to be recurrent to the space programme in this period, their return was delayed due to the lack of availability of a Soyuz launcher to transport their replacement crew.

Major tasks faced the next long-stay crew to Mir—Yuri Onufrienko and Yuri Usachov—launched the day after Mir's anniversary, on 21 February 1996. Their first task was to welcome the first of a rotating set of American astronauts to stay on board Mir as part of an agreement with the United States to pave the way for the International Space Station. Space shuttle veteran and 53-year-old biochemist Shannon Lucid arrived on board Mir from the space shuttle Atlantis on 23 March. She settled in much more quickly than the previous year's visitor, Norman

Saying goodbye at the foot of the launch tower

Thagard. Her enthusiasm for the flight became apparent when she began to mail home accounts of life aboard Mir, her letters from orbit being posted on the internet for the benefit of journalists and spaceflight enthusiasts. They recounted everything from the joy of receiving a parcel of American food on board, reading who-dunnit novels in her spare time, watching the Priroda laboratory arrive ('like a gigantic silver bullet'), to sitting calmly in the airlock with her Russian colleagues as they collected their thoughts before spacewalking. Here, Shannon Lucid describes the arrival of a Progress supply craft:

I saw it first. There were big thunderstorms out in the Atlantic, with a brilliant display of lightening. The cities were strung out like Christmas lights all along the coast—and there was the Progress like a bright morning star skimming along the top. Suddenly, its brightness increased dramatically and Yuri said, 'the engine just fired'. Soon, it was close enough so that we could see the deployed solar arrays. To me, it looked like some alien insect headed toward us. All of a sudden I really did feel like I was in a cosmic outpost anxiously awaiting supplies—and really hoping that my family did remember to send me some books and candy!

The second task of the Onufrienko and Usachov crew was to receive the last of the large Mir modules, Priroda. The 21-tonne Priroda was launched on 23 April. Lacking solar panels, it took a short, three-day rendezvous profile, arriving on schedule despite a battery failure which took out half the module's power supply. Priroda carried an advanced set of scanners, radiometers, imagers, a French lidar (laser radar) and a volume of American research equipment. Mir was now a massive object in orbit—a core block, host to one astrophysical module, four 20-tonne modules its own size, freighter spacecraft, manned Soyuz spacecraft, with solar panels, cranes and arms extending in all directions. Well might arriving cosmonauts and astronauts gasp and call it 'the dragonfly'.

The missions of all three crew members were extended during the summer: Lucid's because of delays in launching the space shuttle due to problems with the solid rocket boosters, and those of her Russian colleagues because of continuing funding problems. The drama intensified on 24 July when the launch of the Progress M-32 supply ship was aborted only 45 seconds before lift-off when the launcher failed to pressurize properly, though it managed to lift-off a week later. A week before the next manned launch, crew commander Gennadiy Manakov was grounded with a heart irregularity. Both he and his colleague, Pavel Vinogradov, were pulled from the mission, to be replaced by Valeri Korzun and Alexander Kaleri who left for the space station on 17 August. Flying with them was France's first woman cosmonaut, Claudie André-Deshays, on a two-week visiting mission. She was a 39-year-old rheumatological doctor and CNES neuro-sensory physiologist who had beaten hundreds of men to become her country's first spacewoman and the 31st woman ever to fly in space. From 1990 she had taken part in the regular zero-gravity tests carried out for intending cosmonauts on board France's old Caravelle airline. Her main experiments were with a cargo of six lizards brought on board to test the effects of zero gravity.

Claudie André-Deshays

As the new Soyuz with Korzun, Kaleri and André-Deshays began to dock, Yuri Usachov went to the window in the Mir base block to watch. 'We managed to capture on video the firing of the attitude control thrusters. They superbly completed a standard profile: hovering turn, thruster burn and reverse turn.' As the new crew members came through the hatch, Yuri Usachov recalled how the station, used to its three occupants for the past number of months, suddenly became a noisy and crowded place and they partied till 2 am to welcome the new arrivals.[8] When they returned to Earth, Claudie André-Deshays flew back to Paris with her lizards and presented the results of her experiments to a symposium presided by the secretary of state for research.

Shannon Lucid's shuttle, further delayed by hurricanes in Florida, did not arrive until 19 September. By the time she returned to Earth the following week, she had spent six months in orbit, becoming the longest-flying woman astronaut from any country and the longest-flying American astronaut. NASA greeted her as a hero and President Clinton was on hand to welcome the returning crew. She was the star of the show and Bill Clinton read out a letter which Lucid had written as a young girl to her teacher expressing her interest in travelling to space—and her teacher's reply telling her not to be so silly.

Her position on Mir was taken by John Blaha, who was in turn replaced in January 1997 by Jerry Linenger. In early spring 1997, the outlook for Mir was upbeat. The station was complete and had never been working so well. The visiting missions and the joint programme of American missions guaranteed it a steady stream of work and, equally important, money. In Moscow, the Yuri Gagarin Cosmonaut Training Centre sent out its first call for new cosmonauts for some time. Two more cosmonaut groups were recruited. A training group was set up for the International Space Station. However, they soon received a reminder of how dangerous flying in space really was: Mir faced its worst crisis and the future of the space programme swung precariously in the balance.

SPRING, SUMMER '97: THE LONG, DESPERATE MIR CRISIS

The troubles began not long after the arrival on 12 February 1997 of the Soyuz TM-25 crew of Vasili Tsibliev and Alexander Lazutkin of Russia and Reinhold Ewald of Germany. They joined Alexander Kaleri, Valeri Korzun and Jerry Linenger to make a six-man crew. Before flying, veteran pilot Vasili Tsibliev had consulted his astrologer and her cards were ominous. She tried her best to warn him.

On 24 February, a flash fire broke out in the Kvant module in the station. The cause of the fire was an oxygen candle, which are small 20-cm-long, 8-cm-diameter objects inside a small oxygen generator and are used by the cosmonauts periodically to renew the air supply—a process which takes 20 minutes. However, instead of

Alexander Lazutkin, Vasili Tsibliev and Reinhold Ewald

releasing additional oxygen into the cabin, the candle on this occasion burst into flames and, acting like a blowtorch, discharged flame and thick smoke which filled the cabins of the station within less than a minute. The crew quickly donned masks and tried to extinguish the fire, but with no effect. The next minute was tumultuous as some cosmonauts raced through Mir to bring extinguishers and others tried to unload the retrorocket schedule from the computer printer. Eventually, after 90 seconds, though it must have seemed much longer, the fire burned out, but smoke continued to fill the station for the next six or seven minutes as the cosmonauts coughed and spluttered and the master alarms sounded. The problem in a space station is, of course, that smoke has nowhere to go—there are no windows to open— although the ventilation system will eventually remove it. Thankfully the smoke was not toxic. The smoke had even spread to the descent craft, making the use of Soyuz for evacuation problematic.

The station was over Africa at the time and once things had calmed down, the crew radioed in news of the incident to ground control over a crackly air-to-ground link. One of the NASA engineers in mission control noticed the sudden flurry of activity among the controllers and tried to find out what had happened. Eventually he picked up the word *pozhar*—the Russian word for fire—being mentioned several times.

This set the scene for a clash of cultures that was a perennial theme of the Shuttle–Mir operation. NASA felt that it was being side-lined, was kept out of the loop when key developments took place and was always the last to hear, which was simply not acceptable when one of its astronauts was on board. The Russians, for their part, took the view that they had the situation under control and could cope. They did not want NASA fussing or interfering. They knew how to run orbital space stations.

THE NEW RENDEZVOUS AND DOCKING SYSTEM, TORU

Alexander Kaleri, Valeri Korzun and the German Reinhold Ewald returned to Earth on 2 March, but then the problems multiplied. The Progress M-33 resupply craft, which had been flying free from the station in order to make available a docking port for Soyuz TM-26's arrival, failed to redock two days later. The reason for this attempted docking manoeuvre was that the guidance system on the Progress space-craft was in the process of replacement.

Hitherto, Progress had used an automatic docking system called *Kurs* made in the Ukraine. As witnessed by more than 70 consecutive and successful dockings, *Kurs* had proved to be highly reliable, but following Ukranian independence the manu-facturing company had raised the price of *Kurs* to a level which the Energiya corporation considered exorbitant. Now a new system called TORU was installed. Essentially, TORU was operated like a child's computer game: a television on Progress was relayed to the commander's station on Mir; this showed its approach to the station and gave a readout of range and direction; on board the space station, the commander operated a left–right forward-and-back joystick to fire

Progress approaching Mir

Progress's thrusters and bring the freighter gently in. TORU cost a fraction of the old *Kurs* system.

The first attempt to use the new TORU rendezvous system went badly wrong. Tsibliev commanded the Progress to come in from a distance of 7 km, whence it should reach the orbiting station in 15 minutes. At a distance of 5 km the television camera on Progress should have operated, beaming in to Tsibliev a picture of Mir. But it did not, which meant that Tsibliev was now flying blind. All he could do at this stage was watch out the window for Progress, try to steer it away and hope there was no collision. Spotting Progress from Mir was extremely difficult, since the windows were designed to look out sideways (much as in an aeroplane cabin or in a train) rather than forwards. The cosmonauts braced themselves for impact. In the event, Progress cruised past. At the time, ground control quoted a distance of 5 km, but subsequent accounts suggest it was dangerously close (200 metres). Now out of fuel, the supply ship was abandoned.

BAD TO WORSE

Days later the station's oxygen generators failed, meaning that the station was no longer able to generate fresh air and depended on canisters. Moreover, there was a glycol leak on the station which created the risk that the water supply had become poisoned. Then on 19 March, the station's omega attitude control system failed. Mir drifted out of alignment, the solar panels lost their bearings on the Sun and the gyrodines had to be shut down. Later that day, the cosmonauts got the station back on line. To complicate things even further, the Russians had, earlier in the year, lost the services of one of the Luch relay satellites, which meant that Mir was in contact with mission control for less than the full day.

By early April, the station's crew had only a week's air supply left. All now depended on the successful launch of a delayed freighter, Progress M-34. Taking off on 6 April, the resupply craft arrived two days later, amidst scenes of considerable relief. Progress brought up fresh air, supplies and repair equipment. The crew soon fixed the glycol leak and repaired the oxygen generation system. A sign that things were getting back to normal was a spacewalk by Jerry Linenger with Vasili Tsibliev. Not only was this the first time an American had spacewalked in a Russian space suit from a Russian space station, but it marked the introduction of a new, much improved spacewalking suit, the Orlan M with new joints and a top visor.

The next shuttle arrived on 17 May. Although the purpose of the mission was to replace Jerry Linenger with British-born American astronaut Michael Foale, the shuttle also ferried up 2 tonnes of dry cargo, a tonne of fresh water, a new oxygen generation system and a gyrodine. Scientific results and failed equipment were transferred to the shuttle for the return journey to Cape Canaveral. As the shuttle dropped away, cameras followed the Mir space station as it tracked across Russia, southeast Asia and the blues of the Pacific Ocean.

Michael Foale was a quite different personality from Jerry Linenger. The rapport between Linenger and his Russian colleagues had been less than ideal. For his part, Linenger had complained vigorously of what he considered to be the mismanagement of the Shuttle–Mir programme (not without good reason, it seems). He took the view that he was a resident researcher on Mir. He was there to get on with his experiments and that is what he did. The more gregarious Michael Foale worked well with his Russian colleagues from the very beginning, made a big effort with the Russian language, asking for and taking responsibility for non-scientific tasks on the station. He quickly won the respect of his Russian colleagues.

Va korabl!

Disaster struck on 25 June. Mission commander Vasili Tsibliev undocked the Progress M-34 freighter for a fresh rendezvous and redocking experiment using the TORU system. Station commander Vasili Tsibliev was again in charge and set to operate the TORU joystick. Later, his colleagues recalled how apprehensive he seemed to be before the manoeuvre. To make things worse subsequently, ground control (called TsUP in Russian) had not explained to NASA the purpose or significance of the manoeuvre, even though it had been the subject of some radio traffic over the previous several days. Western listeners of the Mir radio traffic knew of the manoeuvre, although they attributed no special importance to it, but NASA had not been informed about it at all.[9]

The event seems to have occurred like this. Progress was undocked to a safe distance the previous day and now Tsibliev commanded the freighter to come in. He asked his colleague, Alexander Lazutkin, to position himself near one of the windows to help him to spot the arriving Progress. This time, the television set did work. However, it was extraordinarily difficult for the Progress TV to pick out Mir from a distance against the speckled clouds of Earth underneath. First, Progress seemed to come in too slowly. So, Tsibliev fired its thrusters to make it come in more quickly. Tsibliev and Lazutkin used a set of squares on the screen to measure the distance (a square on the grid meant it was 5 km distant). Using a combination of the grid, a stopwatch and the television camera, Tsibliev steered Progress in. When it was 1 km out, or so he thought, Tsibliev applied a standard braking manoeuvre to slow Progress to a walking pace for the final approach. All this time, he had Lazutkin and Foale watching out. On the screen, Mir was much bigger now, filling four grids on the square. This time, Tsibliev was alarmed at its rapid rate of approach. He fired the thrusters repeatedly to slow the freighter, but to no avail. He and Lazutkin dashed from one window to the other, trying to spot Progress and enlisted the visiting astronaut Michael Foale in the effort. According to the plan, Progress should now have been 400 metres away.

But it was too late. Lazutkin at last spotted the Progress, not at 400 metres as in the worksheet but at 150 metres and closing rapidly. 'It's here already!' yelled Lazutkin. Progress careered into the Spektr module at some speed, bumped along its hull twice, crumpled its solar panels and drifted off to the side. The astronauts then felt a pop in their ears as the pressure in the station began to drop. The space station had been hit, was punctured, and air was hissing out. The master alarm at once rang out. The manual suggested that at most they would have 18 minutes of air left in the station: they would have to either evacuate the station within that time, or seal the leak or they would be dead. Tsibliev yelled to Foale: *Va korabl! Get into the Soyuz lifeboat!*

Foale pushed himself quickly down the tunnel, into the node, into Soyuz at the far end. Once there, he removed the hoses and cables going from the node into Soyuz, in order to prepare it for emergency descent. Michael Foale was soon to realize that it was Spektr that had taken the hit. Spektr was, after all, his module, where he slept and carried out his experiments. According to the book, it was simple enough to close the hatch between Spektr and the node, but the hatchway was full of cables and ducts. They disconnected some, and searched for the switching board to disconnect the rest, but at that rate it would take forever. So Lazutkin and Foale used a knife to cut the cables, with sparks flying, and tried to close the hatch between the node and the module. It would not close—because the escaping air was pulling it outward. However, the two men found a hatch cover on their side which they jammed in its place from the node side, the escaping air sealing it in. The leak was now on the other side of the hatch and they were saved. The job took 14 minutes and the air stopped venting at once. Pressure had fallen from 760 to 693 mm.

Disaster had been averted, but a whole series of knock-on problems were only just beginning. Cutting the cables into Spektr meant that the main part of Mir lost all the solar electric power coming from Spektr. This was considerable, for Spektr had four

Only hope of return—the Soyuz cabin

large panels which supplied 40% of Mir's energy requirements. Tsibliev, Lazutkin and Foale powered down Mir's equipment, abandoned scientific work for the time being and learned to live by torchlight. Experiments and non-essential equipment were powered down.

SILENT, DYING STATION

Gradually, they tried to restore the situation. By 27 June, the pressure on the station was back to 770 mm. The errant Progress M-34 was moved to a safe distance 25 km from Mir and de-orbited. It is possible that the collision was caused, in part, because

Progress was full of rubbish. This had disturbed its equilibrium and given it more mass than had been allowed for either by ground controllers or the hapless Tsibliev.

On the next communication pass, Tsibliev related what had happened to mission control. There, the man in regular contact with the crew was Vladimir Solovyov, who had been on the very first crew to Mir. Chaos broke out. Once again, the American team in Moscow was making the middle-of-the-night phone calls to Houston to tell NASA that an American astronaut had narrowly survived disaster. Solovyov did his best to calm the situation and at once put together a plan to save the station. The next Progress mission, M-35, was delayed. Standard supplies were unpacked: instead, additional cables, repair items and new personal effects for Michael Foale were stacked on board.

Just as they were beginning to return to normal, the station drifted out of alignment from the Sun and lost all its remaining solar power on 3 July. Its gyrodines powered down. Though stable, Mir was now drifting helplessly in Earth orbit, unable to lock on to the Sun and acquire its electricity-giving powers. The ventilation system was silent. The normal clatter and hum of the orbiting station was replaced by dead calm. With vents turned off, Mir became a quiet station, an eerily quiet station, like a ship drifting in the horse latitudes of old.

This was not a situation foreseen by the manual; indeed, Mir did not have a manual. American astronauts training to fly on Mir were shocked by the lack of written documentation and how knowledge of Mir and its systems was passed down the line orally. On Michael Foale's initiative, the crew climbed into the Soyuz cabin. He used its thrusters, puny though they were compared to the mass of the station as a whole, to turn the panels of the station towards the Sun. This was a risky manoeuvre, which was by no means guaranteed to succeed, and used up scarce Soyuz fuel supplies that they would need later for their return to Earth. Whatever he did, it was successful and Mir's panels were able to acquire sufficient sunlight to get a flicker of power back into the system. They locked on and it took a full day to power the station up again. Foale had used up 70 kg of Soyuz's fuel reserves.

SAVING MIR

The Mir crash hit the world headlines at once. Indeed, Mir's calamities received more press coverage in the next week than the sum of all that the mission had attracted in the previous 11 more successful years. NASA responded with a mixture of worry and anger. NASA had not been advised in advance of the Progress manoeuvre and would probably have objected to its astronaut being put in danger in this way. The Progress crash was on the Internet before most people in NASA knew. Mir's critics came into the open, demanding that if Mir not be abandoned then NASA should at least take its astronauts off before one was killed.

The collision and fire brought differences between the Russian and American programmes and their respective approaches and temperaments into sharp relief. NASA worried that its astronauts were forced to take unnecessary risks. Ironically, three years earlier, European astronaut Ulf Merbold on his return to Earth had drawn attention to the way in which the hatches in Mir were always kept open and how cables snaked all around the station. He noted that 'they did not seem concerned about this. Apparently their view is that the risk of fire or depress is so low that the need to use the hatches for cables was of greater priority than the need for a quick closure.' The Russians for their part accepted that there were risks and problems and that they would continue to deal with them. For a nation reared on the 900-day siege of Leningrad and the horrors of the Great Patriotic War, what was fighting the occasional fire?

Progress M-35, unloaded and now reloaded, left Baikonour for the station on 5 July. The supply craft arrived smoothly alongside at 7 a.m. on 7 July and docked safely. The main cargo was equipment that would enable the cosmonauts to reconnect the cables between the node and Spektr, thereby restoring a certain amount of power from the undamaged parts of Spektr's panels. This would require an internal spacewalk. Originally, the Americans had classified spacewalks as either Extra Vehicular Activity (EVA—outside) or Intra Vehicular Activity (IVA—inside). Although the classification had existed for over 30 years, hardly anyone had ever done intravehicular activity before. The cosmonauts would have to work in a very confined space, fully suited, in vacuum, by torchlight—an intimidating prospect. If anything, the risk of a fatal tear was greater than in open space.

Just as they were preparing for the IVA, fresh setbacks struck. On 16 July, Alexander Lazutkin, in the course of normal operations controlling the station, accidentally pulled out the main computer cable in the guidance system. This sparked another master alarm and, worse, shut the system's navigation system down, causing Mir to drift out of alignment and lose power. This time the cosmonauts knew what to do and got Mir back on line again within a day. Once again they had to retreat to the Soyuz, explain the situation to ground control on Soyuz's radio and use the Soyuz thrusters.

HEART ATTACK

By the time the situation was restored, Tsibliev was to suffer a personal setback. As he ran his daily exercises, the doctors at ground control detected an irregularity in his heartbeat. Quite simply, he was suffering the strain of the events of the previous weeks. Ground control decided that it was too risky for him to carry out the upcoming IVA and that this should be delayed until the next crew arrived on board—a decision which greatly upset Tsibliev, but was probably the only one that the doctors could take.

Meanwhile, back in Moscow, the assignment for the crew for the next mission was changed. First, the third member of the crew, Frenchman Leopold Eyharts, was taken off. Given the problems on Mir and the lack of sufficient electrical energy, it was unlikely that he would have been unable to carry out any useful scientific research. His place was taken by much needed repair equipment. The two Russian cosmonauts slated for the next mission, Anatoli Soloviev and Pavel Vinogradov, were assigned to the hydrotank in Star Town for intensive training for the IVA.

Mir commander Vasili Tsibliev (left) with Reinhold Ewald

Anatoli Soloviev was one of Russia's most experienced cosmonauts and no one would have been a better choice to lead the new mission.

After so many difficulties, the next stage of the mission went much more smoothly. Soyuz TM-26 was launched on 5 August, the crew arriving at Mir two days later. Tsibliev and Lazutkin must have been glad to see them. After a week, Tsibliev and Lazutkin boarded their own Soyuz TM-25 for their return to Earth. They had spent 185 days on Mir in what must be rated as the most perilous space mission ever undertaken.

Not that the problems were over. As they came down, the landing rocket failed to fire on touchdown, giving them an unusually bumpy landing. Whisked away to Moscow for debriefing, neither Energiya nor ground control seemed to show much sympathy for them and appeared to hold them responsible for the collision. Michael Foale, who had actually been there, took a different view and was visibly upset when he learned of the shabby treatment of his colleagues.

INTRAVEHICULAR ACTIVITY

Michael Foale was still on board Mir, and the American space shuttle that was to bring him home was not due to arrive for another few weeks. The long-awaited IVA eventually took place on 22 August. Pavel Vinogradov, on his first space mission, successfully entered Spektr and managed to reconnect the cables between the module and Mir's node. This was not simply a matter of reconnecting wires. The pressure in Spektr was still zero, and as there was no prospect of it being repressurized, the wires had to be reconnected to both sides of an electrical plate on the hatch which would then be resealed. Struggling like miners at the coal face, they inserted the new plate, and new electrical current flowed from Spektr into Mir. The cosmonauts were able to get their ventilation systems going again, and Mir was humming back to life. On 6 September, the Russian Anatoli Soloviev and the American Michael Foale made a 6-hour spacewalk alongside the hull of Spektr in an attempt to find the point of the leak that had nearly killed the crew. They were unable to find it, suggesting that the leak must have been quite small, though potentially deadly. Using the Strela crane on the outside of Mir, they removed damaged insulation on Spektr and repositioned its solar wings to better attract the sunlight and thereby increase the flow of electricity.

Mir suffered its third power-out on 9 September with a computer failure, prompting ground control to prepare new computer parts for the next mission. These were hastily flown to the United States for delivery to Mir by the space shuttle, which was nearly ready to go. The American space shuttle Atlantis was duly launched from Cape Canaveral on 25 September, bound for Mir. The original purpose of the mission had been to bring Michael Foale back and replace him with David Wolf. However, because of the strength of objections in the Congress to continued American participation on Mir, NASA chief Dan Goldin delayed a decision on flying up David Wolf until the day before his mission, and not until he had received a safety report from astronaut Tom Stafford.

RECONNECTED

David Wolf was not disappointed and arrived on Mir on 27 September. Atlantis took over responsibility for the orientation of the space station while Mir's unreliable computer was stripped out and new parts inserted. Hopefully, that would be the end of the orientation problems. Even as they did so, congressional hearings were taking place in Washington, DC. Astronaut Frank Culbertson was sent in to bat for NASA, but apart from him, comment on Mir was harshly critical, with portentous warnings than an American would soon die on the orbital wreck. The United States had withdrawn from the monkeys programme (Bion) so they should give their astronauts equal consideration and withdraw from this one too.[10]

Also on board the shuttle was Russian cosmonaut and year-long veteran Vladimir Titov. On 1 October, Titov made an EVA with American astronaut Scott Parazynski to retrieve experimental packages which had been left on the outside of the space station. On 4 October, the Americans and Russians together celebrated the 40th anniversary of the launching of Sputnik 1 in 1957. Atlantis undocked from Mir, flew around the station to inspect and film the damage done by the rogue Progress and headed back to Earth, delivering Michael Foale back to his mightily relieved

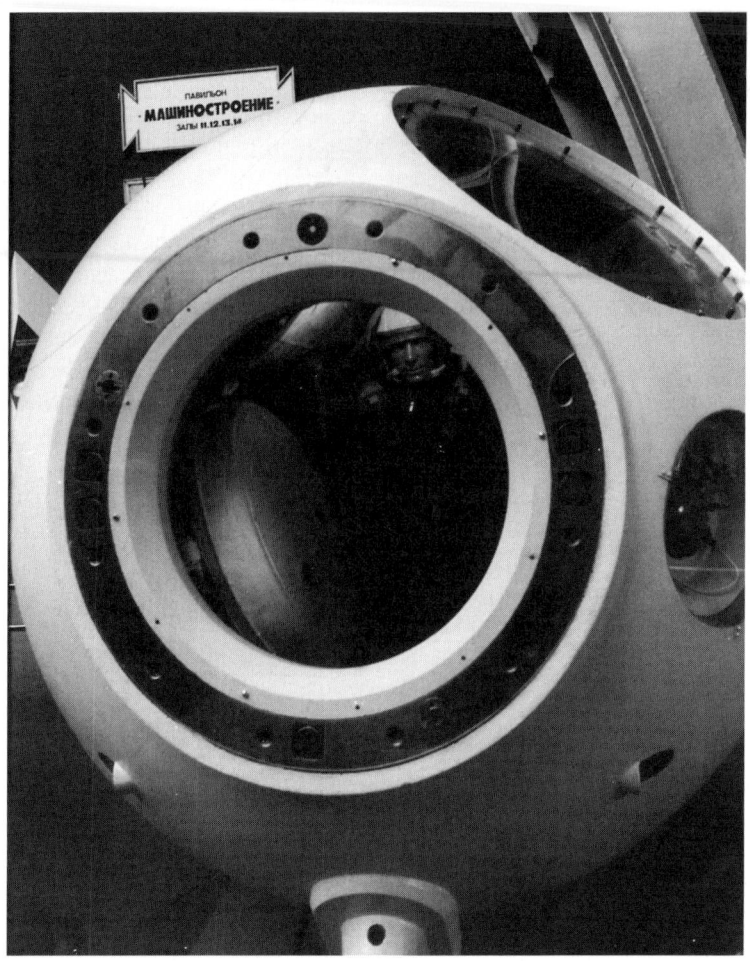

Soyuz's cramped cabin

family. Although he was to make several space journeys, this was one which would not quickly dim in his memory.

Things began to settle down on Mir and, once lives were no longer in danger, faded from the front pages of the world's press. Back on the ground, the causes of that summer's near-disaster were evaluated. The American conclusion, supported eventually by the Russians, was that Tsibliev had been asked to perform a series of hastily-planned manoeuvres for which he had not been properly trained. Several cosmonauts were invited to perform a similar docking manoeuvre on the simulator in Star City: all crashed. In the event, Tsibliev and Lazutkin received their pay that had been stopped and their post-flight medals that had been due to them. They also received the highest medal for fire-fighting, along with complementary helmets. In addition, Michael Foale received a citation for having saved the Mir station—a reward he more than deserved and the one thing that most made him feel it had all been worth while.

A new Progress arrived on 8 October. On 20 October, while Dave Wolf remained in the Soyuz TM-26, Anatoli Soloviev and Pavel Vinogradov made a second IVA in Spektr, reconnecting more cables from the module to the space station. They made further spacewalks on 3 and 6 November, this time outside (EVAs), to reconfigure Mir's solar panels. Mir was now back to 85% of its pre-June levels of power.

Highlight of the 3 November spacewalk was the launch of Sputnik 1! Anatoli Soloviev held in his hands a working model of Sputnik 1, built by Russian and French schoolchildren, transmitting on the same frequencies as the original Sputnik and delivered to Mir by the Progress to mark the 40th anniversary. With a sweep of his arm, he hurled the gleaming silver ball, aerials trailing, into a separate path in orbit. In honour of the occasion, it was named Sputnik 40.

Some earlier problems returned to Mir towards the end of the year. Despite the new computer, there was yet another power-out on 14 November. The station was on the sunny side of the Earth and temperatures rose to 30 °C before it could be rectified. The next week, there was a minor leak in the airlock hatch, so a new sealant was ordered up for delivery by the next Progress. However, on 22 November, the computer crashed again.

SOLOVIEV, SPACEWALKER

Progress M-37, arriving in December, brought up a small fly-around robotic space-craft called Inspektor. The aim was to fly the German-made Inspektor at close range around the station to investigate the collision damage and take continuous high-precision television pictures. Inspecktor was a robot that was prism-shaped, 930 mm long and 560 mm wide, weighed 72 kg, with three solar panels, batteries, antennae, gyros, video cameras and tiny nitrogen gas jets for manoeuvring. However, Inspektor lost its star lock and soon posed a new collision threat to the station! Accordingly, Progress fired its engines to move Mir out of Inspektor's way.

Anatoli Soloviev

In January 1998, Anatoli Soloviev and Pavel Vinogradov spacewalked to remove scientific equipment from Krystall and secure the Kvant 2 airlock. New experimental packages were installed by Anatoli Soloviev and Dave Wolf on 16 January. This was Soloviev's 16th spacewalk, making him the most experienced spacewalker in history. These successful excursions paved the way for the arrival of the final American astronaut on Mir on 24 January. The space shuttle Endeavour came in, bringing up Andy Thomas to replace Dave Wolf (the shuttle also carried cosmonaut Salizan Sharipov). The shuttle brought 3,350 kg of supplies, transferred to Mir a new refrigerator, compressor, computer, gyrodine and 700 kg of water, and brought back 1,500 kg of unwanted cargo.

Only 19 minutes before the shuttle undocked from Mir, a new Soyuz was leaving at Baikonour. Leopold Eyharts, who had been bumped from his mission the previous summer, was eventually on his way, accompanied by veterans Talgat Musabayev and Nikolai Budarin on Soyuz TM-27. The three men reached Mir on 31 January, just as the returning Endeavour was coming in to land in Florida. It was a sign of the restored stability on Mir that this pace of activity excited little media interest. Mir had become routine again. Anatoli Soloviev, Pavel Vinogradov and Leopold Eyharts returned to Earth near Arkalyk on 19 February, leaving Talgat Musabayev, Nikolai Budarin and Andy Thomas manning the outpost.

Andy Thomas moved into the Priroda module (the now abandoned Spektr had been the American outpost until the collision). In advance, NASA had sent up music cassettes, CDs and books for him for his off-duty hours. It was from there that he was able to watch Soloviev, Vinogradov and Eyharts undock from Mir, drift out of sight and then fire their retrorockets to descend to Earth.

WINTRY RETURN

Landing conditions for Soloviev, Vinogradov and Eyharts was as severe as any midwinter landing. The temperature was $-30\,°C$ and blizzards hit the landing site as the cosmonauts re-entered the Earth's atmosphere. For fear of a collision, only one Mil-8 helicopter was dispatched to recover the crew. Due to the poor weather, Kazakh air control was unable to hold a radar image of the descending Soyuz TM-26, but thankfully Mil captain Anatoli Mikhalishev picked them up on his radar as they came through 4,500 metres. Soyuz landed in snow, the Mil close beside an instant later. Normally, the rescue team erects an inflatable field hospital for the crew's medical examinations, but conditions were so poor that Soloviev, Vinogradov and Eyharts were taken by stretcher directly to the floor of the Mil, which had kept its engines revving to stop them freezing in the extreme cold. The medical rescuers, led by Oleg Fyodorov and his assistants in white coats, had to give peremptory physical examinations as the helicopter whisked them away to the nearest landing point at Kustanai.[11]

Musabayev and Budarin made an IVA on 3 March in yet another attempt to fix the Kvant 2 airlock, which was continuing to give trouble. One of its 10 bolts was malfunctioning and causing a small leak. All the cosmonauts managed to do was break three wrenches in an attempt to fix the errant hatch bolt. They asked for new, tougher wrenches to be added to the manifest of the next Progress. These duly arrived with Progress M-38 on 17 March.

Almost unnoticed, the Progress system had begun to work with the safety and reliability with which it was normally associated. With the impending arrival of Soyuz TM-27, all of Mir's docking ports were in use, so Progress M-37 had been undocked for 25 days of independent flight and then brought back to the station once a new port was free. When Progress M-38 arrived on 17 March, Talgat Musabayev steered it in using the new TORU system and there were no problems.

The crew then embarked on what was to be the last intensive period of space-walking on Mir. Talgat Musabayev and Nikolai Budarin went down the hull of Mir on 1 and 6 April. They installed new experiments on the outside of the station and attempted to readjust the solar panels on Spektr to derive more power from them. On 11 April, they took down the old VDU station-keeping engine block, pushing it away to decay in a slightly lower orbit. This was a special box-shaped engine attached to a long boom which regularly fired small engines to stabilize the station. On 17 April, they clambered down to the Progress, on whose hull was attached a new 700-kg VDU block. On 22 April, they used the boom on the outside of Mir to lift the VDU from Progress and then attach it to its proper place on the sofora girder. They then connected up the VDU to the control system of Mir with a number of electrical and command cables. Each spacewalk had lasted in the order of 6 hours.

The American space shuttle took off in blazing June heat from Cape Canaveral on 2 June 1998 for its final visit to the Mir space station. The purpose of the flight was to take off the last American on board, Andy Thomas, bring up 2 tonnes of water and other supplies to Mir and bring down over a tonne of American equipment in use on Mir. On board the shuttle was veteran Salyut 6 cosmonaut Valeri Ryumin. The four-day visit was concluded with a final fly around Mir by Discovery. As it circled Mir, Talgat Musabayev and Nikolai Budarin pumped coloured nitrogen acetone gas into Spektr in the hope that leaking gas, escaping the module, might be spotted by the shuttle crew and give away the leak's location. They saw nothing.

WINDING MIR DOWN

Mir continued in operation, its permanent crew down to two for the first time in three years. At this stage, the Russian Space Agency began to discuss the termination of the Mir operation. The end of American financing coincided with a decline of interest in Western Europe in further manned missions to Mir. German government spending was extremely restrained in the late 1990s and space spending declined sharply. Although one final French mission was still being considered, the new French minister responsible for spaceflight, Claude Allegre, was opposed to a national manned spaceflight programme. This was despite the fact that France had been extraordinarily successful in having its spationautes fly both on Russian

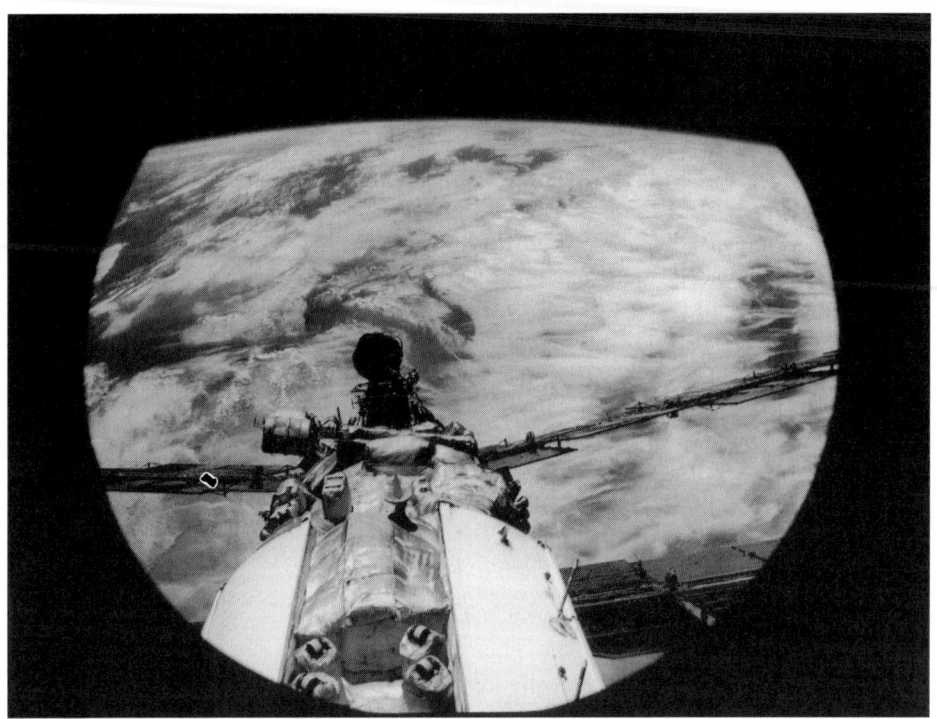

Spacewalker's view down the hull of Mir

space stations and the American shuttle, with seven spationautes flying to Salyut 7
and then Mir. Allegre ended the Mir missions, transferred his remaining astronauts
into the European astronaut team in Cologne and redirected the French space
programme around applications missions and a probe to take samples from Mars.

With the funding drying up and the delayed international space station drawing
nearer, the de-orbiting of Mir came onto the agenda. When Progress M-39 arrived at
Mir on 16 May 1998, the space agency announced that one of its objectives was to
fire its engines to lower Mir's orbit from 400 to 250 km, preparatory to final de-
orbiting in 1999.

Ironically, although the situation on Mir had now completely stabilized (radio
traffic intercepts during the summer suggested a lot of experimental work was going
on), morale on board had plummeted. The cosmonauts knew that, down on the
ground, the money was simply running out. In July 1998, the Energiya Corporation
made a special appeal to the prime minister (then Sergei Kiriyenko) to provide
emergency funding to keep Mir aloft. The next Soyuz mission, the one to replace
Musabayev and Budarin, was then delayed. Instead of two separate missions in
1999—one Slovak and one French—the two would be merged. A date was even
given for the de-orbiting of Mir: June 1999. In a strange move, President Boris
Yeltsin proposed that his adviser on spaceflight, Yuri Baturin, join the next Mir

crew and report back to him on the condition of the station. Although then a 48-year-old lawyer on military reform, Baturin had qualified as a physicist, had worked in RKK Energiya in the 1970s and even applied for the cosmonaut squad then, though he had been rejected on grounds of poor eyesight.

Things went from bad to worse. September 1998 saw the worst financial crisis in the Russian economy since the transition to capitalism had begun. The ruble collapsed, there was a fresh bout of inflation and bank queues lengthened as people deperately tried to protect what was left of their savings and investments. Kiriyenko was replaced by Primakov, not that it made much difference. The Russian Space Agency, for its part, floated off a further 13% of the shares of the Energiya Corporation, delayed the international space station programme again, and slowed down the production line of Soyuz, Progress and rockets to the bare minimum.

A new crew had joined Mir that August. Gennadiy Padalka, Sergei Avdeev and adviser Baturin reached Mir on Soyuz TM-28 on 13 August. Dispelling notions that presidential advisers might not have the right stuff, the unlikely Yuri Baturin relished his new assignment and quickly mastered the Soyuz systems. He even had a schedule of 20 experiments to perform during his eight-day visit to Mir and duly advised his president on his return. He blended into the cosmonaut corps effortlessly, was awarded his flight engineer's certificate in 2000 and was assigned to a new mission the following year.

WHO WILL BE THE LAST TO MIR?

Back on Earth, Baturin made an unequivocal appeal for more money to keep Mir flying. He made little impact and the Russian Space Agency announced that the next crew to Mir would be its last. This was an assignment which no cosmonaut fought for, no one wanting to be on 'the crew that killed off Mir'. Veteran Viktor Afana-sayev was duly given command of the final assignment. His choice of flight engineer was already made for him. Because the Slovak–French mission was merged, there was no return seat available for the existing Mir flight engineer Sergei Avdeev, who would just have to stay on to the end, much as Sergei Krikalev had done seven years earlier. For the same reason, the last Frenchman, Jean-Pierre Hagnieré, who was originally to fly only 35 days, had his mission extended from 35 days to 99 days at no extra cost to the French, who had paid €20m for his mission but saw no reason to pay more. The Slovak mission arose from somewhat more complicated circum-stances. The Slovaks had felt badly treated when a Czech had represented Czecho-slovakia on Salyut 6 in 1978, and the Slovaks had always wanted to have their own cosmonaut. With Russia owing the Slovak republic a substantial amount of debt arising from the break-up of the Eastern bloc in 1989, the Slovaks at last saw their chance to cash in their favours, or $20m of that debt.

Routine operations continued on Mir, despite its approaching demise. Progress freighters arrived and left. Gennadiy Padalka and Sergei Avdeev made an IVA to improve further the electrical supply from Spektr. The crew carried out medical, astrophysical and geophysical experiments, using the Alissa lidar on the side of

Priroda. Progress M-40 arrived on 27 October, bringing fresh supplies, new year gifts in advance, a new *Znamia* space sail test and experiments for the Slovak–French mission scheduled for the spring. On 10 November, Padalka and Avdeev space-walked for 6 hours to put up a micrometeoroid shield to record strikes from the forthcoming Leonid meteor shower and launched another amateur hand-held Sputnik working model into orbit, Sputnik 41. On 24 December, they were ordered to fire the engines of Progress, and, rather than drop the orbit of Mir as had been expected, to raise it instead, albeit by a modest 14 km. This was the first indication that plans to crash Mir into the ocean in a year's time might not be so definite after all. On 1 January, the cosmonauts celebrated the new year in orbit, the ninth consecutive year that orbiting cosmonauts had welcomed in the new year. Several weeks later, new prime minister Yevgeni Primakov signed a decree extending Mir's life to 2002 if a foreign investor could be found. The government had apparently taken a semi-neutral position on Mir—not seeking its termination, but not being prepared to put money in either, yet prepared to indicate one possible way forward.

In February, Padalka began to prepare for his return to Earth (Avdeev was to stay on). On 4 February they had a disappointment when they undocked Progress M-40, carrying the second solar sail experiment. This time, the *Znamia* sail was 25 metres in diameter, a span sure enough to really brighten up the night sky. As before, the idea was for the sail to pop open after the Progress undocked—except that the sail snagged in one of the Progress's antenna and the experiment had to be abandoned.

The crew of Soyuz TM-29 was duly launched on a white rocket on 20 February, exactly 13 years to the day since Mir itself lifted off. On board were 50-year-old veterans Viktor Afanasayev and Frenchman Jean-Pierre Hagnieré, accompanied by 34-year-old Slovak research engineer and fighter pilot, Ivan Bella. They entered an orbit of 195 to 235 km, one which they gradually raised as they climbed to Mir's altitude of 360 km. They reached Mir on the 22 February. The old Soyuz TM-28

Last crew to Mir

cabin returned on 28 February with Ivan Bella and Gennadiy Padalka, the latter having spent 201 days on board Mir. Bella had brought with him a package of six experiments, including one to hatch out quail eggs in orbit.

What was originally advertised as the penultimate Progress mission to Mir took off on 2 April. Besides its normal cargo of fuel, food and water, Progress M-41 carried 18 lizards for a French experiment called Genesis. Viktor Afanasayev and Jean-Pierre Hagnieré made the first of two spacewalks together on 16 April when they installed experiments on the outside of the hull and retrieved the Leonid collector left outside the previous November. They launched the third mini-sputnik: Sputnik 42. Viktor Afanasayev and Sergei Avdeev made a further two spacewalks on 23 and 27 July. They erected a Georgian-built antenna called Reflektor, brought in exobiology experiments and installed an atmospheric measuring unit called Indikator. The crew observed that summer's solar eclipse—a ball-shaped cloud of darkness crossing northwest Europe across to Bulgaria. Progress M-42 arrived in July with computers to control Mir in automatic mode.

'GRIEF IN OUR HEARTS'

In June, the Russian Space Agency confirmed that the current crew would be the last on Mir. The back-up crew, which had been training in the hope of getting a mission to Mir in the autumn, was stood down. The station would be put in automatic mode and de-orbited in February 2000. Thus it was that at 6 p.m. on 27 August Viktor Afanasayev closed the hatch on Mir for the last time. 'We have grief in our hearts' he told television viewers as he did so. The crew collected in as much as they could of scientific results and equipment. They programmed the computer to run the complex without a human presence. They closed off the fans and the ventilators and the station fell silent. Separation took place at 9.17 p.m. and Mir could be seen receding in the distance over Petropavlovsk. Soyuz TM-29 came down two orbits later, 80 km east of Arkalyk. The solid rockets fired, cushioning the final descent but also setting some steppe grass on fire. The cabin turned over, making Jean-Pierre Hagnieré quite sick, but he was pulled out in 15 minutes. He was given water and quickly recovered and was soon greeted by fellow cosmonaut Claudie André-Deshays. Sergei Avdeev had been in orbit 389 days, or a total of 742 days in three missions. Mir continued to circle the Earth between 353 and 359 km, a level gradually falling. In December, the station was taken out of hibernation for computer and control tests. Progress M-42 fired its engines to raise its orbit from time to time.

The return of Afanasayev, Avdeev and Hagnieré brought almost ten years continuous occupation of the station to an end. In a total of 135 missions, 103 individuals had visited the station; these comprised 44 Americans, 41 Russians and 18 others. Of the 11,762 days spent on board the station, 10,726 were Russian (87%), 981 were American (8%) and 565 were others (5%).

MIRCORP RIDES TO THE RESCUE

In a last desperate effort to raise funds to keep Mir aloft, cosmonauts and engineers made a public appeal for funds. Hard-pressed Russian citizens contributed what they could, but the total amount collected was only 485,000 rubles, about $15,000, enough to keep Mir going for only a few days.

It was all over. Or so it seemed until 20 January 2000 when Russian Space Agency officials confounded the critics and announced that a foreign investor had been found for Mir. Such rumours had abounded before and various names had been mentioned, from Rupert Murdoch to the government of the People's Republic of China. The unlikely saviour of the Mir station was an American technology investor called Walt Anderson who ran a company called Golden Appel registered in the British Virgin Islands. He decided to invest $20m in Mir and set up a new company for the operation, MirCorp, registered in the Netherlands. Walt Anderson and his representatives talked to Energiya and they devised a plan for a 45-day mission which would refit Mir, report on its condition, and prepare it for new operations while further investors were found. MirCorp spoke of how Mir could be a base for new experiments in zero gravity and even a space hotel.

Talk aside, Russia asked for a downpayment of $7m. Unlike previous promises of money, this time it actually arrived. Energiya responded quickly by assigning two Progress freighters to the MirCorp project and ordered the Soyuz TM-29 back-up crew quickly back into training for a new mission. To the fury of the Americans, they sequestered the freighters and the Soyuz from the production line of spacecraft destined for the International Space Station. The Russians just shrugged their shoulders: 'We'll make good the production line later', they said.

MIR FLIES HIGH

No sooner was the cash register ringing with the MirCorp cash than Progress M-42 fired its engines again to raise the orbit, to 301–324 km. On 2 February, the new Progress, designated M1-1, took off for Mir with supplies of water, food and fuel for the station's new crew. This new version of Progress, the third of its kind, was able to carry 220 kg more cargo than its predecessor.

Sergei Zalotin and Alexander Kaleri resumed training. They were joined by an actor, Vladimir Steklov, who was to join the crew to make a film and who had beaten off competition from two actresses, 22-year-old Natalia Gromushkina and Olga Kabo. Under the director of film-maker Yuri Kara, he would make the first space drama. The film under consideration was called *The Mark of Cassandra—The Last Journey*, by Kyrgystan-born Chingiz Aitmatov about—appropriately enough—the efforts of a Russian space commander to save a doomed orbital station. The story line tells of a renegade cosmonaut determined to stay on board an ageing space station until the end of its days—until ground control decide on the ultimate means to lure him back to Earth: send up a woman cosmonaut! However,

Alexander Kaleri

although Steklov surprised everyone by passing all his training, Kara's film backers failed to deliver the money. He was stood down.

With the money from MirCorp, Sergei Zalotin and Alexander Kaleri duly took off from Baikonour on 6 April 2000 on Soyuz TM-30 to recommission the Mir station on a mission set to last 45 days or more. On arrival at the orbiting station, which had now completed 14 years aloft, their first task was to track down a number of troublesome air leaks. Pressure in the station had fallen and they burned oxygen candles to bring it up from 628 mm. Next they again fired the Progress engines to raise the orbit of the complex. The recommissioning of Mir took place just as the Sun was reaching the busiest point in its 11-year cycle. At solar maximum, the Sun's rays had the greatest effect in agitating the molecules of the upper atmosphere, increasing air drag, which caused Mir's orbit to dip by about a kilometre each day. Hence the continued need to reboost the orbit.

The cosmonauts managed to locate and seal the small air leak which had been bothering them—there had been a slight break in the seal on the pressure plate connecting Spektr to the node. The second supply craft, Progress M1-2, arrived on 28 April. Zalotin and Kaleri made a 3-hour 31-minute spacewalk on 12 May, in the course of which they found, on the outside of the hull, a burned out cable which had shorted. They restored all Mir's systems to full working order and carried out a full range of experiments with the orbital greenhouse (Svet) as well as medical and other tests. The station had never been in better working order, at least not since before the fire and collision of 1997. After a relatively uneventful 73-day mission, they undocked late on 15 June and made a smooth return to Earth 45 km southeast of Arkalyk early the following morning. Zalotin and Kaleri were removed from the cabin, helicoptered out and flown to Star Town later in the day.

No sooner were they down than MirCorp announced that it had put together a further two missions for Mir. With funding from an American, Santa Monica-based, 59-year-old technology investor called Denis Tito, two long-stay expeditions would fly to Mir in 2001. Denis Tito himself would ride up with the second crew, spend a week on board Mir and return with the first crew, leaving the new, long-stay cosmonauts on board. Paving the way for the time when ordinary citizens will visit space, it was called the Citizen Explorer programme.

In autumn 2000, MirCorp announced that it had managed to put together an agreement with the American *Survivor* television programme for one of 13 or 14 contestants to fly in space as part of a television show. *Survivor* was one of a genre of television shows in which people were put on a deserted island and viewers could watch weekly how they survived, and other series in which the group was expected to expel one of its members weekly (*Big Brother*). Survivor's idea was to send its dozen finalists to Moscow for training at the Yuri Gagarin cosmonaut training centre. Each week, one of the group would be deselected by one of the official trainers, leaving only one lucky survivor who would get a flight into space in the end of 2001—all of this aired on television both live and in instalments. This genre of television has been criticized as voyeuristic and crude entertainment, but it nevertheless had huge pulling power. NASA had turned down a similar proposition from *Survivor* but the free-enterprise-minded Russians had no such problem if it kept Mir aloft. The deal between the *Survivor* series and MirCorp showed that the corporation's talk about imaginative uses for the Mir space station was not just rhetoric.

Mir orbited Earth in what was called 'automatic mode' throughout the autumn of 2000. Ground control in Korolev checked its telemetry from time to time to ensure its stability. A new freighter, Progress M-43, arrived at the station on 21 October. However, with continued financial problems and the importance of supporting the International Space Station, the Russian government came under increasing pressure to abandon Mir in 2001. Although two further Mircorp missions were slated for Mir, it was by no means certain that they would actually take place.

MIR IN RETROSPECT

Mir was the most durable single achievement of the Soviet, now Russian, space programme. Mir became to the Russian space programme the type of achievement that Apollo represented for the Americans. While part of the evolutionary process of building space stations, beginning with the first generation Salyut (Salyuts 1 and 4) and then the second-generation Salyuts (Salyuts 6 and 7), Mir was the first-ever space station which made possible the permanent occupation of near-Earth space. With the exception of a brief period in 1989, Mir was manned permanently ever since its first long-term resident crew arrived in early 1987. Cosmonauts flew continuously on Mir from 1989 to 1999. While less glamorous than landing men on the Moon, it was no less an achievement in its own right. The quality of Mir's design was confirmed when it carried out a mission more than three times longer than originally intended, becoming home to modules, experiments and tasks quite different from those originally planned. At a time when American shuttles could fly for only two weeks at a time, Mir was home to missions which averaged half a year and was the home of cosmonauts who pushed back duration in space to 326 days, 366 days and then a year and a quarter (428 days). These long-duration missions represented the furthest frontiers of medical science and space medicine. By the end of the century, over two-thirds of all time in space had been on Soviet or Russian spacecraft and most of these long hours were clocked up on Mir.

Returning to Earth from Mir

Mir was a huge area of investment by the Russian space programme. Although the number of spacecraft may be less than the numbers committed in other pro-grammes, they are still very considerable. They are also more expensive spacecraft, where, because human life is involved, quality control is of the essence. The following table summarizes the commitment during the years 1992–2000 alone:

Spacecraft committed to the Mir project, 1992–2000

Soyuz	17
Progress	34
Modules	2
	—
Total	53

By summer 2000, Mir had received 104 individuals, comprising 42 Russians (or Kazakhs), 44 Americans and 18 others. There were 28 long-stay resident missions and two visiting flights. There were nine shuttle dockings with 57 visitors (some of whom were Russian). Mir received 70 unmanned dockings, of which five were modules, 65 Progress and one unmanned Soyuz. Its international nature was such that Mir had received 17 sets of visitors from countries as diverse as Afghanistan, Austria, Britain, Bulgaria, France, Germany, Kazakhstan, Syria, Japan, Slovakia, and the United States.

The Russian record in long-duration flight was, by the turn of the century, unchallenged. If one compiled a list of the longest completed spaceflights made by 2000, the only American would appear as the 25th name, and that was on the Mir space station (Shannon Lucid, at 188 days, the longest travelled American astronaut). The seven longest flying American astronauts all achieved their records on Mir. The longest flown Russian is Sergei Avdeev, who spent 742 days in orbit, followed by Valeri Poliakov (678 days), Anatoli Soloviev (651 days), Viktor Afanasayev (545 days), Musa Manarov (541 days), Alexander Viktorenko (489 days) and Sergei Krikalev (483 days). The longest American flight remains 84 days, set by Skylab 4 astronauts Gerry Carr, Ed Gibson and William Pogue in 1974. The longest a shuttle can fly is 17 days, which is still shorter than the longest solo Soyuz flight in 1970.

The table below shows how space experience has built up on Russian orbital stations.

Occupation of Soviet/Russian orbital stations

Station	Period of operation	Flight duration	Manned flight
Salyut 1	19 Apr 1971—11 Oct 1971	170 days	23 days
Salyut 2	2–29 Apr 1973	26 days	
Salyut 3	16 Jun 1974—25 Jan 1975	213 days	15 days
Salyut 4	26 Dec 1974—3 Feb 1977	774 days	93 days
Salyut 5	22 Jun 1976—8 Aug 1977	441 days	62 days
Salyut 6	29 Sep 1977—29 Jul 1982	4 years 10 months	617 days
Salyut 7	19 Apr 1982—7 Feb 1991	8 years 10 months	1,075 days
Mir	20 Feb 1986—	More than 14 years	11,908 days
Zarya	20 Nov 1998—		
Zvezda	12 Jul 2000—		

The volume of science carried out on Mir has been enormous, problems and interruptions notwithstanding.[12] Work done by the cosmonauts included such diverse disciplines as Earth observations, smelting, human and animal biology, medicine, radiation, antibiotics, astrophysics, atmospherics, meteorites and astronomy. Even the biology studies could be subdivided into a further range of cardiovascular studies, blood, urine, the neurovestibular system and musculoskeletal studies. Eighteen different instruments were used to measure radiation. Even some offbeat experiments proved productive: the *Mariya* spectrometer found a means to predict earthquakes by detecting electrostatic disturbances associated with imminent movements in the Earth's tectonic plates.[13] One of the least acknowledged but biologically most interesting experiments was the attempt to grow plants through the entire reproductive cycle. In 1990 the Bulgarians managed to plant, grow and re-seed radish and cabbage through the whole cycle (the *Svet* experiment). In 1997, the cosmonauts grew mustard and rape seed the whole sequence through from seeding to re-seeding. Science operations were hindered by insufficient power (25 kW up to 1997, 15 kW thereafter), excessive temperatures (frequently up to 30 °C), insufficient and cluttered space, limited communications and infrequent opportunities to send research results back to Earth.[14] Despite this, much was done.

Sustaining the permanent occupation of near-Earth space required a considerable industrial and organizational commitment. The regular replacement of crews, the frequent Progress resupply missions, the year-long round-the-clock operation of flight control, the recovery of crews and capsules, the Luch communications network, all represented a commitment to quality, standards and sustained effort. The successful arrival of every manned and unmanned supply mission—101 in all—was proof if it were needed that the Russian space programme could reach the highest standards of reliability.

The amount of spacewalking carried out aboard Mir was enormous. By July 2000 it amounted to 149 EVAs by 36 cosmonauts totalling 691 hour 18 minutes or almost 29 days.[15] One cosmonaut, Anatoli Soloviev, accumulated the extraordinary personal total of 74 hours 41 minutes. Efforts continued to go into the improvement of space suits. The Zvezda company, created as far back as 1952, had designed Gagarin's spacesuit, the suit and life-support system for Leonov's spacewalk (*Berkut*), the subsequent Soyuz 4/5 walk (*Yastreb*) and the unused suit to support a 10-hour EVA on the Moon (*Kretchet*). On Salyut, semi-rigid suits (*Orlan*) had been introduced and these stayed on the space station for whichever cosmonaut might need to use them. Unlike American suits, which were put on piece by piece, the Russian suit was like a spaceship in its own right and one climbed into it through the back. April 1997 saw the introduction of the *Orlan M* suit, its fourth modification, which will be used on the international space station.[16] It was first used by Vasili Tsibliev and Jerry Linenger and won immediate praise from its users.[17]

Modules to Mir

Name	Launch	Arrival
Soviet Union		
Kvant	31 Mar 1987	12 Apr 1987
Kvant 2	26 Nov 1989	6 Dec 1989
Krystall	1 Jun 1990	10 Jun 1990
Russia		
Spektr	20 May 1995	1 Jun 1995
Priroda	23 Apr 1996	26 Apr 1996

Resident crews on board Mir 1986–2000 (Visitors launched with them in brackets)

Soviet Union

1	13 Mar 1986	Soyuz T-15	Leonid Kizim, Vladimir Solovyov
2	5 Feb 1987	Soyuz TM-2	Yuri Romanenko, Alexander Laveikin
3	21 Dec 1987	Soyuz TM-4	Vladimir Titov, Musa Manarov, (Anatoli Levchenko)
4	26 Nov 1988	Soyuz TM-7	Alexander Volkov, Sergei Krikalev, (Jean-Loup Chrétien)
5	5 Sep 1989	Soyuz TM-8	Alexander Viktorenko, Alexander Serebrov
6	11 Feb 1990	Soyuz TM-9	Anatoli Soloviev, Alexander Balandin
7	11 Aug 1990	Soyuz TM-10	Gennadiy Strekhalov, Gennadiy Manakov
8	2 Dec 1990	Soyuz TM-11	Viktor Afanasayev, Musa Manarov, (Toyohiro Akiyama)
9	18 May 1991	Soyuz TM-12	Anatoli Artsebarski, Sergei Krikalev, (Helen Sharman)
10	2 Oct 1991	Soyuz TM-13	Alexander Volkov, (Franz Viehböck, Toktar Aubakirov)

Russia

11	17 Mar 1992	Soyuz TM-14	Alexander Viktorenko, Alexander Kaleri, (Klaus-Dietrich Flade)
12	27 Jul 1992	Soyuz TM-15	Anatoli Soloviev, Sergei Avdeev, (Michel Tognini)
13	24 Jan 1993	Soyuz TM-16	Gennadiy Manakov, Alexander Poleschuk
14	1 Jul 1993	Soyuz TM-17	Vasili Tsibliev, Alexander Serebrov, (Jean-Pierre Hagnieré)
15	8 Jan 1994	Soyuz TM-18	Viktor Afanasayev, Yuri Usachov, Valeri Poliakov
16	1 Jul 1994	Soyuz TM-19	Yuri Malenchenko, Talgat Musabayev
17	3 Oct 1994	Soyuz TM-20	Alexander Viktorenko, Elena Kondakova, (Ulf Merbold)
18	14 Mar 1995	Soyuz TM-21	Vladimir Dezhurov, Gennadiy Strekhalov, Norman Thagard
19	26 Jun 1995	Atlantis	Anatoli Soloviev, Nikolai Budarin
20	3 Sep 1995	Soyuz TM-22	Yuri Gidzenko, Sergei Avdeev, (Thomas Reiter)
21	21 Feb 1996	Soyuz TM-23	Yuri Onufrienko, Yuri Usachov
22	17 Aug 1996	Soyuz TM-24	Valeri Korzun, Alexander Kaleri, (Claudie André-Deshays)
23	10 Feb 1997	Soyuz TM-25	Vasili Tsibliev, Alexander Lazutkin, (Reinhold Ewald)
24	5 Aug 1997	Soyuz TM-26	Anatoli Soloviev, Pavel Vinogradov
25	28 Jan 1998	Soyuz TM-27	Talgat Musabayev, Nikolai Budarin
26	13 Aug 1998	Soyuz TM-28	Gennadiy Padalka, Sergei Avdeev, (Yuri Baturin)
27	20 Feb 1999	Soyuz TM-29	Viktor Afanasayev, (Jean-Pierre Hagnieré, Ivan Bella)
28	6 Apr 2000	Soyuz TM-30	Sergei Zalotin, Alexander Kaleri

Orbital parameters are given in the appendix (see pp. 309–316).

Visiting missions to Mir, 1986–2000
Excluding cooperative programme with United States (see Chapter 3)

Soviet Union

1	22 Jul 1987	Soyuz TM-3	Alexander Viktorenko, Alexander Alexandrov, Mohammed Faris (Syria)
2	7 Jun 1988	Soyuz TM-5	Anatoli Soloviev, Viktor Savinyikh, Alexander Alexandrov (Bulgaria)
3	29 Aug 1988	Soyuz TM-6	Vladimir Lyakhov, Abdul Mohmand (Afghanistan)
4	26 Nov 1988	Soyuz TM-7	Jean-Loup Chrétien (France)
5	2 Dec 1990	Soyuz TM-11	Toyohiro Akiyama (Japan)
6	18 May 1991	Soyuz TM-12	Helen Sharman (Britain)
7	2 Oct 1991	Soyuz TM-13	Franz Viehböck (Austria), Toktar Aubakirov (Kazakhstan)

Russia

8	17 Mar 1992	Soyuz TM-14	Klaus-Dietrich Flade (Germany)
9	27 Jul 1992	Soyuz TM-15	Michel Tognini (France)
10	1 Jul 1993	Soyuz TM-17	Jean-Pierre Hagnieré (France)
11	3 Oct 1994	Soyuz TM-19	Ulf Merbold (ESA)
12	14 Mar 1995	Soyuz TM-20	Norman Thagard (USA)
13	3 Sep 1995	Soyuz TM-22	Thomas Reiter (ESA)
14	17 Aug 1996	Soyuz TM-24	Claudie André-Deshays (France)
15	10 Feb 1997	Soyuz TM-25	Reinhold Ewald (Germany)
16	13 Aug 1998	Soyuz TM-28	Yuri Baturin (presidential advisor)
17	20 Feb 1999	Soyuz TM-29	Jean-Pierre Hagnieré (France), Ivan Bella (Slovakia)

The list also includes Soviet/Russian cosmonauts on visiting missions who did not stay on board for long-duration, resident flight. Some of these visitors flew duration missions. Orbital parameters are given in the appendix (see pp. 309–316).

Progress and unmanned resupply missions to Mir, 1986–2000

Soviet Union		*Russia*	
19 Mar 1986	Progress 25	25 Jan 1992	Progress M-11
23 Apr 1986	Progress 26	19 Apr 1992	Progress M-12
21 May 1986	Soyuz TM-1	30 Jun 1992	Progress M-13
16 Jan 1987	Progress 27	15 Aug 1992	Progress M-14
3 Mar 1987	Progress 28	27 Oct 1992	Progress M-15
21 Apr 1987	Progress 29	21 Feb 1993	Progress M-16
19 May 1987	Progress 30	31 Mar 1993	Progress M-17
3 Aug 1987	Progress 31	22 May 1993	Progress M-18
23 Sep 1987	Progress 32	10 Aug 1993	Progress M-19
20 Nov 1987	Progress 33	11 Oct 1993	Progress M-20
20 Jan 1988	Progress 34	28 Jan 1994	Progress M-21
23 Mar 1988	Progress 35	22 Mar 1994	Progress M-22
13 May 1988	Progress 36	22 May 1994	Progress M-23
18 July 1988	Progress 37	25 Aug 1994	Progress M-24
9 Sep 1988	Progress 38	11 Nov 1994	Progress M-25
25 Dec 1988	Progress 39	15 Feb 1995	Progress M-26
10 Feb 1989	Progress 40	9 Apr 1995	Progress M-27
16 Mar 1989	Progress 41	20 Jul 1995	Progress M-28
5 May 1989	Progress 42	8 Oct 1995	Progress M-29
23 Aug 1989	Progress M-1	18 Dec 1995	Progress M-30
20 Dec 1989	Progress M-2	5 May 1996	Progress M-31
28 Feb 1990	Progress M-3	31 Jul 1996	Progress M-32
15 Aug 1990	Progress M-4	19 Nov 1996	Progress M-33
27 Sep 1990	Progress M-5	6 Apr 1997	Progress M-34
14 Jan 1991	Progress M-6	5 Jul 1997	Progress M-35
19 Mar 1991	Progress M-7	4 Oct 1997	Progress M-36
30 May 1991	Progress M-8	20 Dec 1997	Progress M-37
20 Aug 1991	Progress M-9	14 Mar 1998	Progress M-38
17 Oct 1991	Progress M-10	14 May 1998	Progress M-39
		25 Oct 1998	Progress M-40
		2 Apr 1999	Progress M-41
		16 Jul 1999	Progress M-42
		1 Feb 2000	Progress M1-1
		26 Apr 2000	Progress M1-2
		17 Oct 2000	Progress M-43

All launched by Soyuz from Baikonour cosmodrome. Orbital parameters are given in the appendix (see pp. 309–316).

Mir Spacewalks/EVAs, 1986–2000

Soviet Union				
1987	Romanenko/Laveikin	3	8 h	48 min
1988	Titov/Manarov	3	13 h	46 min
	Volkov/Chrétien	1	5 h	47 min
1990	Viktorenko/Serebrov	5	17 h	39 min
	Soloviev/Balandin	2	10 h	47 min
	Manakov/Strekhalov	1	3 h	48 min
1991	Afanasayev/Manarov	4	20 h	48 min
	Artsebarski/Krikalev	6	32 h	09 min
Russia				
1992	Volkov/Krikalev	1	4 h	12 min
	Viktorenko/Kaleri	1	2 h	03 min
	Soloviev/Avdeev	4	18 h	21 min
1993	Manakov/Poleschuk	2	9 h	58 min
	Tsibliev/Serebrov	5	14 h	13 min
1994	Musabayev/Malenchenko	2	11 h	07 min
1995	Dezhurov/Strekhalov	5	18 h	43 min
	Soloviev/Balandin	3	14 h	32 min
	Gidzenko/Avdeev	1		29 min
	Avdeev/Reiter	1	5 h	16 min
1996	Gidzenko/Reiter	1	3 h	06 min
	Onufrienko/Usachov	6	30 h	31 min
	Korzun/Kaleri	2	12 h	33 min
1997	Tsibliev/Linenger	1	5 h	18 min
	Soloviev/Vinogradov	4	22 h	10 min
	Soloviev/Foale	1	6 h	
1998	Soloviev/Wolf	1	3 h	52 min
	Musabayev/Budarin	5	30 h	32 min
	Padalka/Avdeev	1	6 h	
1999	Afanasayev/Hagnieré	3	17 h	48 min
2000	Zalotin/Kaleri	1	3 h	31 min

Since each EVA was made by two cosmonauts, total EVA time is double the times given.
Intra Vehicular Activities (IVA) not included.
Spacewalks by American astronauts and Russian cosmonauts in the course of shuttle visiting missions not included.

3

The international frontier
Mir 2 and the International Space Station

So, Russia *had* managed to keep the Mir space station in operation throughout the 1990s, eventually evacuating the station in 1999 but then returning the following year. It had done so despite chronic shortages of money, difficulties in using Baikonour cosmodrome, a crash and fire, with missions juggled between cash payments, nosecone shortages and rockets not yet ready.

Ultimately, Mir could not last forever and the new Russian space programme must be built on fresh, post-Soviet foundations. However, anyone who predicted in the early 1990s that the manned space programmes of Russia and the United States would effectively merge would have been sent to see the men or women in the white coats. Yet the political and economic circumstances of both nations converged to the point that Russian–American crews on the shuttle and Soyuz became the norm and the two countries jointly, with other countries, operated the successor to Mir. The new space station frontier was, for the Russians, a joint enterprise with its old adversary.

HISTORY OF ON–OFF COOPERATION

It was not that there had never been cooperation between the United States and the Soviet Union in spaceflight. There had been: in the early 1970s, political and economic circumstances had made possible the first joint flight between the two countries.

In 1973, the Americans realized that there would be a long gap, perhaps four years, between the end of their Skylab space station programme (1974) and the start of their space shuttle programme (1978, though this was to extend to 1981). Not only that, but cancellations in the Apollo programme had led to a surplus, not just of astronauts wanting to fly, but of hardware without a mission. On the Soviet part, the desire to restore their space programme in the eyes of the international community and before their own people in the aftermath of the Soyuz 11 disaster provided the impetus for a high-profile mission. These factors, coupled with a move to detente between the Soviet Union under Brezhnev and the United States under Nixon and Ford, provided the political climate that made a joint mission possible.

So it was that in February 1972 NASA and the Academy of Sciences agreed on an Apollo docking with a Salyut in 1975. It would be the first Salyut with two docking ports—Soyuz would arrive at one end and Apollo at the other. The proposal was ratified in Moscow on 24 May 1972 by President Richard Nixon and Prime Minister Alexei Kosygin. The plan envisaged a joint flight in two years and a second mission after that. Yet suddenly the USSR withdrew the Salyut part of the mission. A two-port Salyut would simply not be ready in time, it said. The link-up would have to be between Apollo and a two-man Soyuz instead, a much less ambitious project.

Crews were duly announced the following year: NASA's commander was Tom Stafford of Apollo and Gemini fame; and the other members were Donald Slayton—the only Mercury astronaut who had never flown—and Vance Brand. The Soviet team were Alexei Leonov and Valeri Kubasov, who had been waiting for a mission together for two years. With all the Salyut and Soyuz mishaps, the USSR felt that success was at a premium in the upcoming Apollo–Soyuz Test Project (ASTP). It must be seen to be at parity with the United States. Nothing must be left to chance. While the Americans allocated one spacecraft and six astronauts, the Russians allocated eight cosmonauts and seven Soyuz spacecraft. Four unmanned tests took place in the Cosmos programme. Of the manned version, one was for a technical test before the flight; one was for the mission itself; and another was to stand by as reserve.

ASTP required new hardware—a docking unit which would enable the two space-craft to link together. Because there was spare weight capacity on the American Saturn IB rocket, but not on the R-7, this was NASA's responsibility. The Americans constructed the docking unit—a box-shaped tunnel, 3.15 metres long and 1.42 metres in diameter, weighing 5,907 kg. As preparations for the mission went ahead over 1973-4, cosmonauts became regular visitors to Cape Canaveral and Houston; American engineers and astronauts were seen in Star Town. Together they worked on the problems of pressures, optical and radar methods of navigation and docking and tracking. The Americans even got used to being bugged in their rooms in Moscow: it amused them more than it angered them. At one stage they complained loudly in their rooms about the lack of coat hangers: next morning they miraculously appeared. Tom Stafford made a visit to Baikonour to inspect preparations—a condition of the mission. His demand was met, but he was flown in and out by dead of night so that he would see as little as possible.

The USSR's pre-mission technical test took place from 2 to 8 December 1974. It was flown by Anatoli Filipchenko and Nikolai Rukavishnikov and designated Soyuz 16. They flew the 142-hour profile planned for the ASTP and carried out 20 experiments, mostly relating to the docking tunnel, atmospheric pressures, and frequencies. Soyuz 16 was tracked by the NASA network and communicated through mission control in Houston.

The Soviet press felt obliged to advertise the whole mission in advance. It made the best of this and announced plans to broadcast the launch live as 'a novel departure from the Soviet tradition'. Well they might. They made a virtue of necessity and ran the show with all the razzmatazz of an American network doing the same thing, even if the style was a little stilted. Full flight details were released some weeks ahead. Some alert State enterprises even marketed Apollo–Soyuz cigarettes and perfume as exclusive once-off brands.

HANDSHAKE IN ORBIT

In the event, the mission went entirely smoothly, at least until the very end. It attracted the greatest world-wide media interest in space travel since the Moon landing six years earlier. Soyuz and Apollo met on the second day of their joint mission, 17 July 1975. Television showed the smiling figure of Alexei Leonov, in his communications soft hat and light coveralls, stretch forward his hand to greet Tom Stafford in the docking tunnel between Apollo and Soyuz. They stayed docked for two days, exchanging pleasantries, conducting experiments, taking calls from Leonid Brezhnev and Gerald Ford. Soviet TV cameras fitted to the large Mil recovery helicopters picked out Soyuz high over the town of Arkalyk just after the parachute opened. A stiff wind blew it across the brown steppes like a glider on a cross-country race. A whoosh of dust blew up as the retrorockets fired. It was over. Flight control applauded wildly. The two cosmonauts emerged, signed the side of their capsule in chalk, chatted to pressmen and were flown off.

The aircraft carrier USS *New Orleans* took the last Apollo out of the Pacific Ocean on 24 July. Only a few minutes after the three astronauts had walked across the flight deck of the aircraft carrier did it transpire that they were in the intensive care unit, poisoned. As soon as the parachutes had opened, astronaut Vance Brand had thrown a switch to dump unused fuel. This was a normal procedure. But instead of venting to the atmosphere, the fumes were blown back into the cabin. All three astronauts began to lose consciousness. Tom Stafford just managed to get his oxygen mask on and helped his colleagues. The three only stopped coughing and spluttering once the hatches were opened up and the fresh sea air of the ocean blew in. The medium-term effects of the toxic poisoning were severe and the astronauts were not declared out of danger until 28 July.

Despite allegations to the contrary, the Russians probably learned little about the American space programme and its technology and nothing they could not have found out through the open technical literature. The Americans, by contrast, got a grandstand view of how the Soviet Union ran its space missions, warts and all. Although the Americans had to design and build a special docking module, the adventure probably cost the USSR more financially. After all, they built no less than seven spacecraft for the flight, flying six of them (the back-up orbited solo as Soyuz 22 the following year). There were numerous advantages to the Americans: it bridged a gap which in the end lasted seven years between Skylab 4 and the first shuttle and it enabled NASA to hold on to many good engineers and astronauts for longer than would otherwise have proved possible. Apollo 18 carried an Earth resources package which would alone have justified the flight. However, American suspicions that the Russians were getting more out of it were a theme of the undertaking—and were to resurface when the two sides came together again 20 years later.

ASTP AFTERMATH: COOLING RELATIONSHIPS

The ASTP had been envisaged by the Americans as the first of two joint flights. However, in October 1973, the Russians made it clear they did not wish to discuss a second flight until the first had been successfully accomplished. Notwithstanding, NASA deputy administrator George Low floated the Russians the idea of a follow-up mission the following year, 1974, suggesting the shuttle fly to a future Salyut in the 1980s and that together the two nations build an international space station in 1990. In May 1975, with the joint mission only two months away, George Low again proposed a further joint mission, an astronaut/cosmonaut swap, space station link-up studies and cooperation in other areas. At this stage, NASA was desperate to keep its ASTP team together. Otherwise, it would face certain disbandment the moment Apollo 18 splashed down.

The Soviets did not respond for over a year. Approval for their own second-generation space station with two docking ports was not given until November 1975. A year later, on 27 November 1976, NASA received a positive response, proposing an American shuttle with spacelab module dock with their second-generation Salyut. In May 1977, NASA and the USSR signed a second agreement for a joint flight, replacing and updating the 1972 texts. Two working groups were set up—one for experiments, the other for operations. The experiments working group was soon considering proposals for the future mission: the USSR put forward 11 possible experiments, the Americans four. The task of the operations working group, which met in November 1977, was to prepare a joint flight for 1981. NASA was confident that it could carry out such a mission even within its existing shuttle development budget and without having to ask for extra money from Congress. The working group discussed joint experiments and devised communication arrangements.

The second working group meeting was scheduled for spring 1978, but never took place as the political climate worsened sharply during this period. It appears that President Carter cancelled scientific visits to the USSR because of the treatment of dissidents there. NASA tried again for another meeting in October 1979 to revive the flagging project, but the State department again objected. The mission was never officially cancelled: it just died.[18]

RESUMPTION OF JOINT FLIGHTS

The resumption of joint Russian–American spaceflights 20 years later was the combination of different factors at work in both Russia and the United States. The renewal of the relationship took place in two phases: the resumption of joint missions, first; and the building of an international space station, second.

The resumption of joint flights was the outcome of the much improved political relationship between East and West. From 1989 onwards, the Soviet Union began to withdraw its troops from Eastern Europe, and the Berlin wall was torn down. With the collapse of communism and the subsequent transition to democratic government and a market economy, the principal reasons for not collaborating with Russia had disappeared. With financial shortages affecting its armed forces, Russia became less and less of a military threat and both countries removed the procedural systems which targeted their nuclear missiles at one another.

On 15 July 1992, following approval by both their governments, the first agreements were signed between Russia and the United States providing for a cosmonaut to fly on the shuttle, an American to fly on Mir and for the shuttle and Mir to link up together in 1995. The signatories were Yuri Koptev for the Russian Space Agency and Dan Goldin for NASA. The purpose of these missions was the sharing of expertise and experience. Ironically, few people remembered that the agreement differed little from what George Low had proposed nearly 20 years before.

AMERICA'S SPACE STATION

Coinciding with this agreement, the United States was experiencing considerable difficulties in developing its own space station. Although announced in 1984 by President Reagan, their space station Freedom was by this stage seriously over-budget and still many years from flying. Indeed, Freedom had achieved the distinction of having spent its budget but without having actually built any hardware at all! In February 1993, within a month of taking office, President Bill Clinton ordered a complete redesign of Freedom. He directed its costs to be cut from $30bn to $10bn, for a shorter lifetime, fewer assembly and resupply missions and this review to be completed by 1 June.

In the event, NASA came back with three redesigns, popularly called A (labelled austere), B (labelled basic) and C (which some simply called a 'can'), all budgeted in a range of $12bn to $13.8bn. President Clinton proposed that the station be kept in the budget while he considered the options. Just how little room for manoeuvre he had became evident later that summer when the station survived a House of Representatives budget vote by 216 votes to 215. Clinton ordered NASA to come back to him in mid-September with a scaled-down version of option B, but to come in at between $10.5bn and $11bn. He also gave permission for negotiations to take place with Russia to see if cooperation would be possible and if that would reduce costs.

These negotiations came to a head sooner than expected, and on 5 September Vice-President Al Gore and Russian Prime Minister Viktor Chernomyrdin signed an agreement. There were two parts. In the first, called phase I, the United States would pay Russia $400m for joint operations on Mir for the period to 1997. In return, Russia would provide two years of astronaut flying time on Mir. In effect, the 1992 agreement was radically extended in a purposeful way to lay the ground for a bigger project. In phase II, the American space station and the projected Russian Mir 2 space station would merge in a joint international space station, which would have common controls and environmental systems. The new station would be jointly assembled by the United States and Russia using the space shuttle and Proton rocket respectively. The United States also relaxed some of its restrictions on the commercial uses of the Russian Proton rocket on the world satellite market.

EVOLUTION OF MIR 2

On the Russian side, the Mir 2 space station had already run into difficulty, although its problems were financial rather than technical or design. Originally, Mir 2 would have followed Mir 1 much like Salyut 7 had followed Salyut 6 and would follow a similar profile. Indeed, the Mir 2 service module was the back-up to Mir itself, since it was normal to build such modules in pairs (it still carries hull number #12801). When the Americans announced Freedom, Mir 2 was predictably scaled up into something bigger and more ambitious than Freedom. The new Mir 2 would still use the original Mir module, but it would be transformed into what was called OSETS or *Orbitalny Shorochno Eksploratsionnyy Tsentre* (Orbital Assembly and Operations Centre). To the service module would be added a 90-tonne module launched by Energiya and huge truss structures and solar arrays. The new station would be serviced by a new 5-man spacecraft launched by Zenit called Zarya and by a large 13-tonne Zenit-launched supply craft, Progress M2.[19] Zarya was a 13-tonne spacecraft, to be launched on the Zenit, able to carry eight cosmonauts or 4 tonnes of supplies, or various combinations in between. It was shaped just like the Soyuz, but with lightweight Buran tiles as a heat shield and the downward-firing touchdown rocket moved to the sides.

   In the austere, post-Soviet financial climate, these plans were revised and scaled back by the Council of Chief Designers in November 1992. The new Mir 2 in effect became the original, pre-1986 Mir 1 design: a core block with four 7-tonne science modules delivered by Soyuz spacecraft, a 'newer, smaller, better' Mir. It would have an external truss structure for its solar arrays, to try to address the problems of power shortages experienced on Mir 1, making it appear much larger in scale. It received the hybrid, temporary nickname of 'Mir 1.5'. The new Mir 2 attracted the interest of the European Space Agency in spring 1993 and the project went through further evolutions to take account of European research interests. When the Buran programme was cancelled in June 1993, the continuation of Mir 2 was reaffirmed and a flight date was given as 1997, about the same time as Mir 1's termination was scheduled.

   However, Mir 2 lasted as a project for only a few more months. By late summer, the financial situation in Russia had tightened further. Russia's eagerness to do a deal quickly matched that of the Americans after the narrow congressional vote. Circumstances conspired in which both countries could offer a lifeline to one another. Neither side needed much further persuasion, hence the quick deal in early autumn.

PHASE I: NEAR-MIR

Phase I of the September 1993 Gore–Chernomyrdin agreement involved four years of preparatory missions in which the two nations learned how to work together in space on board the Mir station.

The first of the joint missions took place on 3 February 1994, when veteran Soviet cosmonaut Sergei Krikalev joined the crew of STS-60, the space shuttle Discovery, which carried the Wake Shield experiment, Spacehab 2 and a German free-flying satellite. The second occurred just over a year later on 3 February 1995, when the shuttle Discovery, on mission STS-63, flew a rendezvous profile with Mir (inevitably, it was popularly named 'mission near-Mir'). The Russian cosmonaut on board was Vladimir Titov and the purpose of the mission was to test the ability of the shuttle to close in on the space station. At 11.55 a.m. on the morning of 6 February, Vladimir Titov spotted his old home in space from the shuttle window from a distance of 338 km. Seven hours later, shuttle commander James Weatherbee and pilot Eileen Collins let the shuttle drift in to only 10 metres, ever so careful that the two did not collide. For the next hour, the shuttle and Mir flew in a gentle space ballet, crew members waving to each other through the portholes, exchanging greetings in Russian and English. Television pictures of the link-up were stunning, showing the shuttle closing in on the Russian space complex as it glided slowly over the blue and white background of the planet Earth. With the rendezvous over, the two spaceships separated, Discovery pursuing an independent mission before returning to Cape Canaveral on 11 February. A month later, the exchange was reciprocated when American doctor Norman Thagard flew to Mir for three months of experiments on board Soyuz TM-21.

CLASH OF SPACE CULTURES

Phase I had a number of overlapping aims. Formally, the purpose was to provide 600 American days on Mir in the course of which the astronauts would carry out a number of dedicated experiments in the areas of engineering, microgravity, physiology, neuroscience, the cardiovascular system and human performance in orbit. The subtext was to learn how two quite different space programmes could integrate their operations (or not) and to anticipate the risks that might arise in the construction of the international space station.

American astronauts were divided in their reaction to collaboration with the Russians. Some indicated that they wanted no part in the programme. By contrast, others were genuinely enthusiastic about the prospect of working with a quite different programme and approach. Several astronauts saw phase I as an opportunity to get a spaceflight for which they might otherwise have to wait a long time. Some relished the opportunity for a long flight, which would never be possible on the shuttle alone. Those most interested made a point of letting the mission selectors know that they were learning Russian in their spare time.

Black Sea water training

The Russian language was the first hurdle. American astronauts flying to Mir began their language training while still in the United States. All found the language difficult and, in their own words, *bezkonechni* ('endless'). The next challenge for the Americans was learning about Mir and Soyuz systems and how the space station functioned. This they had to learn through lectures in Russian and memorize. They could not read the manual, because there wasn't one, and although the Russians used simulators, they were a much less important aspect of training than in the United States. Next, the American trainees were brought off for water and survival training. The first involved exiting from the Soyuz cabin 2 km off Sochi in the Black Sea, in case of a sea landing. For survival training, they were dumped in Tiksi on the shore of the Arctic Ocean and had to find their way out. To complete the test at the other climatic extreme, they were then left in the Turkmenian desert for three days with only two litres of water and told to find their own way home.

Although we know much about the American experience on Mir, we know little about the experience of the Russian cosmonauts who flew on the space shuttle and how they adapted to the American way of life. Just as Americans came to live and train in Star Town, so too did Russians, in reverse, settle in the Manned Space Flight Centre in Houston, Texas, learn English, get to know the shuttle systems and adapt to a different way of flying into space. The media image showed cosmonauts joining bar-b-queues in the neat back gardens of astronaut homes in Houston, and, presumably, they too experienced similar problems of cultural adjustment.

Although the press interest was on the Americans who flew to Mir, the missions of Russian cosmonauts on board the space shuttle were largely footnoted. In the course of 1994-7, 11 cosmonauts were to fly on the shuttle. One can only guess their feelings about flying on the American shuttle while their own Buran was grounded. The 11 included one Ukranian, Leonid Kadenyuk (trained originally in the Russian cosmonaut corps), two Russians who flew up to Mir on the shuttle, and two colleagues who took their place back down. Newcomer Salizan Sharipov made his first spaceflight on the shuttle. Vladimir Titov flew on the shuttle twice. The deputy head of Energiya, Valeri Ryumin, although 58 and towards the end of his cosmonaut career, was selected for the last shuttle mission to Mir which brought back the last American on Mir, Andy Thomas. As 17 years had passed since his previous space-flight, there were comments that Ryumin had used his senior position in the space programmes to select himself: either way, he lost 25 kg to qualify and passed his medical. A year earlier, Elena Kondakova, coincidentally married to Valeri Ryumin, had also flown to Mir.

SECOND HANDSHAKE

The February 1995 rendezvous was a rehearsal for the first of seven docking missions. The heart of the system was a new docking unit called the Orbiter Docking System (ODS), designed for attachment to the APAS-89 Buran port on Krystall. The ODS weighed 1.7 tonnes and included a capture ring made in Moscow by RKK Energiya and sold to Rockwell for $18m. The ODS was to be the basic system for the International Space Station.

Atlantis was the first shuttle to test the new docking system, eventually leaving Cape Canaveral on 26 June after two cancellations due to summer thunderstorms in Florida. Veteran shuttle commander Robert 'Hoot' Gibson brought the Atlantis close in to the Krystall docking unit on Mir, the two large space objects meeting slowly but firmly and exactly on schedule. Gibson brought Atlantis in just 1 km under Mir, with astronaut and Mir trainee Bonnie Dunbar calling out the closing rates in Russian to the Mir crew. Flying over the Caspian Sea, they were 10 metres apart. Five minutes later, over Lake Baikal, they had capture. It took two hours to equalize the pressure between the two spacecraft. 'Hoot' Gibson opened the hatch and floated through to be greeted by Mir commander Vladimir Dezhurov, his Russian companion Gennadiy Strekhalov and fellow American Norman Thagard, who had been on board Mir for the previous three months. Besides gifts of chocolate, sweets and flowers, 'Hoot' Gibson brought with him the new Mir crew, Anatoli Soloviev and Nikolai Budarin, the first time the Mir crew had ever been changed by courtesy of an American spacecraft. The combined assembly of Mir/Atlantis weighed a record 230 tonnes, the largest single object ever assembled in space. Atlantis brought Mir new supplies of water and air.

The American space shuttle docks with Mir

When the 10 astronauts and cosmonauts posed for their group photograph in the Mir cabin on the afternoon of 29 June 1995, that moment could be said to mark the true end of the space race between the two nations. Unlike Apollo–Soyuz, this rendezvous was the beginning of a long-term relationship. The shuttle undocked from Mir on 4 July, spending 90 minutes flying around Mir at a distance of 1 km and going on to make three days of independent flight before returning to Cape Canaveral. Thagard returned from Mir to Cape Canaveral in the shuttle with a new American endurance record in space, 115 days. The most serious problem the mission encountered was bureaucratic. No one had thought to get permission for returning Mir crew members Vladimir Dezhurov and Gennadiy Strekhalov to enter the United States and the US Customs Service, considering itself now in the business of inspecting arriving space shuttles at Cape Canaveral, insisted that the Russians have visas. These were duly brought up to them in time so that they could present them at Cape Canaveral when they came down from space, duly classified as aliens.

The missions of phase I became routine surprisingly quickly. Atlantis met with Mir for a second time on 15 November 1995, delivering a 5-tonne Russian-made docking module, 2.2 metres in diameter and 4.7 metres long. Hitherto, the Krystall docking port provided insufficient clearance for the arriving shuttle and there was a danger of collision with Mir's extensive array of solar panels. For the first docking, the 20-tonne Krystall had been moved to the longitudinal, axial docking port to facilitate an arriving shuttle, but it was deemed impractical to carry out such a manoeuvre each time. The new docking module, built quickly by Energiya, provided the additional space required. A new set of solar panels was included on the module, so once it had delivered its cargo, the shuttle had spare capacity for its return leg. The cosmonauts took advantage of this by getting the shuttle to take back to Earth as much rubbish, broken and unwanted equipment as possible. Indeed, with the availability of the shuttle for the ensuing years, the Russians discontinued use of their own Raduga capsules which had been introduced with Progress M-5 in 1991 and finished with Progress M-25 in February 1995.

The third docking mission the following year began a period of permanent American presence on board Mir. From then on, the standard crewing pattern for Mir was for two Russian cosmonauts and one American astronaut to be on board. For the Americans, Mir was able to provide their astronauts with their first long-duration space experience since their Skylab missions of 1973-4. The first such astronaut was Shannon Lucid who arrived on Mir on 23 March 1996. She was replaced by John Blaha and he in turn by Jerry Linenger, followed by Michael Foale. As an example of just how international and diverse these missions had become, STS-84, which met with Mir in May 1997, included Russian cosmonaut and Mir veteran Elena Kondakova, Jean-François Clervoy of France (representing the European Space Agency) and Americans of Chinese, Peruvian and British descent. The shuttle was piloted by a woman, space veteran Eileen Collins.

Americans had a mixed experience on board Mir. Shannon Lucid brought an infectious enthusiasm to bear to her mission, one captured by the superb IMAX film *Mission to Mir*. Michael Foale made a supreme effort to mix in with his Russian colleagues, which they undoubtedly appreciated. Norman Thagard seems to have become homesick from an early stage, while Jerry Linenger found the management of the Mir programme to be bureaucratic and even corrupt. The Americans did different things with their leisure time: Shannon Lucid was an avid reader; John Blaha watched *Star Trek* tapes and even had the film *Apollo 13* sent up (was he tempting fate?); Michael Foale's video favourite was the zany British comedy *Monty Python*; and Jerry Linenger wrote letters for his newly born son to read when he was older. Shannon Lucid sent her thoughts down to space fans by e-mail.

Although American politicians complained, often vociferously, at the costs of collaboration with Russia, a permanent presence in space was a capability they lacked themselves and would not be available to them until the International Space Station was built. The Americans got not only the 600 days given to them under the programme, but substantially more, 908 days, and their first long-duration experience in over 20 years. The Russians, for their part, benefited from the substantial cargo-carrying capacity of the shuttle, which was able to carry equipment and supplies substantially in excess of their own Progress freighters. They were able to reduce, though not eliminate, the number of their Progress missions at a time of acute economic pressure. Whatever the external political pressures, relationships between the two nations on board Mir itself were always good, the single American on board finding himself in command of the Russian station whenever his colleagues ventured out for spacewalks. Who could have imagined such a situation ten years earlier?

Phase I of Russian–American cooperation on board the Mir space station

14 Mar 1995	Soyuz TM-21	Norman Thagard	105 days
22 Mar 1996	STS-76 Atlantis	Shannon Lucid	175 days
16 Sep 1996	STS-79 Atlantis	John Blaha	118 days
12 Jan 1997	STS-81 Atlantis	Jerry Linenger	122 days
15 May 1997	STS-84 Atlantis	Michael Foale	134 days
25 Sep 1997	STS-86 Atlantis	Dave Wolf	119 days
22 Jan 1998	STS-89 Endeavour	Andy Thomas	135 days
Total			**908 days**

Russians on the space shuttle

3 Feb 1994	STS-60	Discovery	Sergei Krikalev
4 Feb 1995	STS-63	Discovery	Vladimir Titov
26 Jun 1995	STS-71	Atlantis	Soloviev and Budarin up; Strekhalov and Dezhurov down
15 May 1997	STS-84	Atlantis	Elena Kondakova
25 Sep 1997	STS-86	Atlantis	Vladimir Titov
19 Nov 1997	STS-87	Columbia	Leonid Kadenyuk (Ukraine)
22 Jan 1998	STS-89	Endeavour	Salizan Sharipov
2 Jun 1998	STS-91	Discovery	Valeri Ryumin

PHASE II: DESIGNING THE INTERNATIONAL SPACE STATION

Turning to phase II of the American–Russian agreement, NASA resubmitted its new space station design in late 1993. The four basic elements in its design were two American nodes, Mir 2 (renamed the service module) and a Russian functional block called the FGB (*Funkstionalii Gruzovoi Blok*) which would be built on contract for the Americans for $215m.[20]

The FGB was the first element of the station to fly. Although it dated to Vladimir Chelomei's *Almaz* space station designs of the 1960s, it was the most advanced space station of its kind when it eventually flew. The FGB was a 23-tonne module, 12.8 metres long, 4 metres in diameter with a volume of $55\,m^3$. It had two solar panels of $28\,m^2$, able to supply 6 kW of electricity. It had four docking ports, three at one end, one at the other. The rear end was designed to take Russia's service module, the front end America's node 1. The FGB carried two 417-kg main engines, 20 rendezvous motors of 40-kg thrust and 16 stabilization motors of 1.3-kg thrust. It had 16 fuel tanks holding 6.1 tonnes of propellants, 42 manoeuvring engines and a tonne of micrometeoroid shielding. Its design lifetime was a minimum of 13 years (to

2010), with at least 15 years expected. Its principal function was to stabilize the station and provide it with electricity until the station's main power system was constructed. Although buried by his rivals during his lifetime, it was Chelomei's ship which became the core of the world's largest scientific project ever undertaken.

The FGB was the first component to launch and was followed by America's node 1, Unity. The FGB, renamed Zarya, was shipped from the Khrunichev factory in Moscow for the 1,500-km rail journey to Baikonour in mid-winter, in January 1998. American engineers were puzzled when they learned of the extraordinarily low temperatures ($-50\,^\circ$C) and high g forces ($50g$) that the FGB had been built to withstand—until it was explained to them that they were based on the conditions the FGB must survive in the unheated Russian railcars and on its rail and road network, rather than the more benign conditions of space.

Zarya was originally to have flown in 1997 but dates repeatedly slipped towards the end of 1998. The station was eventually powered up on 7 October 1998 at Baikonour, an irrevocable action committing the station to launch in the near future. NASA chief Dan Goldin stood beside Russian Space Agency leader Yuri Koptev and 160 other invited guests under a corrugated iron shelter as the Proton counted down on the morning of 20 November. Television pictures beamed the launch on the Eurovision network, showing the black-tipped Zarya against the brown steppe and a grey Kazakh sky. As usual, Russian transmissions did not show countdown clocks and because the Proton used storable fuels there were none of the telltale wisps of cold fuels burning off to indicate that a launch was imminent. Eventually, acrid yellow and orange clouds billowed out of the base of the rocket which rose with quickly accelerating speed, heading into clouds and shortly being lost to sight. At 43 km, 2 minutes into the mission, the large first stage dropped off. At 78 km, the outer fairing came off, exposing Zarya to the now airless

The FGB, the Functional Control Block

atmosphere. The second stage was discarded at 138 km, 334 seconds into the mission, and the third stage then burned until 587 seconds. The most crucial stage then followed: would the solar panels deploy? If not, the mission would be in serious trouble. After 15 minutes, ground control confirmed that Zarya had safely reached orbit and, to cheers, that the two 30-metre solar panels had deployed.

As champagne corks popped below, Zarya was in serious trouble. The computer system on board was not responding to commands, which meant that its motors could not be fired to raise its orbit and it would decay within a few days. This problem was first picked up in the secret military tracking centre in Golitsyno west of Moscow. There chief test engineer Nazarov wrote, in a frantic 90 minutes, a computer programme to bypass the fault on Zarya and within a short period the station was responding. Zarya's initial orbit of 179–341 km was duly raised three days into its mission to 385–396 km. As for Nazarev, he received a medal from the commander-in-chief, a bonus, and four years of delayed backpay!

MARKING TIME, 1999: THREE SHUTTLE MISSIONS

The next most crucial single element in the construction of the international space station was the Russian service module. Until it was ready, three shuttle servicing missions flew up to Zarya to maintain the station and extend the first building blocks.

The first such mission was shuttle mission STS-88, Endeavour, launching on 4 December, two weeks later. Besides the American crew led by Robert Cabana, Russian cosmonaut Sergei Krikalev was on board, doubly appropriate as he was scheduled to fly on the first resident Russian–American long-stay crew on the station. The first task of Endeavour was to attach to Zarya the first American module, called Unity. Endeavour used its telescopic arm to pull the 14-tonne

Space station ground assembly

Unity out of its payload bay, and closed in on the gleaming space station on the 6 December, approaching it gingerly from above. Unity was attached and construction of the International Space Station (ISS) had begun.

The first of three scheduled spacewalks began on 7 December when astronauts Jerry Ross and Jim Newman left the porch of Endeavour, clambered onto the hull of Zarya, and began attaching 40 power cables between the two modules. Soon thereafter, ground control commanded electrical power to flow from Zarya's arrays into Unity. The astronauts installed sockets and rails to facilitate future EVAs. On a second spacewalk the next day, Ross and Newman installed two S-band relay antennas to enable the ISS to communicate to mission control in Houston via the TDRS relay system.

In true international spirit, Robert Cabana and Sergei Krikalev entered the International Space Station together on 10 December, switched on the lights and checked that all systems were in order (they were). Compared to Mir, Zarya looked fresh and uncluttered (a situation bound to change). On 12 December, Ross and Newman made their third spacewalk, pre-positioning tools and other equipment on the outside of Zarya and Unity and making a photographic record of the conditions of both hulls. Cabana undocked Endeavour from the ISS on 13 December, made a lengthy fly around and dropped away for a night-time landing on the runway at Cape Canaveral.

The second shuttle mission up to the International Space Station took place on 27 May 1999. Discovery was commanded on mission STS-96 by Kent Rominger, with new cosmonaut Valeri Tokarev as the Russian visitor on board. Discovery reached the station on the 29th and a day later, astronauts Dan Barry and Tamara Jernigan made a 7 hour 55 minute spacewalk to attach Russian and American cranes to the outside of the ISS. Discovery's crew entered the station, transferring almost 2 tonnes of supplies into the station—a mixed group of items including water, clothing, sleeping bags and medical equipment. Rominger fired the engines of Discovery to raise the orbit of ISS by 9 km. After being linked to the station for 80 hours, he brought the shuttle back for a night landing. Later, it transpired that the visiting astronauts had suffered serious discomfort and nausea in the course of their visit to Zarya. Initially attributed to the air control systems in the Russian module, it turned out that the explanation lay in disrupted patterns of air flow between the station and the arriving shuttle, a space-borne equivalent to sick building syndrome.[21]

Originally, the third shuttle mission to the ISS was to have flown after the service module launch, but the Americans decided to bring it forward. The main reason for doing so was that four batteries on Zarya had accidentally discharged and required repair. The shuttle Atlantis left Cape Canaveral on mission STS-101 just before dawn local time on 19 May 2000 after a long period of countdown delays. The mission was commanded by Jim Halsell and included, as its Russian crew member, Mir veteran Yuri Usachov.

Atlantis closed in on the International Space Station two days later. Astronauts Jim Voss and Jeff Williams soon suited up for a 6-hour spacewalk, attaching a new Russian Strela crane and reattaching some external equipment that had come loose. They placed a range of external equipment on the outside of the station. The astronauts opened the hatches into Unity and Zarya on 22 May, becoming the first visitors to the complex in a year. They rigged up a new system of air supplies, recharged the FGB batteries and then used the Atlantis engines in 27 pulses of 130 seconds each to raise the station's orbit by 30 km.

Russians on the space shuttle to the ISS

STS-88	4 Dec 1998	Sergei Krikalev
STS-96	27 May 1999	Valeri Tokarev
STS-101	19 May 2000	Yuri Usachov

DELAY, DELAY AND DELAY

Next to fly was the Russian service module. Although the American-financed FGB was completed in 1997 on schedule (though launched late), the Russian service module languished due to lack of funding and ground to a halt. The service module was actually the back-up hardware for the Mir space station. The hull had been completed as far back as February 1985 and the module fitted out by October 1986. However, its internal systems had to be completely rebuilt and modernized to enable it to operate with the FGB and the other components of the ISS.

Repeated American visits to the Energiya plant in Moscow produced renewed commitments to resume work, but nothing ever seemed to happen. Frustration grew and NASA began to consider contingency plans should Russia drop out of the project. Not until May 1997, following the personal intervention of President Yeltsin, did funding resume on the Russian side with a presidential commitment of $260m. Within days, engineers were back at work on the project. However, this did not lead to the expected improvement, possibly because President Yeltsin's new funding turned out to be a complex set of bank loans rather than on-the-spot cash. Dates again slipped and slipped. The situation worsened with inflation and the slide of the ruble on foreign exchanges. In 1998, as launch dates came and went, the only progress was that the service module acquired a name, Zvezda, or Star. Reluctantly, in late 1998, the Americans provided more money, between $60m and $100m in order to try to resume some momentum.

This was an advance on a planned series of annual $150m subsidies over four years. The new funding was linked to plan whereby the assembly of the station would be stretched over another year, to 2005 and the shuttle would make up for some Russian resupply missions. This was a tricky issue for NASA: many in Congress and the press were unhappy about the links with the Russians and were quick to exploit these bail-outs. Even those who favoured the subsidies worried about the way in which money seemed to disappear when it arrived in Russia. There were allegations that top officials were pocketing the money and building fancy private homes for themselves in Star Town. NASA did not wish to buy more Russian parts of the ISS outright, like the FGB and indeed Congress might well have prevented this. Instead, NASA made contributions towards operational costs and rented future Russian facilities on board the station, even though they had not yet flown. NASA spent 1998 investigating a number of scenarios whereby it could substitute its own module for the service module and proceed without Russia. An Interim Control Module (ICM) was even designed. Reluctantly, NASA concluded that a substitute module would cause further delays, do much less and ultimately cost more. It was cheaper, albeit frustrating and even exasperating, to continue to subsidize the Russians.

By 1999, these additional injections of cash had their effect. By then too, the financial crisis had eased a little and the service module was at last shipped to Baikonour. It might have flown that autumn had not the Proton rocket that was due to launch it suffered a double launch failure in July and October. The Russians insisted (as did Kazakhstan, which was likely to suffer the most from further disasters) that there be a careful, step-by-step requalification programme. Zvezda was slipped to May 2000, then July. The Proton returned to flight on 12 February 2000 and Zvezda was finally powered up. A surge of Proton launches in the early summer cleared the way for the crucial service module launch.

For the United States, these delays were testing, although the Americans glossed over problems on their own side which would certainly have led to some (though much less serious) delays to the project. Many Americans attributed these delays to the costs and energy involved in keeping Mir aloft. The sooner the old station was de-orbited and sunk in the Pacific Ocean, the sooner the ISS project would be under way, they argued. When the Russians recommissioned Mir in spring 2000, the Americans responded by announcing that they would build their interim control module if the Russians failed to orbit Zvezda by the autumn. There was even an implied threat in some quarters to kick the Russians out of the ISS project. Relationships between the two sides was poor in spring 2000. The General Accounting Office presented a report listing the defects of the Russian FGB, such as loud ventilators, poor air supplies and insufficient protection against depressurization—even though the module had been built to American specifications and so certified before launch.

ZVEZDA, 12 JULY 2000

Zvezda counted down in the scorching summer of the Baikonour morning heat of 12 July 2000. Just to remind people of the cash shortage, advertising space on the side was sold to Pizza Hut, who paid €1m to Energiya for the privilege of relaying its logo to the world's watching television companies. Against a light blue sky, brown clouds billowed out beside the lightning towers on the Proton pads as Zvezda sped skyward. Although unmanned, it was probably the most important single launch of the programme, for the prospects of replacing a failed Zvezda were financially not worth thinking about. At 80 km, halfway through the second stage burn, the payload fairing was jettisoned, and Zvezda started to transmit its own data.

Just less than 10 minutes after take-off, Zvezda was in orbit 184 km high and 3 minutes later its 30-metre solar panels sprung open, beginning a 13-day rendezvous profile. NASA's Dan Goldin put his previous comments to one side and spoke of all the difficulties the Russians had overcome to launch the station 'and they came through'. Ground control in Korolev then steered Zvezda through a series of manoeuvres, bringing it into an ever-higher orbit to FGB. At an 80-km distance from one another, the approach and docking manoeuvres were begun, with Zvezda being the passive ship. The FGB was commanded to yaw and made a radar sweep for its sister ship. The whole process was flawless and by early morning on the 25 July the two craft were 200 metres apart, 366 km over Kazakhstan. The FGB moved gently in at 0.2 metres per second and they clunked together.

Over the next week, ground controllers in Korolev and Houston began reconfiguring the software of the FGB, Unity and Zvezda in order to achieve a unified control system. On 6 August, the first unmanned supply craft for the ISS, Progress M1-3, left Baikonour to take up equipment to prepare the station for permanent occupation, arriving smoothly two days later. By this stage, the station comprised three modules and one supply craft, weighed 60 tonnes and was orbiting between 351.6 and 380 km.

ISS: THE NEXT STEPS

Just as Zarya was the essential block that controlled the International Space Station, Zvezda was the module that would make it possible for astronauts and cosmonauts to live and work there. Like its predecessor, Mir, it had no scientific equipment, but it hosted bedrooms, a bathroom, running track and exercise galley, cooking galley, refrigerator and even an entertainment rack with videos, CDs and books. The men and women on board could observe the Earth through 13 portholes. It carried a set of computer controls called the Data Management System, developed by the European Space Agency. The new Zvezda telemetry system, called *Regul*, had flown only once before, on Buran, and was now making its operational début. Many of the bulky consoles of Mir had been replaced by new slimline, portable computer systems. In a small but welcome home improvement, the Russians brought a space-age samovar on board—a dispenser able to serve hot tea at any time, day or night.

The first expedition to the International Space Station comprised Mir veterans Yuri Gidzenko and Sergei Krikalev and shuttle pilot William Shepherd. Professionally, they crossed a range of disciplines from MiG pilot (Gidzenko) to underwater commando (Shepherd) to Energiya engineer and computer genius (Krikalev). Their first scheduled task was to get the toilet and air systems in working order, unpack the scientific and other equipment on the space station, unload new supplies arriving on Progress M1 spacecraft and welcome two shuttle missions—one to deliver the second American module, Destiny, and the other to bring up the solar array.[22]

Later in the sequence there will be three small Italian logistics modules (named Leonardo, Rafaelo and Donatello), a European module (Columbus) and a Japanese experimental module (Kibo). In all, 13 Russian and 16 American missions will be required to build the station, not to mention up to five cargo flights a year.

In August 2000, the first Progress M1 freighter resupplied the ISS

Construction will begin of the large power truss and turbogenerators that will power the station. The crew of the station will later be expanded from three to six. A corps of nine European astronauts was selected, of whom six were allocated to train with NASA and three with the Russian Space Agency. In 1999, the Japanese astronaut Soichi Noguchi arrived for training at the Yuri Gagarin Space Centre. Construction will take until 2005.

SOYUZ: NEW LEASE OF LIFE

A little publicized aspect of the space station was the important role played by the Soyuz spacecraft in emergencies. From the beginning, both countries defined a need to be able to evacuate their crews speedily should an emergency take place—a well-identified need in the light of the Mir fire and collision in 1997. The United States eventually came up with the X-38, a small spaceplane that looked very like one of the Russian BOR mini-shuttles tested in the 1970s. The X-38 would, in an emergency, detach from the station with its crew, fly automatically, fire retrorockets, head into re-entry, open a parasail and glide onto the desert floor in the southern United States. However, the X-38 would not be ready until 2004. In the meantime, it was agreed that a spare Soyuz would be docked to the station at all times in order to make possible an emergency evacuation. Since a Soyuz's systems can be guaranteed in orbit for only six months at a time, it would need to be replaced twice a year, requiring a number of what are called taxi missions.

The first taxi will mark the introduction of the Soyuz TMA, the fourth main version of the Soyuz spacecraft to be introduced since 1966 (after Soyuz, Soyuz T and Soyuz TM). The TMA will feature improved computers, controls and life-support systems. Most important, however, from the ISS perspective is that it will have more space. During phase I, two American astronauts—Scott Parazynski and Wendy Lawrence—had been sent for training to Star Town but when they arrived the Russians found that Parazynski was too tall to fit into the Soyuz TM, which has a height restriction of 182 cm and that Lawrence, at 152 cm, was too short to fit into their spacesuit in case she needed to wear it in an emergency. Indeed, about half of America's astronauts were probably too tall for the old Soyuz—a question of being the right stuff, but not the right size. Miniaturization on Soyuz TMA will provide the extra few centimetres to accommodate the remainder, just in case they have to take the Soyuz taxi in an emergency ride back to Earth.

For the first years of occupation of the ISS, as mentioned above, a Soyuz will be permanently stationed on Mir, ready to offer the crew a quick evacuation in an

Safe landing in Kazakhstan—but the Russians had also trained for other landings

emergency. Soyuz was almost used as an escape vehicle during the crash of 1997, although the cosmonauts stopped the air leak just in time. It came to light, thanks to the work of a Swedish researcher in 1999, that the Soyuz always had a procedure for making a quick return to Earth. As far back as 1979, the Kettering group of amateur radio sleuths in England had listened in to transmissions from Soyuz 33 and heard the crew referring repeatedly to the *ugol pasadki* or landing angle. Later, cosmonaut Vladimir Shatalov confirmed that every night, before a crew went to sleep, they set the landing angle so they could make an emergency descent at any time. The *ugol pasadki* was, to be precise, the angle between retrofire, the Earth's centre and the landing spot. Analysis of Soyuz landing angles found that the Russians had designated a series of emergency landing zones world wide. Within the former USSR itself, there were several emergency return spots: the Sea of Okhotsk, Poltava and near Kiev in the Ukraine. However, the most intriguing landing areas are those outside the Russian Federation. These are worth listing, since the inhabitants of these areas have probably been unaware that, for years, they have been in a Soyuz emergency recovery zone! In North America, these are Manitoba, Saskatchewan, North Dakota, Forth Worth Texas, Oklahoma, presumably selected because of their flatness and similarity to the steppes of Kazakhstan. In Europe there are two selected zones: the plain of Flanders (Belgium) and Picardy (France). Therefore, in the event of the crew of the International Space Station having to leave, these are the areas in which they may come down.[23]

THE COMPLETED SPACE STATION

The International Space Station, when completed, will be the largest object ever assembled in the sky. Indeed, down below, people on Earth living between 51°N and 51°S (most people) will be able to spot the station crossing the night sky at its set height of 444 km as a bright and steady object, its solar panels glinting in the Sun. They may be able to see the regular procession of shuttles, Soyuz and other ferries flying up to the station, bringing up supplies and men and women and their equipment.

The total airtight volume of the International Space Station will reach $1,200 \text{ m}^3$. The mass of the station, when complete, will be over 415 tonnes in weight, 74 metres wide and 108 metres long—as big as two football fields end to end. The array of solar panels along its truss will generate up to 110 kW of electrical power. The overall cost of the project, from 1993 through to 2012, is likely to be in the order of €40bn, making it the largest international scientific project ever undertaken. Although led by the United States and the Russian Federation, there is a substantial investment by Europe, Canada and Japan. Europe and Japan are supplying modules and unmanned cargo ferries, while Canada is supplying mobile equipment and manoeuvring arms.

The Unites States will develop a number of experimental and habitation modules for the station, though similar Russian plans remain hazy due to lack of funding. The Institute for Medical and Biological Problems has proposed no less than 50 experiments for the prospective Russian module, whatever it is, with an artificial gravity machine and an on-board X-ray machine to measures changes in cosmonauts' bone structures. Ukraine gave a commitment to supply a module to the station by 2003, but this target date slipped, as the financial situation in the Ukraine was at least as bad as in Russia. Russia had the advantage of having built a back-up version of the FGB, called FGB-2. By summer 2000, plans had been developed by both Khrunichev and the American Boeing company to try out the FGB-2—to be launched in 2002—as a commercial module offering $20 \, m^3$ for scientific experiments.

The principal Russian commitment is called the Universal Docking Module (UDM). Design of the module began in 1997 in RKK Energiya but was halted a year later owing to lack of money. Work was resumed in October 1998, when NASA paid \$24m for services which cosmonauts would provide for their American colleagues on the ISS. Manufacture of this module began in 1999 and was due to finish in 2000, with electrical fitting out and equipping in 2001 and shipment to Baikonour for a launch in mid-2002. The title 'docking module' suggests a minor interconnecting node, but it is a much more ambitious project than its title suggests. The UDM will be a 20-tonne module, 14 metres long and 4 metres wide, to be launched by Proton. It will include $80 \, m^3$ of habitable space and will act as an airlock, a docking port for modules and as the segment connecting the Russian part of the station to the rest. It will have its own power supply, life-support system and some research equipment. It will, ultimately, host Russian modules when these are available. At one stage, the blueprints also showed additional comfort facilities for the crew of the ISS, including a sauna and showers.

The size of the Russian contribution to the ISS diminished as Russia's economic situation deteriorated in the late 1990s. Original manifests committed Russia to 122 launches between 1997 and 2012.[24] This was made up principally of Progress resupply missions (73) and manned Soyuz spacecraft (32), and, as such, was a much higher level of commitment than that given by the United States (79 launches) which was portrayed internationally as the leader of the project. By 2000, revised manifests showed that Russia had committed 25 Soyuz to the International Space Station, of which 17 would be for long-term 140-day missions and the

Soyuz TM will be phased out by the new Soyuz TMA

remainder would be what are termed taxi missions to bring up fresh spacecraft or crews. Assembly of the station would involve 10 Proton assembly missions and 30 Progress M, 65 spacecraft in all. Of the station's total baseline cost of $40m, at least half will come from the United States, $10bn from Russia and the balance from the other international partners (Europe, Canada and Japan). For an economy in Russia's state, this was nevertheless a huge and stressful investment.

COSMONAUT SQUAD

Who would fly the missions? Since 1960, the Soviet Union and then Russia had recruited over 30 groups of cosmonauts. The first, historic selection had been of 20 cosmonauts, generally young test pilots. The emphasis had been on physical fitness and the ability of candidates to react quickly to difficult situations. In the second large intake of cosmonauts, in 1963, older and more experienced air force officers were recruited. Following this, different groups of cosmonauts were recruited at irregular intervals: shuttle pilots, journalists, scientists, women and specialist engineers. About 150 cosmonauts were recruited during the Soviet period, of whom about 80 actually made it into orbit. At some times during the Soviet period, the cosmonaut squad comprised as many as 80 men and women, but these often included older men who were on the books but had no realistic chance of making a mission.

Essentially, there have been three strands to the cosmonaut groups. Military officers and pilots have always been the dominant element of the squad. Almost all Soyuz missions are commanded by a military pilot, as is the case with the American space shuttle. The second type of group to be selected is the flight engineer. Originally, this was drawn from the OKB-1 Korolev design bureau—indeed, it was once called the Korolev kindergarten. Even today, the second largest part of the cosmonaut group consists of OKB-1, now called RKK Energiya engineers and their number reflects the continued dominance of Energiya in the Russian space programme. As often as not, the engineers are older than their military commanders. These engineers have not only a scientific interest but also know the ins-and-outs of a space station's systems intimately and are expected to fix things that go wrong. In one sense, the engineer, although junior in rank, may be more crucial to the success of a mission than the commander.

The third part of the cosmonaut squad might be termed miscellaneous and consists of doctors drawn from the Institute for Medical and Biological Problems (IMBP) and engineers and specialists selected from design bureaux other than Energiya. Despite the strong tradition of space medicine and long-duration flights in the USSR and Russia, remarkably few doctors have flown in space. By 1989, no less than 19 doctors had been recruited for spaceflight, but only three—Boris Yegorov (1964), Oleg Atkov (1984) and Valeri Poliakov (1988–9, 1994–5)—had actually flown. The most recent group of doctors was recruited in 1989—Vladimir Karashtin, Vasili Lukyanyuk and Boris Morukov.[25] Following the problems of nausea on the second visiting shuttle to Zarya, Boris Morukov was assigned to

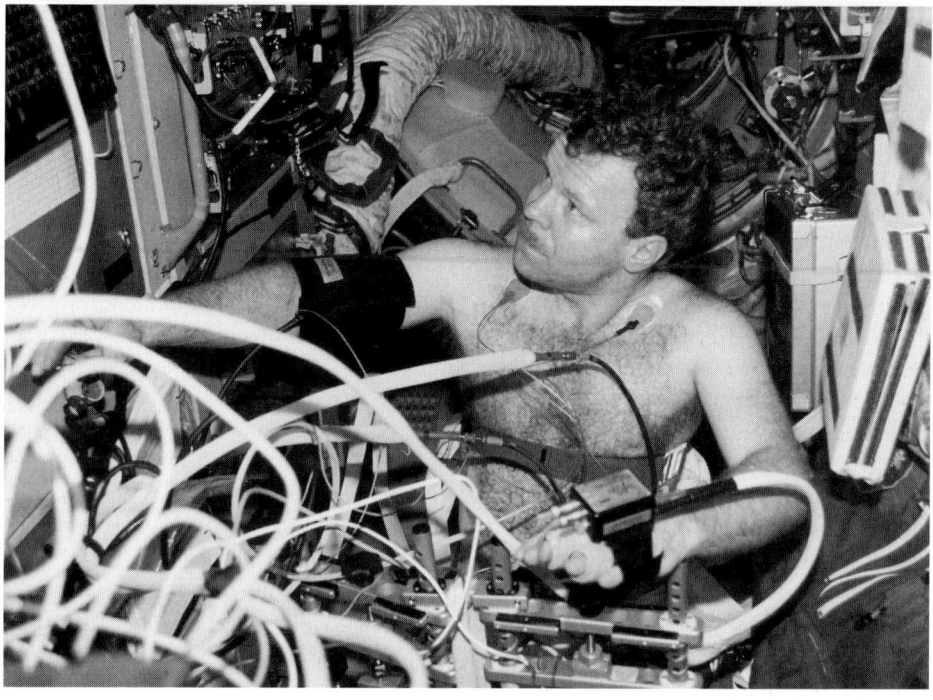

Only four Russian doctors have flown in space. Here a visiting cosmonaut has a medical test.

the September 2000 shuttle mission to visit the ISS in order to be on hand in the event of any further air problems, but he was only the fourth Soviet or Russian doctor to orbit the Earth.

The early cosmonauts were household names. By the time of the long Salyut missions in the late 1970s, much of the glamour had worn off the profession and cosmonauts were less and less likely to be stopped in the street for an autograph or

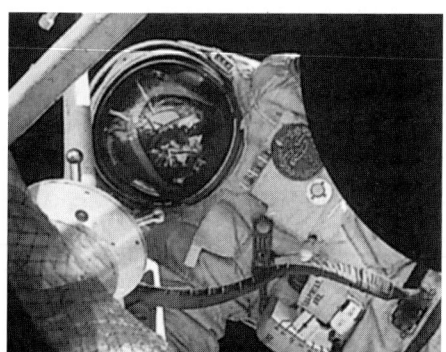

Spacewalking from Mir: $1,000 bonus

quick conversation. For those joining the cosmonaut squad in the 1980s, it was just another, albeit unusual, profession and most valued the falling off in external media attention. The changes in the 1990s led to a significant alteration in the way in which cosmonauts were paid, one which amazed the Americans (and still does). American astronauts receive a standard military paycheck or, as civilians, a standard NASA one, regardless of whether they are on or off the planet. By contrast, following 1991, cosmonauts were put on an incentive system mimicking the worst practices of capitalism. Cosmonauts received a contract for each space mission, for which they were paid at a rate of $100 a day (civilians were paid $80). They got a bonus of $1,000 for a spacewalk. If particular aspects of the mission were not accomplished, fines were applied afterwards and deducted from the contract. For some peculiar reason, a bonus was applied for carrying out a manual docking. From the moment it was introduced, cosmonauts unfailingly informed ground control, about 10 metres before a smooth automatic docking, that unspecified problems had arisen and they had urgently to take over manual control!

For the 1990s, the manned Russian space programme relied on pilots and engineers selected in the course of the 1980s. The occupation of Mir required two, at the most three, manned flights a year. Granted that one crew member must always be an experienced cosmonaut, new flight opportunities for cosmonauts in the 1990s were limited to two or three a year—which was not very many. Accordingly, recruitment was slow. The first group of post-Soviet cosmonauts comprised three civilians selected from RKK Energiya in March 1992: Alexander Lazutkin, Sergei Treshchev and Pavel Vinogradov. Lazutkin and Vinogradov learned about space travel the hard way, for Lazutkin was on Mir during the fire and the collision, while Vinogradov took the next Soyuz up for repair work. Because of financial pressures, the third seat in the Soyuz was nearly always taken by a paying, visiting cosmonaut, which meant that although several doctors were listed as members of the squad, their actual chances of flying remained very slim. None of the new doctors was given a flight in the 1990s and none had been slated for a Russian mission by 2000.

Two more civilians were selected from RKK Energiya in April 1994: Nadezhda Kuzhelnaya and Mikhail Tyurin. Nadezhda Kuzhelnaya was one of a small number of Russian women cosmonauts, as equal opportunities is not a strong feature of the Russian cosmonaut squad. Although the Soviet Union flew the first woman in space (Valentina Terreskhova, 1963) and twice flew Svetlana Savitskaya up to the Salyut 7 space station (1982, 1984), only one Russian woman had flown since (Elena Kondakova, 1994). The cosmonaut squad for 2000 included only one woman, Nadezhda Kuzhelnaya, who was slated for an ISS taxi mission. By contrast, 29 American women had flown in space by then and Eileen Collins had even commanded a shuttle.

In the immediate post-Soviet period, many of the older, experienced cosmonauts who had no chance of making a further mission were retired from the squad. In 1993–4, when Buran was cancelled, the shuttle pilots were taken off the payroll. The group had already been depleted by retirement—its members moving on to other professional opportunities—and by a series of fatal aircraft accidents. By 1994, the cosmonaut squad had contracted to 17 pilots, 12 Energiya engineers and five IMBP doctors (34 people altogether)—probably its lowest number since the early 1960s—but an accurate reflection of the numbers of people available for missions.

Selection of Russian cosmonauts, 1992–2000

Civilians selected from RKK Energiya, March 1992

Alexander Lazutkin	Sergei Treshchev	Pavel Vinogradov

Civilians selected from RKK Energiya, April 1994

Nadezhda Kuzhelnaya	Mikhail Tyurin

Civilian selection, spring 1996

Konstantin Kozeyev	Oleg Kotov	Sergei Revin
Yuri Shargin	Oleg Kononenko	Oleg Skripotchka

Pilot selection, summer 1997

Dmitri Kondratyev	Yuri Lonchakov	Oleg Moshkin
Roman Romanenko	Alexander Skvortsov	Maxim Surayev
Konstantin Valkov	Sergei Volkov	Sergei Moshenko
Fedor Yurchikin		

This situation lasted only a few years. Expansion was prompted by the need to have pilots and engineers available, not immediately, but for International Space Station missions in the 2000–2010 period. Accordingly, a call for new cosmonauts went out in 1996. For the pilot selection of 1997, 55 made it through to the final selection process and, of these, 7 were chosen. Four engineers were selected at around the same time and, on 12 January 2000, pilots Konstantin Valkov, Dmitri Kondratyev, Yuri Lonchakov, Alexander Skvortsov, Maxim Surayev, Roman Romanenko and Sergei Volkov all received their cosmonaut badges, along with RKK Energiya engineers Fedor Yurchikin, Oleg Skripotchka, Mikhail Kornienko and Khrunichev engineer Sergei Moshenko. They posed to have their photograph taken beside the stone bust of Sergei Korolev on cosmonauts' avenue at the Exhibition for Economic Achievements. These men had completed their training and five months later were assigned to training groups. The group was distinguished by including two men who were the sons of cosmonauts: Sergei Volkov, son of Alexander; and Roman Romanenko, son of Yuri. The latter had left the cosmonaut squad after his record—breaking mission in 1987, while Alexander Volkov had only just left, ruining the chances of the first ever father-and-son team to fly into orbit. Alexander Skvortsov was the son of A.A. Skvortsov, who was a cosmonaut from the 1965 group but had never flown.

In 2000, the Russian cosmonaut squad comprised 18 pilots, 15 Energiya engineers, 4 doctors and 4 others (total 41). It had returned to the more typical size of the squad during the Soviet period. In addition to these 41, several American, European and Japanese (and sometimes even Chinese) astronauts were training at the Yuri Gagarin Space Centre . The other astronaut squads in the world comprised Europe (16), Japan (8) and the United States (up to 160). The present group is listed in 'Russia's cosmonaut squad'.

Russia's cosmonaut squad

Air Force pilots	RKK Energiya engineers	Others
Yuri Gidzenko	Sergei Krikalev	*IMBP*
Vladimir Dezhurov	Yuri Usachov	Boris Morukov
Yuri Onufrienko	Mikhail Tyurin	Vladimir Karashtin
Valeri Korzun	Sergei Treshchev	Vasili Lukyanyuk
Yuri Malenchenko	Sergei Treshchev	
Viktor Afanasayev	Sergei Revin	*Star Town doctor*
Valeri Tokarev	Nikolai Budarin	Oleg Kotov
Gennadiy Padalka	Konstantin Kozeyev	
Yuri Lonchakov	Alexander Poleschuk	*Space forces engineer*
Dmitri Kondratyev	Fedor Yurchikin	Yuri Shargin
Konstantin Valkov	Oleg Skripotchka	
Sergei Volkov	Pavel Vinogradov	*TsSKB Samara engineer*
Alexander Skvortsov	Alexander Kaleri	Oleg Kononenko
Roman Romanenko	Sergei Avdeev	
Maxim Surayev	Mikhail Kornienko	*Salyut KB engineer*
Talgat Musabayev		Sergei Moshenko
Salizan Sharipov		
Sergei Zalotin		*Presidential adviser*
		Yuri Baturin

In addition, three foreign cosmonauts also hold cosmonaut certificates: Claudie André-Deshays (France), Thomas Reiter (Germany) and Christer Fuglesang (Sweden) who took a course in piloting Soyuz for emergency missions.

The allocation of assignments within the cosmonaut squad in 2000 can be seen in the table on page 104.

Sadly, the 1990s also saw the passing of a number of Russia's older, senior cosmonauts: Boris Yegorov (1994), Georgi Beregovoi (1995), Georgi Shonin (1997), Yuri Artyukhin (1998), Lev Demin (1998), Yuri Malashev (1999) and Gherman Titov and Yevgeni Khrunov (2000). Many died long before reaching old age: Yuri Malashev was only 58, Georgi Shonin was 61, Yevgeni Khrunov was 64 while Boris Yegorov suffered a fatal heart attack at 59. Yegorov had experienced the loss of his son, a 20-year-old medical student, who died in the course of the 1993 shoot-out at the Ostankino television centre during the crushing of the white house revolt. One cosmonaut died during training: in July 1993, 35-year-old novice Sergei Vozovikov, only three years after his selection, was caught in a fishing net while swimming in the Black Sea and drowned.

THE SPACE STATION FRONTIER

For Russia, the International Space Station was, initially at least, a mixed blessing. On the positive side, it offered an opportunity whereby Russia could continue to

Cosmonaut squad assignments

Mission	Cosmonaut
ISS 1	Yuri Gidzenko, Sergei Krikalev (Shepherd)
ISS 2	Yuri Usachov (Voss, Helms)
ISS 3	Vladimir Dezhurov, Mikhail Tyurin (Culbertson)
ISS 4	Yuri Onufrienko (Bursch, Walz)
ISS 5	Sergei Korzun, Valeri Treshchev (Whitson)
ISS 6	Yuri Malenchenko (Robinson, Fincke)
Taxi 1	Viktor Afanasayev, Nadezhda Kuzhelnaya
Taxi 2	Valeri Tokarev, Sergei Revin
ISS emergency	Gennadiy Padalka, Nikolai Budarin
	Back-ups: Valeri Korzun, Sergei Treshchev
Shuttle	Yuri Malenchenko, Boris Morukov
Training	(1) Oleg Kotov, Yuri Shargin, Oleg Kononenko, Konstantin Kozeyev, Alexander Poleschuk; (2) Yuri Lonchakov, Dmitri Kondratyev, Konstantin Valkov, Sergei Volkov, Fedor Yurchikin, Sergei Moshenko; (3) Alexander Skvortsov, Roman Romanenko, Oleg Skripotchka, Maxim Surayev, Mikhail Kornienko
Mir	EO-29: Salizan Sharipov, Pavel Vinogradov
	EO-30: Talgat Musabayev, Yuri Baturin
Unassigned	Vladimir Karashtin, Vasili Lukyanyuk, Sergei Zalotin, Alexander Kaleri, Alexander Lazutkin, Sergei Avdeev

Soyuz crew prepares to leave Earth

maintain its commitment to keeping cosmonauts in space. Without the ISS it is very doubtful if Russia could have afforded on its own to launch the Mir 2 space station, man it and resupply it. American cash provided a means whereby Mir 2 could be kitted out, the FGB built and the production line for the Soyuz and Progress kept rolling.

On the other hand, there was a debit side to the space station. Financial dependence on the United States for funding at key points created an unhealthy relationship in which Russia came to see itself, and was seen internationally, as the junior partner. Indeed, the ISS came to be projected in some of the Western media as an American project in which the Russians were either minor players or were being given a minimal role out of charity. American financial dominance meant that money talked whenever there were, as indeed there were, arguments about such diverse matters as crew composition and the measurement system to be used during the programme (metric or imperial). This did less than justice to the enormous legacy of orbital construction which the Russians were bringing to the project, the very real and tough level of Russian financial investment (€10bn over its 12-year base lifetime) and the considerable number of launches and resupply missions to which the Russians were committed. The ISS was heavily dependent on Russian expertise, its three large modules (Zarya, Zvezda and the Universal Docking Module) and the fact that it was providing an average of half the permanent crew, either two or three cosmonauts at any given time. Although the Russians had no option but to participate in the ISS in 1993, it was often forgotten that the Americans had been equally obliged to do so, for political support for an independent go-it-alone American station had ebbed equally by the same time.

Some American comment on the Russian commitment to the ISS has been harsh to the point of unfairness. It has frequently been implied that the Russian commitment has depended exclusively on American financing and that Russia has put none of its own resources into the project. This is not true. In 1997, for example, R1.8 trillion out of the space budget of R5.1 trillion requested was for the ISS. Of the $21bn *assembly* costs (not the total cost) of the ISS for 1993–2005, $13bn is borne by the United States, Russia is paying $2.51bn, Europe $2.46bn, Japan $2.36bn, and Canada $68m. The Russian contribution is small compared to the United States, but it is nevertheless the second largest.

Russia's participation in the ISS is likely to be more evident as this huge enterprise gets fully under way. With regular Soyuz crew replacement and taxi flights, the arrival of the Universal Docking Module, the presence of the emergency Soyuz lifeboat and the frequent Progress supply missions, the Russian role in the construction of the ISS may be better appreciated and understood. With an economic recovery, Russia may be able to build and fly its long-promised scientific modules to the station and its presence will be ever more visible.

But so much for the tribulations of the manned Russian space programme, from Mir to the International Space Station. Russia had kept Mir going, against all the odds, until 2000, by which time the new International Space Station was ready. What about its unmanned space programmes?

The Soyuz rocket—mainstay of Russian flights to the ISS

4

The military frontier
Russia's military space programme from 1992

Throughout this period, Russia ran an unmanned programme that was, and still is, the second biggest in the world. It is not well known in the West, although it has many intriguing and unusual features. Chapter 4 describes the main aspects of the Russian unmanned military programme from 1992 to 2000 and Chapter 5 will discuss the civilian unmanned programme.

THE UNMANNED SPACE PROGRAMMES

Although the unmanned space programme was much less well known than the high-profile manned, lunar and interplanetary missions, it was nonetheless a large programme. It was largely a military programme flown under the Cosmos label and this passed the 2373 mark in late 2000.* Russia inherited a huge unmanned military space programme from the Soviet Union. Within the military space programme, there were a number of important subprogrammes. The most important of these was photoreconnaisance, followed by electronic intelligence. Other important military programmes have been concerned with missile early warning, military communications and navigation.

With *glasnost*, parts of the Cosmos programme were declassified. The Russians began to admit that not all the two thousand or so missions launched since 1962 were, as was originally claimed, purely scientific in nature. With *glasnost*, it also became standard practice for civilian parts of the Cosmos programme to receive civilian designations (e.g. Nadezhda). Thus many of the 'new' programmes that appeared in the late 1980s and early 1990s had been running within the Cosmos

---

* A number of important military programmes had become defunct by the end of the Soviet period and these are not described here. These were the system of hunter–killer satellites (1967–82), the fractional orbital bombardment system (1967–9) and the nuclear-powered radar ocean reconnaissance satellites (RORSATS, 1974–88).

programme for years. The Cosmos label became used less and less and when it was, it applied exclusively to military missions, which was largely why the label had really been created in the first place. Even when the military Cosmos label was still used, the Russians were more forthcoming about the type of military mission it was. By the late 1990s, Russia was more open about some aspects of its military programme than the United States. Despite these changes, welcome from the point of view of external analysts, the division between Russian military and civilian programmes is still not always clinically clear, either operationally or by nomenclature. Some military and civilian programmes run alongside one another (e.g. the Nadezhda/Tsikada navigation satellite series, Strela/Gonetz communications programme). Several programmes are also half military and half civilian (e.g. GLONASS).

Soviet and Russian military space programmes (by classification, Russian names and dates)

Photoreconnaissance			*Military communications and relays*		
1st generation	Zenit	1962–7	Relays	Strela 1	1970–92
2nd generation	Zenit	1962–78	Store-dump	Strela 2	1970–94
3rd generation	Zenit	1978–94	Relays	Strela 3	1986–
4th generation	Yantar	1974–	24-hour relay	Potok	1974
5th generation	Terilen/Neman	1982–			
6th generation	Orlets 1	1989–	*Navigation satellites*		
7th generation	Orlets 2	1994–	High altitude	GLONASS	1982–
8th generation	Arkon	1997–	Low altitude	Tsikada	1997–
			Low altitiude	Parus	1974–
			Rescue	Nedezhda	1989–
Electronic intelligence					
3rd generation	Tselina D	1977–93	*Missile early warning*		
4th generation	Tselina 2	1985–	Molniya orbit	Oko	1972
			24-hour orbit	Prognoz	1988
Military Ocean electronic intelligence					
EORSAT	US-P	1967–	*Radar calibration*		
			Yug		1977
			Vektor		1974
			Romb		1976

PHOTORECONNAISSANCE

With the start of the cold war in the late 1940s, both sides realized the importance of obtaining high-quality photoreconnaissance photographs of the other's military facilities. In the early 1950s, high-speed British Canberra bombers fitted with cameras repeatedly intruded into western Russia and the Ukraine to take images of Soviet military facilities. In the late 1950s, the Americans introduced the high-altitude U-2 spyplane which flew deep into Soviet territory from covert American Air Force bases in Turkey. When a U-2 was shot down in May 1960, this was no longer an option and the United States became exclusively dependent on satellite imagery. For the Soviet Union, the possibilities of satellite-based photo-reconnaissance had been realized from the early 1950s. Unlike prying spyplanes, satellites could not be shot down, at least, not legally, and it was difficult technically in any case. The USSR did not have bases close to the United States and was not in a position to overfly North America easily: prowling Bear spyplanes flying across the Greenland Gap were picked up far out to sea and easily intercepted. For the USSR, spy satellites had always been the only option. In the event, both sides found that photoreconnaissance satellites greatly exceeded their expectations and their cameras were able to film a quality of detail neither side had ever expected. By the 1980s, the Americans claimed they could pick out the ranks on the shoulder labels of Soviet military officers from 200 km above. Even if this was an exaggeration (we should not assume it was), the story told something of the capability of the eye-in-the-sky.

The Soviet photoreconnaissance programme has always presented difficulties of analysis and interpretation. Officially, the programme did not exist until the late 1980s, though the amount of official documentation of the early period has now improved radically. Western analysts have applied to the Soviet photo-reconnaissance programme the terms first, second, third and fourth generation and so on, based on the type of spacecraft used, its performance, the orbital paths followed and mission characteristics. The USSR and Russia have applied series names (Zenit, Yantar, Orlets, etc.) and both are used here. Between them, the Soviet Union and Russia have launched over 780 photoreconnaissance satellites, presenting a mammoth task of cataloguing and interpretation.

The decline in Russian military space capabilities was dramatically apparent when war broke out in the Balkans at the end of March 1999. In previous wars (for example, the middle east in 1967 and 1973, the Falklands in 1982 and the Gulf in 1991), the USSR had been quick to get photoreconnaissance satellites over the conflict zone to spot real military operations as they took place. This time, Russia had only one photoreconnaissance satellite and one electronic intelligence satellite aloft to watch and listen to the Balkan conflict unfold. By contrast, the United States had two 15-tonne Lacrosse radar-imaging satellites, three KH-11 photo-reconnaissance satellites and three smaller imaging spacecraft.

Russian space reconnaissance deteriorated to the point that there were several occasions on which Russian military intelligence was 'blind' during this period, though never 'deaf' at the same time. Between the return of Cosmos 2320 on 28 September 1996 and the launch of Cosmos 2343 on 15 May 1997, Russia had no orbiting operational photoreconnaissance satellites. There was a second gap two years later when the Cosmos 2365 Yantar 4K2 Kobalt returned on 15 December 1999; this was not filled until the Cosmos 2370 Yantar 4KS2 Neman launched on 3 May 2000.

ZENIT

The Soviet photoreconnaissance satellite programme was ordered as early as 1956–7 and Sergei Korolev was the chief designer.[26] The 2.3-metre-diameter spherical satellite he designed, called Zenit 2, became two years later the basis of the Vostok manned spaceship. Many people assumed that the spy satellites were derived from the manned programme, but in fact it was the other way around, such was the urgency of the USSR to get spy photographs.

The first attempt to orbit a Zenit was 11 November 1961 and the first partly successful one was called Cosmos 4 on April 1962. Zenit carried four cameras, three 1,000 mm and one 200 mm. The swath was 180 km wide. The cameras could be turned for either downward-looking or oblique views of the Earth's surface. The aim of each mission was to cover an area equivalent to the United States, 10 million km^2. Each camera could take up to 1,500 frames. This, the first series of photoreconnaissance satellites, operated from 1962 to 1967. On average the payload flew for eight days, later models for 14 days, before the cabin returned to Earth. Once it landed, the film was whisked away for analysis. The advantage of Zenit was that high-quality black-and-white images could be obtained systematically and repeatedly of most parts of the Earth's surface. In the event, the quality of pictures was better than expected. Resolution was 7 metres, better than the 10 metres expected. The disadvantage was that this information became available only after the Zenit landed, by which time it was already dated.

The West had many difficulties interpreting the Zenit series. The Cosmos programme had been introduced in 1962 as a scientific research programme. The rapid pace of Cosmos launches, weekly by the 1970s, aroused Western suspicions, for no country could have a commitment to science that justified weekly launches. However, all Western experts had to go on were the launch site, the orbital paths followed and dates when cabins dropped out of orbit. In fact, most Cosmos launches fell into patterns and eventually became quite predictable. Of course, some from time to time defied the pattern and these came to be labelled 'obscure' Cosmos missions, with an obdurate subset called 'even more obscure' Cosmos. Western analysis found that there were three main variants of the early photoreconnaissance Cosmos and these were labelled the first, second and third generation of Soviet photo-reconnaissance satellites. The term 'Zenit', which the Soviets themselves used, was not revealed until *glasnost*. In reality, there were closer to 10 variants of Zenit, which the designers identified as Zenit 2, 4, 2M, 4M, 4MK, MT, 6 and 8. Within the Zenit programme, they were also given the subclassifications of Hector, Hermes, Heracles, Fram, Orion, Argon, Rotor and Oblik.

Only one of these variants flew into the Russian period, Zenit 8 (factory designation 17F116), called Oblik, introduced with Cosmos 1571 and operationally with 1584 in 1984. Typically, Oblik orbited at 70 degrees flying between 350 and 420 km.[27] It was a relatively small part of a large programme and only eight or nine Obliks were ever flown, the last category of the Zenit series. During the period of the Russian space programme, two Obliks were put in orbit, Cosmos 2207 and Cosmos 2281, the last. The Zenit series seems simply to have run to conclusion. The production run of cabins may well have ended and the series had long since been superseded by the more efficient Yantar system.

In both cases, the Oblik circled the Earth, its Vostok-shaped cabin pointing downward towards Earth as its cameras snapped away at military targets. After

22 days in the case of Cosmos 2281, retrorockets fired and the cabin parachuted back to the standard recovery zone in Kazakhstan. Once landed, the precious film was extracted and flown immediately to GRU military intelligence headquarters in Moscow for analysis and interpretation.

The return of Cosmos 2281 did not mean the absolute end of the Zenit series. Zenit capsules were converted for civilian use and called Resurs (see Chapter 5). The characteristics of Resurs flights were not very different from Zenit and probably carried similar equipment. In addition, one more Cosmos satellite, 2260, was also given the label of Resurs T, and may well have been an Oblik with a non-military assignment.

Oblik, third-generation photoreconnaissance satellite, Zenit 8 series

30 Jul 1992	Cosmos 2207	Soyuz	Plesetsk	237–355	90.49	82.33
7 Jun 1994	Cosmos 2281	Soyuz	Plesetsk	237–296	89.84	82.58

YANTAR

The principal shortcoming of the third generation was that military ground controllers had to wait until the satellite returned to Earth before they could analyse the film. This problem was partly solved by the fourth-generation series, Yantar. These were much more sophisticated satellites, able to settle in a parking orbit, but swoop down to skim the upper atmosphere as they zoomed in on targets of special interest before retreating back to higher orbit. Small capsules were ejected for re-entry, up to two on each mission. Pictures could then be analysed when the main satellite was still aloft and the satellite could then be ordered to concentrate on new areas. The main cabins were often reused, up to three times. Long before the 'reusable' space shuttle was invented, the Russians were already reusing their cabins.

This new satellite was commissioned by Soviet military intelligence, the GRU, in 1964 and the project was given the code name Yantar, the Russian word for Amber.[28] The assignment was given to the Kozlov design bureau in Kyubyshev, now TsSKB of Samara.

Originally it was thought by Western experts that just as Soyuz had followed Vostok, so Yantar had followed Zenit and was based on the Soyuz design, but this was incorrect. Yantar was an original design, though it is true that it borrowed concepts from a Soyuz military version then under consideration. Yantar was a fibreglass spacecraft, cone shaped, over 8 metres tall, with a long lense at the tapering end and an overall weight of 6.6 tonnes. It had a demanding specification: the ability to orbit for at least a month (more than three times as long as Zenit at the time), perform up to 50 manoeuvres in orbit, send down two film capsules during its mission and come back to Earth itself for reuse. Power was supplied by two large solar panels at the base of the system. Yantar had an unusually powerful camera lense, with a resolution of 50 cm, sufficient to identify individual tanks, aircraft and rockets. The film was wound either into the main re-entry vehicle or into two 80-cm spherical capsules on the side of the Yantar. These descent capsules, called SpKs, had a solid fuel retrorocket and could be parachuted down as commanded, with a VHF beacon attracting the attention of the recovery forces. When its mission was over, the main Yantar retracted its lense, burned its retrorockets and descended to Russia. A small retrorocket under the parachute cushioned the final stage of the descent to prevent damage to the delicate and expensive camera and optical system. The main orbital manoeuvring engine weighed 375 kg, used UDMH and N_2O_4 and had an impulse of 1,060 kN.

Yantar was a complex satellite which took some time to design and build. The first Yantar appeared in December 1974 with Cosmos 697, but it took six test missions to achieve a fully successful profile. Yantar was declared operational three years later. The first Yantars were launched from both Baikonour and Plesetsk into typical orbits of 180 by 370 km. Once parked in orbit, they awaited commands to inspect individual military targets. To do this, they would fire their engines, descend to 160 km for a close pass and then move back into a higher orbit. The original model, Yantar 2K Feniks, introduced in the 1970s, was flown about 30 times and was replaced in the early 1980s by an improved version, the Yantar 4K1 Oktan, able to stay in orbit twice as long, 45 days. Twelve Oktans flew before they were replaced by the two versions used in the final period of the Soviet/Russian programme.

KOBALT: CLOSE LOOK ANALYSIS

A new version was introduced in 1982 and became operational in 1984, the Yantar 4K2 Kobalt (code number 11F695), was able to stay in orbit for 60 days and take a close look at targets of particular interest. Typically, Kobalts were launched into orbits of 180 by 370 km, orbiting the Earth at 62.8 or 67.2 degrees from Plesetsk. Once in orbit, however, they would move up and down, according to the close-look target selected. Such satellites were ideal for checking on the observance of arms agreements, following conflicts in progress and watching other targets which justified military curiosity. Kobalt was a substantial programme: 83 had been launched by 2000.

The potential of Kobalt as a means of looking closely at military targets was first in evidence during the Gulf War in 1991. Three Kobalts were used to dip low over the battlefield and survey the struggle under way in the desert below—Cosmos 2108, 2124 and 2138.[29]

Russia launched 17 Kobalt missions from 1992 to 2000. This was a much reduced launch rate from the 1980s. Partly, this could be attributed to diminished funding, but also to reduced military tension requiring a lower launch rate. Most importantly, however, the 60-day mission limit gradually became extended. First, it was gradually extended to 65 days in 1993 and to 71 days in 1994 until 120 days became the norm. As a result, Russia was able to obtain twice the value for each Kobalt mission compared to the 1980s.

Despite funding problems, Russia was able to maintain its Kobalt programme throughout the period. In the early part, a new Kobalt was launched almost as soon as the previous one returned. At least one was launched every year, with as many as three in 1993. By the end of the decade, Kobalt missions were reduced to an annual mission. Cosmos 2348, 2358 and 2365 each made 120-day missions in 1997, 1998 and 1999 respectively. The last, Cosmos 2365, returned to Earth on 15 December 1999 after 120 days and no replacement mission had been flown by the following autumn. There was one apparent failure during the period: Cosmos 2259, which should have stayed 60 days aloft, was brought back down after only two weeks.

KOMETA: MILITARY IMAGES FOR SALE

A military mapping version of Yantar was introduced in 1981, the Yantar 1KFT Kometa, starting with Cosmos 1246. This series was much less prolific—once done, a mapping exercise normally lasts for some time and does not need frequent rechecking. Kometa carried two cameras—a main mapping camera taking frames of 200 by 300 km and a precision mapping camera for frames of 180 by 40 km. Kometa flew higher than the Feniks or the Kobalt, at 200–270 km. Kometa used a laser altimeter to measure its precise height over the ground—essential for accurate mapping.

Sixteen Kometas were flown under the Soviet programme from 1981 to 1991. Only two Kometas were flown under the Russian space programme, Cosmos 2185, 2284, and one failure, 2243. Cosmos 2243 malfunctioned on its last stage when there was an explosion, entered an unstable orbit and decayed after 10 days.

In an initiative going back to Soviet times, from 1992 the Russians started to make Resurs and Kometa photographs available to the West on a commercial basis through the state company Soyuzkarta. Whether the pictures in question came from Resurs or Kometa was never entirely clear and they probably came from both. Either way, the razor-sharp picture of the Congress building in Washington, DC, was unmistakable and nailed the notion that, somehow, Russians cameras were not very good. The United States Air Force (USAF), duly intrigued at this development, used commercial and what were described charmingly as 'other' intermediaries to buy a considerable quantity of Kometa photographs in the course of 1992–3. For the Americans, this provided an opportunity to obtain images of their own country, something which their own spy satellites had, by definition, not been doing. More importantly, it gave them the opportunity to assess Russian capabilities. During the October 1993 funeral of General Jimmy Dolittle in Washington, DC, Russian images were used by the USAF to plan the route of the cortege through the national capital.

Making Kometa available on a commercial basis involved the declassification of the two camera systems of Kometa—the topographic camera TK-350, with a resolution of 10 metres for 1:660,000 mapping and the high-resolution KVR-1000 with a resolution of 2 metres and able to make maps on a 1:220,000 scale. Soyuzkarta made an arrangement with Aerial Images Inc. of Raleigh North Carolina for the Kometa pictures for a price of between $13m and $20m, done in cooperation with

Kodak and Microsoft. The idea was to build a topographic library of the United States and other Western countries.

The mission was called SPIN-2 (because of its 2-metre resolution). However, the first SPIN suffered an unusual Soyuz launch failure. The first commercial Kometa failed with the Soyuz rocket explosion of 14 May 1996, but the second flew as Cosmos 2349 from 17 February to 3 April 1998. Aerial Images sold its pictures, with 2 metres resolution, once the Cosmos returned to Earth. The SPIN-2 made maps of 1 million km^2 of US territory which were subsequently sold on the Internet for $4m. The SPIN-2 photographs were staggering in their detail. Their definition turned out to be closer to 1 metre than the 2 metres advertised. Aerial Images put them on the web (view for free, download for fee) and invited people to zoom in on their home or workplace. The commercialization of Russian space images led, against some military advice, President Clinton to permit American companies to make available satellite images on a commercial basis, leading to a famous exchange in which a defence department intelligence analyst noticed an unknown person handling a spy photo of North Korea. He asked: 'Where did you get that? It's a secret!' 'No', the man replied. 'It's for sale!'

A new Kometa, for domestic use, Cosmos 2373, was launched in September 2000.

Yantar Kobalt and Kometa series

21 Jan 1992	Cosmos 2175	Soyuz U Plesetsk	184–337	89.74	67.14	Kobalt
1 Apr 1992	Cosmos 2182	Soyuz U Plesetsk	166–315	89.33	67.15	Kobalt
29 Apr 1992	Cosmos 2185	Soyuz U Baikonour	211–279	89.43	69.97	Kometa
28 May 1992	Cosmos 2186	Soyuz U Plesetsk	182–352	89.94	62.85	Kobalt
24 Jul 1992	Cosmos 2203	Soyuz U Plesetsk	190–312	89.53	62.81	Kobalt
22 Sep 1992	Cosmos 2210	Soyuz U Plesetsk	150–349	89.76	67.26	Kobalt
20 Nov 1992	Cosmos 2220	Soyuz U Plesetsk	167–342	89.62	67.14	Kobalt
19 Jan 1993	Cosmos 2231	Soyuz U Plesetsk	179–376	90.06	67.13	Kobalt
2 Apr 1993	Cosmos 2240	Soyuz U Plesetsk	190–322	89.62	62.85	Kobalt
27 Apr 1993	Cosmos 2243	Soyuz U Baikonour	192–236	88.77	70.35	Kometa failure
14 Jul 1993	Cosmos 2259	Soyuz U Plesetsk	170–366	89.99	67.12	Kobalt failure
17 Mar 1994	Cosmos 2274	Soyuz U Plesetsk	163–350	89.64	67.13	Kobalt
20 Jul 1994	Cosmos 2283	Soyuz U Plesetsk	169–330	89.5	67.11	Kobalt
29 Jul 1994	Cosmos 2284	Soyuz U Baikonour	214–277	89.41	70.38	Kometa
22 Mar 1995	Cosmos 2311	Soyuz U Plesetsk	168–336	89.54	67.19	Kobalt
28 Jun 1995	Cosmos 2314	Soyuz U Plesetsk	166–340	89.56	67.13	Kobalt
14 Mar 1996	Cosmos 2331	Soyuz U Plesetsk	164–358	89.73	67.14	Kobalt
14 May 1996	Cosmos	Soyuz U Baikonour				Kometa failure
15 Dec 1997	Cosmos 2348	Soyuz U Plesetsk	165–345	89.61	67.15	Kobalt
17 Feb 1998	Cosmos 2349	Soyuz U Baikonour	212–278	89.41	70.37	Kometa/SPIN
24 Jun 1998	Cosmos 2358	Soyuz U Plesetsk	167–334	89.52	67.13	Kobalt
18 Aug 1999	Cosmos 2365	Soyuz U Plesetsk	167–343	89.6	67.14	Kobalt

GLOBAL COVERAGE: YANTAR 4KS1/TERILEN/NEMAN

The fifth generation of photoreconnaissance satellites, Yantar 3KS1, was introduced in December 1982 (Cosmos 1426) as project 11F117. Although the Yantar Feniks, Oktan, Kobalt and Kometa satellites were able to send capsules with film down for analysis, they still presented the problem that data could only be analysed afterwards. What was needed was a satellite that was able to transmit data which could be interpreted in real time, before any cabin returned to Earth. The Yantar KS1 series was therefore devised to return imaging data continuously in real time.

The Kozlov bureau sought to overcome the problem by returning images electronically, on-board devices scanning pictures and transmitting them back to Earth almost immediately. This method had first been demonstrated, albeit primitively, when Luna 3's pictures of the far side of the Moon were developed on board, scanned electronically, and transmitted back to Earth. This breakthrough was made possible by computer advances, charge-coupled devices and digital means of storing information. The Americans mastered this technology first, introducing it with the KH-11 satellite in 1976.

The first 14 launches in the series, lasting in the order of 6–8 months, were called Terilen (4KS1) and the later missions, beginning in 1992 and generally of 10 months duration or more, were called Neman (4KS2). Launch is from Baikonour on a Soyuz U into a nearly circular orbit of 225 km with an inclinations of 65 or 70 degrees. Regular engine boosts maintain this altitude quite strictly. The resolution of its cameras is thought to be in the order of 2 metres. Unlike the previous models, the Yantar does not return to Earth. Instead, after a year's work, it is de-orbited to destruction over the Pacific Ocean.[30]

Yantar Terilen and Neman have the same conical shape as the previous Yantar, its long lens peeping downward towards Earth, solar panels at the top. The principal difference is two data relay antennas at the mid-point, directed not downward to mission control near Golitsyno but outward towards 24-hour orbit. No signals have ever been picked up from this generation of satellites and it is thought that they beam their digital information continuously up to Potok relay satellites whence they are relayed back down to the ground. The Potok system, also called Geyser (also spelt Geizer), uses satellites in 24-hour orbit which then retransmit, in code, the Terilen images to ground stations in Konakovo, Golitsyno in Moscow and Nahodka in the far east. Thus the establishment of the Terilen/Neman system also involves a commitment to maintaining a network of Potok relays in 24-hour orbit using the Proton rocket. It is therefore an expensive system at a time of scarce resources.

At the end of the Soviet Union, there was only Terilen in orbit, Cosmos 2153, launched in July 1991 and de-orbited on 13 March 1992. There was then a gap of a month before the launching of its successor, Cosmos 2183, the first in the longer-duration Neman series. Cosmos 2183 made over 40 orbital manoeuvres before it de-orbited on 16 February 1993 after a record 314 days circling the Earth. Typically, it changed its orbit five times a month. Cosmos 2223 was sent aloft on 9 December 1993 while Cosmos 2183 was still in orbit, but nearing the end of its mission and it, in turn, extended its mission duration to more than a year, 372 days. Cosmos 2267 went even further, 419 days, and flew at an angle of 70 degrees, extending the coverage of the programme to latitudes above the arctic circle and being joined in a paired operation by Cosmos 2280 in April 1994. There was a gap of only one day between the de-orbiting of Cosmos 2267 (28 December 1995) and the launch of Cosmos 2305 (29 December 1995).

There was a long gap between the return of Cosmos 2320 (September 1996) after 364 days and Cosmos 2359 (June 1998). This fell back in the atmosphere on 12 July 1999 after 381 days, to be replaced by Cosmos 2370 in May 2000.

Twice it looked as if this series had come to an end: after Cosmos 2320 and 2359. Cosmos 2320 and its predecessors used the Soyuz U2 launcher. With sintine fuel no longer available, it was thought that the Soyuz U would lack the power to launch the satellite, but this did not prove to be the case, or possibly the payload was lightened. Although there was then a second long gap, the series was renewed with Cosmos 2370 and, two months later, there was a further launch of the Potok system of relays (2371). This was a clear indication that Russia intends to continue the series.

Yantar 4KS2 Neman series

8 Apr 1992	Cosmos 2183	Soyuz U2	Baikonour	240–294	89.85	64.87
9 Dec 1992	Cosmos 2223	Soyuz U2	Baikonour	241–293	89.85	64.66
5 Nov 1993	Cosmos 2267	Soyuz U2	Baikonour	240–304	89.95	70.39
28 Apr 1994	Cosmos 2280	Soyuz U2	Baikonour	241–306	89.98	70.38
29 Dec 1994	Cosmos 2305	Soyuz U2	Baikonour	240–298	89.89	64.91
29 Sep 1995	Cosmos 2320	Soyuz U2	Baikonour	242–302	89.95	64.92
25 Jun 1998	Cosmos 2359	Soyuz U	Baikonour	240–303	89.94	64.71
3 May 2000	Cosmos 2370	Soyuz U	Baikonour	241–303	89.95	64.76

ORLETS/DON

What was considered to be the sixth generation of Soviet photoreconnaissance satellites was introduced with Cosmos 2031 in 1989. It was followed by Cosmos 2101 (1990), 2163 (1991) and 2225, 2262 (1993) and then a gap of four years. The standard flight path was an orbit of 230–300 km, with missions being in the order of 60 to 124 days. This is the Orlets 1 programme, design code 17F12, also known as the Yantar FR6, and the Don.

Orlets has proved to be a puzzle. No pictures have been released or hinted at, though if it is a Yantar-derived system, one may make a reasonable guess. Their means of returning data is uncertain. They are believed to carry eight to ten capsules to return to Earth with film. Orlets may also use digital imaging and unload the data electronically over Soviet territory rather than through the Potok system.

The Orlets series have the unmistakable characteristic of exploding at an altitude of 200 km at the end of their missions. Sometimes the débris spreads far out in orbit—up to 1,100 km high. There is as yet no explanation why this is done. Russia may lack a means to recover them or for some other reason prefers to self-destruct them (fear of advanced equipment falling into foreign hands). The débris may reach as far out as 1,100 km, and such deliberate destruction runs counter to growing international concerns about the problem of space débris. All previous Cosmos reconnaissance satellites have carried explosive devices, and these have been used when previous missions have run into difficulty, but this is the first example of explosive destruction being used as a norm.[31]

Orlets1/Yantar FR6/Don

22 Dec 1992	Cosmos 2225	Soyuz U	Baikonour	214–309	89.73	64.91
7 Sep 1993	Cosmos 2262	Soyuz U	Baikonour	207–325	89.83	64.89
15 May 1997	Cosmos 2343	Soyuz U	Baikonour	197–292	89.39	64.86

Orlets cannot be regarded as a successful series of photoreconnaissance satellites. Only six have flown: three before the Russian period, and three after 1991. Of these, there were two in 1993 and then a long gap. None had long lifetimes, certainly not compared to the Neman series. Cosmos 2225 blew up on 18 February 1993, after 57 days in orbit, 2262 after 102 days and 2343 after 124 days. It would be surprising if this series were to resume.

ORLETS 2/COSMOS 2290

The mainstay of the Russian photoreconnaissance programme has clearly been the Yantar Kobalt and the Yantar Neman, with the apparently less than successful Orlets 1 series. One might have thought that these programmes would have been enough to satisfy the needs of the GRU. Far from it, for the 1990s saw the introduction of two new models of photoreconnaissance satellites, Orlets 2 and Arkon. Although only one of each was introduced, they paved the way for future satellites in the series.

In May 1994, Russian television unexpectedly reported that the TsSKB Kozlov Progress works in Samara had created a new large spaceship 'capable of spotting matchsticks from orbit', but that it was so secret that, even in conditions of *glasnost*, no further information could possibly be revealed. What is considered to be the seventh-generation photoreconnaissance satellite appeared not long afterwards on 26 August 1994 with the launch of Cosmos 2290. It was orbited by Zenit from Baikonour, the first photoreconnaissance satellite to take such a powerful rocket. Zenit can carry a payload of up to 13 tonnes into an orbit, but that does not mean that Cosmos 2290 was necessarily that heavy. In the event, it entered an orbit of 220–315 km, 89.5 minutes, 64.8 degrees. Cosmos 2290 made a series of engine firings to maintain its perigee at 200 km but several times raised its apogee, first to 350 km, then to 450 km and finally to 550 km, presumably to test its matchstick-finding abilities from greater heights. It de-orbited over the Pacific Ocean northeast of New Zealand on 4 April 1995 after a mission of 221 days.[32] The first operational version, Cosmos 2372, was launched in September 2000.

The name Orlets 2 has been attached to the series. There are reports that it has up to 22 recoverable capsules. However, little more is known about it and the Kozlov plant has guarded its secrets remarkably well.

Orlets 2 series

26 Aug 1994	Cosmos 2290 Zenit	Baikonour	220–315	89.5	64.8

ARKON

On 6 June 1997, a Proton rocket put a military satellite, Cosmos 2344, into an unusual orbit of 2 hours 10 minutes, 63.3 degrees, 1,516 by 2,749 km. The Russian news services—admitting that it was a spy satellite for the Main Intelligence Directorate, the GRU—referred to it tantalisingly as 'Project 11F664' and said it had been 10 years in preparation. Just as some Western observers thought the Russian military space programme was in terminal decline, it had sprung another surprise with the launch of the biggest spycraft ever.

The failure or the refusal of the Russians to give further details was grist to the mill of Western sleuths and space watchers. The indefatigable Dr Geoff Perry, startled by its orbital parameters, spotted Cosmos 2344 and its Proton launcher flying over Bude later that evening and was able to recalculate its orbit, in the event even more accurately than the official announcement. Later, Dr Perry was able to identify Arkon by putting bits and pieces together from different news and information items given out by Russian companies over previous years.

First, a search through literature given out by Russian optical companies at air shows found reference to a 'Project Arkon'. This was a plan for a satellite flying at 3,000 km, with a reflecting telescope and a 27-metre lense able to zoom in on military targets with a resolution of 2.5 metres. Second, he found a drawing from the Lavotchkin design bureau which referred in the same context both to Arkon and Lomonosov, a project for a stellar mapping satellite. Both featured satellites in the shape of a telescope, with a shade on top and solar panels at the side. Finally, a Lavotchkin souvenir brochure given out at the Moscow Air Show used the same illustration, captioning it Cosmos 2344, a spacecraft capable of high-resolution photography from highly elliptical and circular orbits.

According to official reports, Arkon had to be launched in 1997, otherwise its electronic systems could no longer be guaranteed. Construction of the spacecraft had begun ten years earlier, but had been delayed due to financial shortages. The opportunity for a launch arose because Khrunichev was anxious to test out the new Block

DM2M upper stage which would be used in a number of forthcoming launches of American payloads. The military intelligence directorate, the GRU, need not have worried, for both the block D and the Arkon performed perfectly and within a week the GRU analysts were reportedly 'in seventh heaven' with their new toy.

Arkon series

7 Jun 1997	Cosmos 2344	Proton	Baikonour	1,509–2,747	130	63.42

Analysis of Arkon suggests that it is able to take frames of 1,800 to 3,000 km up to an angle of 45 degrees. The telescope is thought to use a Cassegrain-type optical system, to be 6.89 metres long and indeed to have a focal length of 27 metres. The diameter of the mirror is estimated at 2 metres. From its high orbit, its telescope can swing in many directions and identify objects over 1,000 km off its ground track, enabling it to spy over large areas. Ironically, after the Gulf War, the Americans identified the need for such a spy satellite. Their view was that a slow-moving high-flying satellite with a powerful imaging system might be of more value than close-look satellites. The problem was that the Russians built it instead.

In effect, Arkon was a space telescope pointed not outward to the skies but back towards the Earth. Its more distant orbit gave it a superb vantage point for identifying, revisiting and surveilling a variety of military targets, both from directly above and from slant angles.[33] It is not known if or when the mission concluded.

RUSSIAN MILITARY PHOTORECONNAISSANCE SATELLITES IN RETROSPECT

In a space programme of rapidly falling resources, spy satellites remained a priority. During the 1990s, Russia continued to maintain its Kobalt and Neman programmes, even introducing two new series, Orlets 2 and Arkon, making 36 launches altogether. The Zenit programme was run to conclusion and six satellites were orbited in the Orlets1 series. Russian space-based photography has traditionally been described as inferior to American spy satellites, but there does not appear to be an objective basis for this judgement. The Kometa/SPIN-2 mission suggested that Russian imaging is at least at the same level as American photoreconnaissance.

At present, Russia effectively operates only three satellite types in its photoreconnaissance programme—the Yantar Kobalt for 120-day missions, the Yantar Terilen/Neman for year-long digital imaging missions, and the Orlets 2 series, operational in 2000.

There were two periods when Russia was blind, when it had neither a Kobalt nor a Neman in orbit. The first period lasted for almost eight months, the second for less than five months. This was a clear, and from a military point of view, undesirable gap in the country's intelligence-gathering capacities, one brought about by financial rather than technical reasons. Indeed, the Orlets 2 and Arkon series showed that Russia's ability to innovate in the field of photoreconnaissance was undiminished.

Gaps when Russia was blind

28 Sep 1996–15 May 1997
15 Dec 1999–3 May 2000

Summary of Russian photoreconnaissance programme, 1992–2000

Generation	Name	No of launches	Cosmos
3	Zenit/Oblik	2	2207, 2281
4	Yantar		
	Kometa	4	2185, 2243, 2284, 2349
	Kobalt	17	2175, 2182, 2186, 2203, 2210, 2220, 2231, 2240, 2259, 2274, 2283, 2311, 2314, 2331, 2348, 2358, 2365
5	Neman	8	2183, 2223, 2267, 2280, 2305, 2320, 2359, 2370
6	Orlets 1	3	2225, 2262, 2343
7	Orlets 2	1	2290
8	Arkon	1	2344
		36	

ELECTRONIC INTELLIGENCE

Observing the enemy by photographs is perhaps the most visible aspect of spying. Equally important is observing the other side's capacity and movements through its use of radio and radar. Satellites which do this are termed ferrets or elints (electronic intelligence). Ferrets do two things: first, they listen in to enemy radio traffic and pick up conversations; and, second, they pick up enemy radar and radio frequencies and facilities, enabling their electronic equipment to be located and classified, such as anti-aircraft and ship radars. Each type has a distinctive signature according to type, size and purpose. From the information collected by ferrets, it is possible to follow enemy military manoeuvres and deployments, or, as it is put in military parlance, the Electronic Order of Battle (EOB), just as effectively as visual imaging. Just as it was important for Russia not to be 'blind' and without photoreconnaissance satellites in the 1990s, it was equally important not to be 'deaf' and without electronic intelligence. Ferrets are referred to by the Russians as the Tselina programme, although it has a number of variants (Tselina O, OM, D, 2). Tselina payloads were originally designed and built by the Yuzhnoye design bureau in Dnepropetrovsk in the Ukraine, though some were later also built by KB Arsenal in St Petersburg.

The first generation of Soviet ferrets appeared in 1967 (Cosmos 148) and 48 were launched in the series (Tselina O). Typically, they orbited between 400 and 850 km at 74, 65 and 81.2 degrees. The second generation, a larger satellite, began with Cosmos 189 in the same year, the first of 40. The third generation, called Tselina D, began in 1978, using a Tsyklon launcher, weighing 4 tonnes and operating in constellations of six. The last Soviet-era constellation of Tselina D elints was begun in 1986 and the Russians in effect brought this series to a conclusion with the launching of the last three Tselina D elints in 1992 and 1993, Cosmos 2221, 2228 and 2242. This marked the end of the series and the end of the first three generations of elints.

Tselina D series

24 Nov 1992	Cosmos 2221	Tsyklon	Plesetsk	636–665	97.8	82.51
25 Dec 1992	Cosmos 2228	Tsyklon	Plesetsk	633–669	97.75	82.53
16 Apr 1993	Cosmos 2242	Tsyklon	Plesetsk	634–668	97.74	82.53

TSELINA 2

Tselina 2 was approved in 1973 as a new type of more advanced electronic intelligence satellite. Designed for the new Zenit rocket, it was instead introduced on a Proton rocket in 1984 (Cosmos 1603 and later 1656) because the Zenit was not yet ready. Cosmos 1603 manoeuvred extensively in orbit, drawing attention to itself and sparking speculation in the West—correct in the event—that a new type of spacecraft had appeared.

The first Zenit-launched Tselina appeared with Cosmos 1697 in 1985, intended to be the start of a new four-satellite constellation. With a constellation of four, the entire globe is covered for electronic signals once a day. However, for most of the period of Tselina 2, the Russians have had to content themselves with a lower rate of coverage. The use of the Zenit rocket enabled very large ferrets up to 9,000 kg to be flown, although they may not actually be as heavy as that (indeed, they may be as small as 3,250 kg). The series was declared operational in 1990 and seven Tselina 2 satellites flew during the Soviet period.

Electronic intelligence has been an important aspect of the Russian military space programme throughout this period. Ten Tselina 2 elints flew as part of the Russian space programme, averaging over one a year. Assuming that each has a lifetime of several years, it is reasonable to presume that at no stage were the Russians deaf during this period. The launch of Cosmos 2369 in February 2000 underlined the continued priority being given to the series.

Tselina 2 elints are easy to identify, all using the Zenit from Baikonour, entering very precise orbits. The altitude must be tight in order to obtain precise measurements of the objects surveyed below. No photographs have been published of Tselina D operations and little more is known about them or their capabilities.

The early Tselina D programme suffered badly from the unreliability of the Zenit launcher and three were lost—in 1985, 1990 and 1991. The Russian period also began badly with a Zenit failure on 5 February 1992. Progress with the series was again halted in May 1997 when an elint was lost on a Zenit launcher which exploded 48 seconds into its mission in May 1997. The gap was not made good until Cosmos 2360 in July the following year.

Tselina 2 series

17 Nov 1992	Cosmos 2219	Zenit	Baikonour	849–855	102	71.01
25 Dec 1992	Cosmos 2227	Zenit	Baikonour	849–854	101.96	71.02
26 Mar 1993	Cosmos 2237	Zenit	Baikonour	849–853	101.95	71.02
16 Sep 1993	Cosmos 2263	Zenit	Baikonour	849–855	101.96	71
23 Apr 1994	Cosmos 2278	Zenit	Baikonour	849–855	101.97	71.02
24 Nov 1994	Cosmos 2297	Zenit	Baikonour	849–854	101.97	71
31 Oct 1995	Cosmos 2322	Zenit	Baikonour	849–852	101.94	71.02
4 Sep 1996	Cosmos 2333	Zenit	Baikonour	849–852	101.94	71.01
28 Jul 1998	Cosmos 2360	Zenit	Baikonour	848–854	101.95	71.02
3 Feb 2000	Cosmos 2369	Zenit	Baikonour	848–854	101.95	71.01

US-P/EORSAT

EORSATs are a particularly important subset of ferret. The role of the EORSAT, or Elint Ocean Reconnaissance SATellite, is to track the electronic signals of enemy fleets at sea. The Soviet Union originally devised two quite different systems to track American and NATO fleets. These were RORSATS and EORSATS. RORSATs, or the US-A programme (US being Russian for 'controlled sputnik', A for active), used radars to locate and identify American ships. Radars gulped electrical power and required nuclear energy sources. After several highly-publicized reactor accidents and possibly disappointing performance, the RORSAT programme was abandoned near the end of the Soviet period (1988).

The EORSAT or US-P (P for passive) programme, although based on passive eavesdropping, was no less important. Typically, EORSATS orbit from Baikonour at 65 degrees, 404–417 km. Like Tselina 2, they must operate in precise, unchanging orbits, so as to precisely triangulate the objects they are tracking: to do so, they use ion microthrusters which frequently readjust their flight paths. With the end of the RORSAT programme, the EORSAT became a priority military programme and Russia's prime source of naval electronic intelligence. The programme continued, despite the deep financial problems affecting the Russian military. Traditionally, the length of operation of US-Ps in the 1980s was in the order of 200–300 days. As was the case with the photoreconnaissance programme, this was extended and even doubled in the 1990s, with, for example, Cosmos 2313 running for 683 days and Cosmos 2326 for 659 days.

The EORSAT US-Ps were 3-tonne main spacecraft in the shape of a long thin torch 17 metres long and 1.3 metres in diameter, with two huge solar panels on either side and a large X-shaped antenna for picking up the electronic traffic. The US-P is

US-P EORSAT

launched by Tsyklon into a tight orbit where its orientation is maintained by four 10-kg UDMH and nitrogen tetroxide engines fed by eight 60-litre tanks.[34] The function of the X-antenna is to pick up the signal and identify its precise location, while ground analysts determine the type of ship involved. The launcher used is the Yuzhnoye-built Tsyklon 2, introduced 1969 and later improved as the Tsyklon M, and the US-P programme was its only use during the 1990s.

Just as photoreconnaissance satellites zoomed in on ground wars, so too do EORSATs closely follow fleet movements during times of conflicts or tension. EORSATs tracked the British task force as it sailed into the south Atlantic in summer 1982 to retake the Malvinas Islands during the Falklands War. Presumably, they will have followed American fleet movements in the Gulf War and in that area ever since.

The programme began with Cosmos 699 in 1974 and 46 EORSATS were launched by 2000, of which 10 were during the Russian period. Originally US-Ps operated in pairs, which was probably adequate for routine peacetime operation. Even still, a pattern of paired operation is in evidence: for example, 2258 with 2244, 2264 with 2293 and so on. When the RORSAT programme closed down, resources were transferred to the EORSAT programme and there was a surge of EORSAT launches in the late 1980s. As a result, the USSR was able to build up the system to a peak level of global efficiency.

There remains the question of how this electronic intelligence is relayed back to the military intelligence, the GRU and the Russian navy. Like the Neman series, Tselina 2 uses a series of relay satellites. Signals picked up by the US-Ps are transmitted by a series of relay satellites, Parus. Through Parus, US-P information is passed on to submarines at sea, guided missile cruisers and a dedicated control centre in Noginsk, built specially to handle the ocean surveillance programme. At least 20 Soviet naval vessels were equipped with systems to handle incoming data from US-Ps. The aim of the system is to provide targeting data that can be fed directly into anti-ship missiles. Just like Neman, maintaining the US-P system has required Russia to maintain another satellite programme alongside.

Unusually, for a military satellite, Cosmos 2326 carried a civilian scientific experiment. Even more unusual, it was a joint experiment with the United States to make a two-year study of gamma-ray outbursts. Called Konus, this equipment was developed by the St Petersburg Ioffe Physical and Engineering Institute of the Russian Academy of Sciences. The 130-kg experiment was attached to the top of the US-P and carried a gamma spectrometer, gamma detector and small astrophysical steerable telescope.

The EORSAT programme has been a priority of the Russian space programme. Indeed, the EORSAT constellation was in poor shape when the Soviet period ended and there was even a gap in March 1993 when Cosmos 2122 concluded its 775-day mission (the longest ever). It is possible that in expanding the system in the late 1980s the USSR had got ahead of the production line. However, the Russians rapidly rebuilt the EORSAT system. Several days later Cosmos 2238 was launched and operations were resumed. By 1994, no less than four satellites were operating simultaneously. In mid-1997, the Russians had six EORSATS operating in a constellation—Cosmos 2335, 2326, 2313, 2293, 2264 and 2258. However, this level of operation slipped in the 1990s.

Cosmos 2347 exploded in orbit on 22 November 1999, leaving more than 130 pieces of débris in orbit, almost two years after its launch and towards the end of its mission. The reasons for the explosion, which was presumed to be deliberate, are not known as it was not normal practice to destroy US-Ps at the end of their mission (the previous time was in 1987). Shortly afterwards, on 12 December 1999, Cosmos 2335 fell back into the atmosphere. After a gap of two weeks when Russia was, at least in terms of naval surveillance, deaf, it was replaced in orbit by Cosmos 2367. This made it the only US-P still apparently operational, compared to the situation where the satellites had worked in pairs. It is possible that the problem was due to a shortage of Tsyklon M rockets rather than satellites. The M series was believed to be near the end of its production run and reserve rockets were taken out of the military arsenal.

Gaps when Russia was 'deaf' in naval surveillance

28–30 March 1993
12–26 December 1999

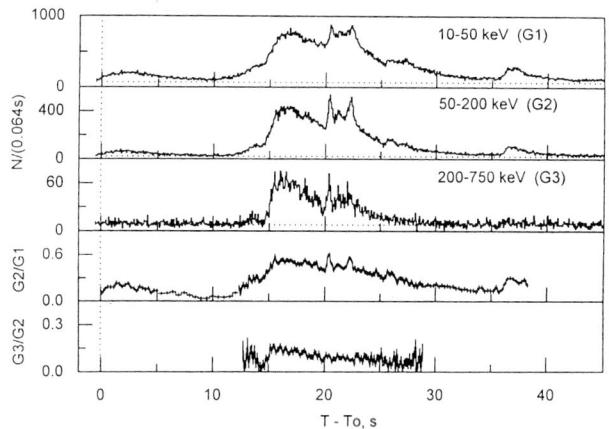

Printout of the results of the Konus experiment

US-P EORSATs

30 Mar 1993	Cosmos 2238	Tsyklon M	Baikonour	404–418	92.78	65
28 Apr 1993	Cosmos 2244	Tsyklon M	Baikonour	404–418	92.78	65.03
7 Jul 1993	Cosmos 2258	Tsyklon M	Baikonour	404–418	92.78	65.05
17 Sep 1993	Cosmos 2264	Tsyklon M	Baikonour	403–418	92.77	65.03
2 Nov 1994	Cosmos 2293	Tsyklon M	Baikonour	404–417	92.78	65.03
8 Jun 1995	Cosmos 2313	Tsyklon M	Baikonour	403–418	92.77	65.04
20 Dec 1995	Cosmos 2326	Tsyklon M	Baikonour	407–415	92.78	65.02
11 Dec 1996	Cosmos 2335	Tsyklon M	Baikonour	403–419	92.79	65.05
9 Dec 1997	Cosmos 2347	Tsyklon M	Baikonour	403–419	92.79	65.04
26 Dec 1999	Cosmos 2367	Tsyklon M	Baikonour	404–418	92.78	65.04

MILITARY COMMUNICATIONS: STRELA, GONETZ, POTOK

Russia operates three military or military-derived communications systems: Strela, Gonetz and Potok. From the early 1960s, the Soviet Union defined a need to provide state-of-the art communications between its military commanders, its navy on the high seas and other far-flung outposts (Strela). One of these programmes was subsequently civilianized (Gonetz). The Potok system was a series of relays in 24-hour orbit to broadcast the digital photoreconnaissance images filmed by the Neman spy satellites.

Strela

Development of the first military communications system began in the summer of 1964 (18 August) when the USSR put three satellites into orbit on one rocket—Cosmos 38, 39 and 40. As they entered orbit, a spring was fired, pushing each of the three satellites into the required orbit in turn. As a result, they were spaced out. The idea was that a ground radio could use one, then another, then another satellite as each came over the horizon. The principle was that messages could be relayed from abroad to one satellite for retransmissions when passing over a control centre in Russia itself several hours later. The system was called Strela, or arrow, to symbolize the speed with which messages could now be communicated globally. The Cosmos 38–40 trio was followed by 5-in-1s and then 6-in-1s (called the Strela 3 series), 8-in-1s (called the Strela 1 series) and single satellites (Strela 2).

During the Russian period, only one launch of the Strela 1 series took place (Cosmos 2187–2194), and this concluded the series. Strela 1 satellites, octets, went back to 1970—indeed the first set was launched the day after China astonished the world with its first satellite launch. These Strela were small, about 61 kg and 100 cm in diameter and operated in constellations of 24 satellites. Typically, they orbited at 1,500 km every 115 minutes at 74 degrees from Plesetsk on a Cosmos 3M rocket.

The second Strela system is called in the West the 'store-dump' or the 'store-and-forward' system (Strela 2). This programme started in 1970 and uses a single satellite at a time. Strela 2 was thought to be linked to the work of KGB agents in the field. Using attic transmitters or even hand-held devices, an agent could send a message up to a Strela travelling overhead. Once over a ground station some time later in the orbit, the vital intelligence is beamed down to Russia. Strela 2s generally worked in pairs and two operated during the Russian period, Cosmos 2251 (June 1993) and Cosmos 2298 (December 1994), the latter apparently marking the end of the series. These single satellites weighed about a tonne and flew at an altitude of around 800 km on a Cosmos 3M from Plesetsk.

Strela 3 sextet satellites are larger, weighing about 225 kg; they fly at 1,400 km at 82.6 degrees and take a Tsyklon 3 rocket from Plesetsk. They have short lifetimes, in the order of two years. Strela are cylindrical-shaped, 230 kg, 1.6 metres tall and 80 cm in diameter with solar cells on the outside and a boom at either end. Inverted cone-shaped communications aerials are attached to the back. The Strela 3 series began in 1986 (Cosmos 1617–1622). There were nine Strela 3 launchings in the Russian period, averaging one a year, though the current rate of launching has slowed, reflecting either financial shortages or longer operating times. It is possible that many of the functions of the old Strela 1 and 2 systems were taken over by the Strela 3 series, so that the end of Strela 1 and 2 may not necessarily mean a loss of capacity. The last 1990's launch in the Strela 3 series was Cosmos 2352–2357 in 1998.

Generally, orbital insertions in the Strela series have been quite precise and must be so if the system is to operate effectively. However, in the launch of 16 June 1998, the final stage misfired, with excess thrust from the upper stage of the Tsyklon putting the satellites into a 1,300 by 1,900 km orbit instead of the planned 1,400 circular orbit, sparking a commission of inquiry and a rescue plan to ensure that the scarce satellites could still be used operationally. Builders of the Strela 2 and 3 systems are the NPO Polyot of Omsk.

Gonetz

Although Strela was and is a military programme, part of it was civilianized in the 1990s. This is called Gonetz, the Russian word for messenger. Gonetz is the same as Strela, but with civilian rather than military transmitters. They were first tested out on Cosmos 2199 and 2201 in July 1993 but started to use the civilian designation

Gonetz

Cosmos 3M at Plesetsk

from the next launch. From 1996, the two types have been launched at the same time, with three Gonetz accompanying three Strela 3. Precisely how the division of functions operates between the two is unclear. By 1998, there were between 4,000 and 10,000 users of the civilian Gonetz system. Gonetz are able to transmit text, picture, voice, paging, internet, mobile phone calls and e-mail all over the world, from equatorial to polar regions.

From 2003, Gonetz will be replaced by three constellations each of four Gonetz put in orbit by the new Rockot launcher. These Gonetz, called Gonetz D1 and D1M, will be smaller, rectangular box-shaped, 200 kg, with solar panels able to provide 500 W and a lifetime of seven years. The intention is to push the number of users of the system up to 100,000. In addition, a number of low-flying electronic communications satellites programmes have been studied for the future years, named Signal, Forport, Rostelcom, Polar Star, Marafon and Molniya-Zond, but their appearance will depend on the state of the Russian economy and the level of Western investment.[35]

Potok

Cosmos satellites have been used in geosynchronous orbit since 1974 both for general military communications purposes and more specifically to relay signals from the Terilen/Neman digital photoreconnaissance satellites. Each spacecraft weighs about 2.5 tonnes, is built by NPO-PM in Krasnoyarsk and requires the expensive Proton launcher to reach 24-hour orbit. No diagram of Potok has ever been published, making the system one of the more impenetrable within the Russian intelligence-gathering system.

The Potok system, also called Geizer, was introduced with Cosmos 1366 in 1982 and operationally in 1987 as Cosmos 1883. Since then, there have been seven launchings, of which three have been during the Russian period (Cosmos 2291, 2319 and 2371). The Potok system comprises three satellite slots, located at 190°E, 80°E and 346°E. The Russians have rarely had more than two operational at any given time, generally moving one into position before its predecessor reaches the end of its life.

Strela/Gonetz/Potok series

Strela 1						
3 Jun 1992	Cosmos 2187–2194	Cosmos 3M	Plesetsk	1,402–1,480	114.93	74.01
Strela 2						
16 Jun 1993	Cosmos 2251	Cosmos 3M	Plesetsk	781–806	100.74	74.04
20 Dec 1994	Cosmos 2298	Cosmos 3M	Plesetsk	786–810	100.83	74.03
Strela 3/Gonetz						
13 Jul 1992	Cosmos 2197–2202	Tsyklon 3	Plesetsk	1,397–1,1416	113.99	82.59
20 Oct 1992	Cosmos 2211–2216	Tsyklon 3	Plesetsk	1,400–1,415	114	82.59
11 May 1993	Cosmos 2245–2250	Tsyklon 3	Plesetsk	1,398–1,417	113.95	82.59
24 Jun 1993	Cosmos 2252–2257	Tsyklon 3	Plesetsk	1,409–1,481	114.77	82.59
12 Feb 1994	Cosmos 2268–2273	Tsyklon 3	Plesetsk	1,412–1,426	114.2	82.58
28 Dec 1994	Cosmos 2299–2304	Tsyklon 3	Plesetsk	1,402-1,416	113.97	82.57
19 Feb 1996	Gonetz D1,2,3 Cosmos 2328–2330	Tsyklon 3	Plesetsk	1,400–1,414	113.94	82.58
14 Feb 1997	Gonetz D4,5,6 Cosmos 2337–2339	Tsyklon 3	Plesetsk	1,413–1,423	96.88	82.6
15 Jun 1998	Cosmos 2352–2357	Tsyklon 3	Plesetsk	1,310–1,875	118.03	82.59
Potok/Geizer						
21 Sep 1994	Cosmos 2291	Proton	Baikonour	35,758–35,817	1,436	1.53
31 Aug 1995	Cosmos 2319	Proton	Baikonour	35,806–35,960	1,441	1.48
5 Jul 2000	Cosmos 2371	Proton	Baikonour			

This may well have been sufficient for the satisfactory operation of the system. With gaps between Neman launches growing longer, and an unusually long period of quiet after the launch of the Cosmos 2319 Potok, it might have been presumed that Russia was no longer able to maintain the Neman/Potok system. This was not the case, for summer 2000 saw the renewal of both systems with the launch, two months apart, of a new Terilen/Neman (May) and then a Potok/Geizer (July): Cosmos 2370 and 2371 respectively.

NAVIGATION SATELLITES: PARUS, NADEZHDA, TSIKADA

The USSR began its first launches of navigation satellites in 1967. Working on a similar principle to the military communications satellites, constellations of communications satellites can provide extremely accurate coordination and reference points for ships at sea, submarines underneath it and aircraft in the air.

The USSR operated a low-altitude navigation system and a high-altitude system for global positioning. All launches used to take place within the Cosmos label. Subsequently, the low-level system was divided into three: civilian (Nadezhda, the Russian word for 'hope'); a semi-military system, Tsikada (Russia word for a chirping cricket); and Parus, a fully military system, which still operates within the Cosmos label.

Parus

The first generation of low-altitude military navigation satellites was phased out in 1978 and replaced by the current Parus system (Russian word for 'sail') which had originally been introduced with Cosmos 700 in 1974. Parus operates in a constellation of six satellites to ensure global coverage, each satellite being replaced as it reaches the end of its lifetime.

Parus comprises small satellites of about 800 kg in weight. They are cylindrical in shape, about 2 metres in diameter, 2.1 metres in length and are equipped with a gravity boom. The navigational accuracy of Parus is estimated to be in the order of 180 metres. Their main role is to provide accurate navigational fixes for the overseas submarine fleet of the Russian navy, but they are also considered to have a key function in transmitting data from US-P EORSATS both to maritime ground control in Noginsk and to ships and submarines. Granted the low level of sorties of the Russian navy in the 1990s, the relay role of Parus may well now have become the dominant one. The priority which Russia has given to maintaining the US-P system in the 1990s has of necessity required the Parus system to be kept operational.[36]

Russia launched 17 Parus, meaning an average of almost two a year. Although most were during the earlier period, launches continued into the late 1990s and suggest that this is a priority programme. There was one Parus failure. Cosmos 2321 suffered a rare Cosmos 3M rocket failure: with the second stage valve becoming stuck, placing the satellite in a highly elliptical orbit.

Nadezhda

Nadezhda is a civilian navigation system originally introduced within the Cosmos programme. The Nadezhda system began under that title in 1989, and consists of a constellation of four satellites each of about 810 kg orbiting at 83 degrees, 1,000-km altitude. It also carries a search-and-rescue satellite system called COSPAS/ SARSAT. Three more followed, with one more within the Cosmos programme. The COSPAS/SARSAT system is one of the quiet and least publicized triumphs of the space age.

The COSPAS/SARSAT beacon is a transponder system which ships in distress may use to summon help. It was established jointly by the Soviet Union, United States, France and Canada at the end of the 1970s. Since 1982 it has saved more than 5,500 lives on more than 1,800 rescue calls and is widely used in the West as well. Over half a million beacons are now operationally linked to the system. A ship in distress will emit signals every 50 seconds to be picked up in real time by an over-flying satellite. COSPAS/SARSAT is most prominently associated with the rescue of round-the-world yachtsmen and trekking climbers in the Himalayas, but most of its work is more mundane but no less critical: merchant shipping in difficulty. Because Nadezhda and the Western satellites overfly most of the world, except some of the extreme polar regions, distress signals can be picked up within less time than one orbit. The average waiting time for receipt of a distress signal is about half that time, about 44 minutes. Generally, a Nadezhda satellite will get a fix on the distressed ship to an accuracy of less than 5 km, which should be sufficient for searching planes and ships.

The series was maintained into the 1990s and beyond. The Nadezhda launch of 28 June 2000 carried two more small satellites built by Surrey Satellite Technology, the British company which specialized in the construction of small Earth satellites. The first, Tsinghua 1, was a microsatellite built for the Tsinghua University in Beijing, China, with a 39-metre resolution multispectral imaging camera designed to help in disaster monitoring and with a store-and-forward communications system. SNAP 1, a €1.5m project, which was even smaller and weighed only 6.5 kg, was designed to test communications and rendezvous capacities of very small satellites. Its tiny size packed a Global Positioning System navigation system, camera, computer, propulsion and attitude control system. The camera was designed to look through clouds and take clear pictures of the Earth from 500 km. The computer and propulsion system was intended to enable SNAP to rendezvous and fly in formation with its companion, Tsinghua, and test out the possibility of building nanosatellite constellations.

Tsikada

Tsikada is a semi-military system operating distinctly from Parus, though probably using the same type of satellite and operating in conjunction with the Nadezhda wholly civilian system. The system was declassified on 24 January 1995 with the launch of 825-kg Tsikada 1 from Plesetsk on a Cosmos rocket and it was

COSPAS/SARSAT in assembly

Navigation satellites

Parus						
17 Feb 1992	Cosmos 2180	Cosmos 3M	Plesetsk	962–1,016	104.94	82.93
15 Apr 1992	Cosmos 2184	Cosmos 3M	Plesetsk	967–1,014	104.89	82.94
1 Jul 1992	Cosmos 2195	Cosmos 3M	Plesetsk	958–1,011	104.85	82.93
29 Oct 1992	Cosmos 2218	Cosmos 3M	Plesetsk	968–1,015	105	89.92
9 Feb 1993	Cosmos 2233	Cosmos 3M	Plesetsk	954–1,009	104.72	82.94
1 Apr 1993	Cosmos 2239	Cosmos 3M	Plesetsk	967–999	104.75	82.93
2 Nov 1993	Cosmos 2266	Cosmos 3M	Plesetsk	950–1,019	104.79	82.95
26 Apr 1994	Cosmos 2279	Cosmos 3M	Plesetsk	957–1,007	104.73	82.95
22 Mar 1995	Cosmos 2310	Cosmos 3M	Plesetsk	980–1,011	105.02	82.94
6 Oct 1995	Cosmos 2321 (fail)	Cosmos 3M	Plesetsk	258–793	95.14	82.94
16 Jan 1996	Cosmos 2327	Cosmos 3M	Plesetsk	952–1,021	104.82	82.98
5 Sep 1996	Cosmos 2334	Cosmos 3M	Plesetsk	970–1,000	104.8	82.94
20 Dec 1996	Cosmos 2336	Cosmos 3M	Plesetsk	979–1,012	105.03	97.83
17 Apr 1997	Cosmos 2341	Cosmos 3M	Plesetsk	978–1,014	105.03	82.92
23 Sep 1997	Cosmos 2346	Cosmos 3M	Plesetsk	939–996	104.92	82.92
24 Dec 1998	Cosmos 2361	Cosmos 3M	Plesetsk	969–1,013	104.91	82.94
26 Aug 1999	Cosmos 2366	Cosmos 3M	Plesetsk	963–1,008	104.81	82.93
Nadezhda						
9 Mar 1992	Cosmos 2181	Cosmos 3M	Plesetsk	973–1,013	105.03	82.95
14 Jul 1994	Nadezhda 4	Cosmos 3M	Plesetsk	954–1,005	104.68	82.95
10 Dec 1998	Nadezhda 5	Cosmos 3M	Plesetsk	977–1,013	105.01	82.95
	Astrid 2					
28 Jun 2000	Nadezhda 6	Cosmos 3M	Plesetsk	684–708		83
	Tsinghua 1					
	SNAP-1					
Tsikada						
12 Jan 1993	Cosmos 2230	Cosmos 3M	Plesetsk	973–1,007	104.91	82.94
9 Feb 1993	Cosmos 2233	Cosmos 3M	Plesetsk	954–1,009	104.72	82.94
1 Apr 1993	Cosmos 2239	Cosmos 3M	Plesetsk	967–999	104.75	82.93
24 Jan 1995	Tsikada 1	Cosmos 3M	Plesetsk	965–1,021	104.97	82.97
5 Jul 1995	Cosmos 2315	Cosmos 3M	Plesetsk	970–1,014	104.94	82.91

announced that it would be part of a seven-satellite constellation. It went into an orbit of 965–1,021 km, 104.97 minute, 82.97 degrees and was co-planar with the already orbiting Cosmos 2123. Russia launched five Tsikada satellites. A new version of Tsikada, called Tsikada M, will be introduced in the future.

GLONASS

In 1982, the Soviet Union introduced a single high-altitude navigation system, GLONASS, which paralleled the American Global Positioning System (GPS) introduced in 1978. GLONASS stands for *Globalnaya Navigatsionnaya Sputnikovaya*

GLONASS series

27 Jan 1992	Cosmos 2177–2178	Proton	Baikonour	19,111–19,149	675	64.81
30 Jul 1992	Cosmos 2204–2206	Proton	Baikonour	19,121–19,140	675	64.83
17 Feb 1993	Cosmos 2234–2236	Proton	Baikonour	19,115–19,138	675	64.85
11 Apr 1994	Cosmos 2275–2277	Proton	Baikonour	19,114–19,143	675	64.82
11 Aug 1994	Cosmos 2287–2289	Proton	Baikonour	19,112–19,134	675	64.88
20 Nov 1994	Cosmos 2294–2296	Proton	Baikonour	18,782–19,135	669	64.89
7 Mar 1995	Cosmos 2307–2309	Proton	Baikonour	19,113–19,149	675	64.8
24 Jul 1995	Cosmos 2316–2318	Proton	Baikonour	19,104–19,131	675	64.85
14 Dec 1995	Cosmos 2323–2325	Proton	Baikonour	18,679–19,133	666	64.83
30 Dec 1998	Cosmos 2362–2364	Proton	Baikonour	19,125–19,129	675	64.8

Sistema, or global navigational sputnik system. GLONASS uses the Proton rocket and is a valuable but expensive system to maintain.

Like the GPS, signals from the high-orbiting satellites are able to give system holders very precise estimates of their position. Officially, the accuracy was supposed to be 65 metres, but in fact GLONASS provides an accuracy of about 20 metres in its civil version and 10 metres on its military signal (the military signal code has been quite easy to crack, as demonstrated by a hacking professor in Leeds University in England). Their main users are ships in the Soviet navy and mercantile marine, and military and civilian aircraft. By 1990, over 1,200 ships had been equipped with GLONASS receivers to enable them to fix their position accurately. A ship, or aircraft, must be able to receive such signals from three satellites, either simultaneously or in succession. GLONASS operates in a constellation of 24 satellites (21 plus three spares) in 64.8-degree orbits out to 19,000 km with a period of 675 minutes. Its orbit makes it attractive for ships and aircraft operating in northerly latitudes, an important consideration for Russia, especially for the cross-polar and northern Siberian air routes which were opened up in the 1990s. GLONASS are large satellites, 7.8 metres tall and 7.2 metres across their solar panels, each of which weighs 1,260 kg. The satellites are cylindrical, with a signal box array at the bottom pointing down towards Earth.

The GLONASS system is launched in threes on Proton rockets. The present GLONASS military and civilian aircraft navigation system was inaugurated with Cosmos 1413–1415 on 12 October 1982. After several years of testing, the first fully operational GLONASS system was commenced in 1989 and completed on 7 March 1995 when a Proton put Cosmos 2307, 2308 and 2309 into space, entering orbits of 64.8 degrees, 11 hour 15 minutes, 19,132 km.[37] The number of users in Russia has been small, compared to the West: about 500 compared to many tens of thousands in the West. The GLONASS system is of similar standard to the Western GPS, but has the advantage of providing better coverage at higher latitudes. Responsibility for the programme falls to NPO-PM in Krasnoyarsk, though they are actually built at NPO Polyot in Omsk.

The GLONASS satellite has a design life of about three years, which means that the system requires fairly regular renewal. With financial shortages in the 1990s, the

Two GLONASS almost completed

Russians struggled to maintain the GLONASS system. Two GLONASS launches were required each year to maintain the system and as this was a launch rate the Russians were unable to maintain: there were no GLONASS launches in 1997 or 1999. The system fell into decline in the mid-1990s. By 1998, only 16 of the 24-satellite system were actually operating (some say only 13).[38] In an effort to retrieve the situation, three new GLONASS were orbited at the end of that year. Cosmos 2362–2364 were the 72nd, 73rd and 74th GLONASS satellites, giving an indication of the considerable investment put into the service. In 1998, the Russian government transferred the GLONASS system from the defence ministry to the Russian Space Agency as a prelude to civilianization. Russia also entertained hopes that GLONASS could be integrated into the new European Union-funded Galileo GPS system approved in 2000. Improved systems, called GLONASS M1, GLONASS M2 and PROPNASS have been promised with improved lifetimes of seven years or more.

MILITARY EARLY WARNING SYSTEMS: OKO, PROGNOZ

When Iraq launched its Scud missiles against Saudi Arabia and Israel during the Gulf War of 1991, the only possible advance notice of an impending attack was a satellite picking out the hot gas plumes of the Scud as it left its mobile launch pad. Although the Gulf War was the first to actively involve the detection and shooting down of missiles fired in anger, the use of infrared detectors to give early warning of impending missile attack was the focus of large efforts by both the United States and the Soviet Union during the cold war. The United States operated two systems— Midas, from 1960 to 1966; and the Defence Support Programme (DSP) satellite series, from 1970. The Soviet system was called Oko. It had the additional benefit of monitoring nuclear tests and civilian rocket launches.

Oko

The Soviet early warning system was introduced in 1972 (Cosmos 520) and became operational in 1976 (Cosmos 862), expanding into a nine-satellite constellation by 1980. Called Oko (appropriately the Russian word for eye), it used a constellation of nine satellites in 12-hour orbits. These operated in what are called Molniya orbits which have low perigees, generally in the southern hemisphere, but high apogees, generally in the northern hemisphere. The climb to, and descent from, apogee is a slow, lazy one and could be set in such a way that Oko could curve high over either the continental United States land mass or the American missile bases in Western Europe, keeping both under constant observation. In their 12-hour orbits, Okos would swing slowly over the northern hemisphere twice a day. Their telescopes could spot the ascending flame of a rising missile against the stellar background within 20 to 30 seconds, enough time to alert the anti-missile forces.

No models or replicas of Oko have been presented publicly or seen by Western experts, although some fuzzy pictures have been published showing an Oko in assembly. The satellite appears to be drum shaped, with a shaded telescope pointing towards Earth and two solar panels at the bottom. The 1,900-kg Oko satellite, built by the Lavotchkin bureau, has a 350-kg, 0.5-metre diameter, Earth-pointing infrared telescope, quick-deploy 4-metre conical sunshield, topped by an instrument bus and solar panels which deliver 2.8 kW of power. From 1976 to 1983, it was standard Soviet practice to destroy Oko satellites at the end of their mission, though on some occasions the explosive charge went off prematurely.

Over 80 Oko satellites have been launched since the series began. Although the constellation was supposed to comprise nine satellites, it is doubtful if Russia had more than three operational for most periods during the 1990s. Indeed, the only period when a complete constellation was fully operational was the early 1980s. The

Satellite assembly

average lifetime of each Oko was about three to four years. Generally, this is a reliable, well-established system, using proven technology and the relatively inexpensive Molniya M launcher. Russia launched 13 Oko satellites. Although the rate of Oko launchings fell, there is no indication that Russia intends to abandon the programme and it must be considered a continuing priority. Expressed negatively, Russia would not like to be in the position of *not* being able to get early warning of a missile attack, from whatever source.

Prognoz

A drawback with the Oko system was that it did not provide global coverage. The Molniya orbits of Oko swung over the main land mass of the United States and Western Europe, the two most likely traditional locations of prospective NATO missile attacks on the USSR. They were not as effective in catering for American west coast or Pacific-based launches, nor indeed, for other potential enemies or what American intelligence analysts called 'rogue states'. For the Russians, a system based in 24-hour orbit offered the best hope of global early warning. In 1981, the USSR requested and was allocated seven satellite slots in 24-hour orbit for Earth observations.

It was and is still not clear if this system, called Prognoz, was intended to supplement or fully replace the Oko system. In the event, it has supplemented Oko, which has continued. Prognoz has only ever used four of the seven slots allocated (336°E, the main one, but also 12°E, 80°E and 35°E) and these missions have been far from trouble-free. Requiring a heavier, Proton booster, it was a much more expensive system than the tried and reliable Oko, which is probably why both systems continued to operate.

Cosmos designations were always given for this series. The first was Cosmos 1940 (1988), followed by Cosmos 2133. Cosmos 1940 was originally announced as a geostationary Earth resources programme, which it was not, and the true nature

Military early warning systems

Oko						
24 Jan 1992	Cosmos 2176	Molniya M	Plesetsk	604–39.747	717.68	62.8
8 Jul 1992	Cosmos 2196	Molniya M	Plesetsk	590–39,733	717.14	62.95
21 Oct 1992	Cosmos 2217	Molniya M	Plesetsk	599–39,757	717.79	62.95
25 Nov 1992	Cosmos 2222	Molniya M	Plesetsk	591–32,288	708.15	62.88
26 Jan 1993	Cosmos 2232	Molniya M	Plesetsk	591–39,363	709.66	62.78
6 Apr 1993	Cosmos 2241	Molniya M	Plesetsk	641–39,772	718.95	62.88
10 Aug 1993	Cosmos 2261	Molniya M	Plesetsk	582–39,760	717.52	62.89
5 Aug 1994	Cosmos 2286	Molniya M	Plesetsk	569–39,767	717	62.89
24 May 1995	Cosmos 2312	Molniya M	Plesetsk	604–39,545	717	62.89
9 Apr 1997	Cosmos 2340	Molniya M	Plesetsk	541–39,815	717	62.94
14 May 1997	Cosmos 2342	Molniya M	Plesetsk	513–39,381	708	62.83
7 May 1998	Cosmos 2351	Molniya M	Plesetsk	525–39,829	717	62.96
27 Dec 1999	Cosmos 2368	Molniya M	Plesetsk	554–39,720	716	62.83
Prognoz						
10 Sep 1992	Cosmos 2209	Proton	Baikonour	35,764–35,806	1,435	1.32
17 Dec 1992	Cosmos 2224	Proton	Baikonour	35,877–36,179	1,448	2.3
7 Jul 1994	Cosmos 2282	Proton	Baikonour	35,775–35,813	1,435	2.29
14 Aug 1997	Cosmos 2345	Proton	Baikonour	34,295–37,274	1,435	1.3
29 Apr 1998	Cosmos 2350	Proton	Baikonour	35,958–36,089	1,448	2.3

of the system was not formally revealed until 1993. Prognoz is made by the NPO Lavotchkin, weighs 3 tonnes and has a 600-kg Cassegrain optical assembly with aluminium-coated beryllium mirrors which scan the Earth's surface every 7 seconds and are so sensitive that they can pick out the afterburner of a jet fighter.[39] It has sunshades to protect it against laser attacks. Prognoz is believed to be a handle-shaped rectangular box with solar panels on the front and may be the basis for the forthcoming Spektr X-ray astronomy satellite project.

Russia made five Prognoz launches. Ideally, each satellite should last five years, but the programme has suffered from satellites failing long before their time and has not, therefore, been a successful programme. The first Prognoz lasted less than a year and Cosmos 2155 nine months. The problem of longevity persisted into the Russian period, for Cosmos 2282 failed after 16 months. Cosmos 2350 lasted only two months, ceased operation in July 1998 and was not replaced. By 2000, none appeared to be operational. The Oko series, introduced long before, continued in operation.

MINOR MILITARY COSMOS SATELLITES

Within the Cosmos programme, a small number of satellites have always been difficult to classify. These have often been small satellites, taking unusual orbits. In the West, these were classified as 'minor military' by some and 'obscure Cosmos' by others.

In 1997, some of these mysteries were identified as the Yug, Vektor and Romb programmes. Yug and Vektor were used to measure atmospheric density and to calibrate Soviet and Russian radars, at least in part for the antiballistic missile system. Yug was a series of passive satellites, Vektor being active. Cosmos 2332, a Vektor, was a 500-kg spherical satellite launched on a Cosmos 3M from Plesetsk.[40] Three Yugs were launched by the USSR, two by Russia. Ten Vektors were launched by the USSR, one by Russia. The Romb (Russian word for 'Rhombus') series dropped off mini-satellites in orbit, presumably for calibration purposes. One of the functions of the Romb sub-satellites may be to test the ability of military radars to pick out simulated multiply-targeted independent re-entry warheads. The USSR put up 29 Rombs, and Russia one. Cosmos 2306 deployed no less than 10 subsatellites. These programmes would no longer appear to be a priority, reflecting either declining budgets or reduced pressure for antiballistic missile systems or both.

Minor satellites

26 Oct 1993	Cosmos 2265	Cosmos 3M	Plesetsk	291–1,474	103.68 82.94	Yug
27 Sep 1994	Cosmos 2292	Cosmos 3M	Plesetsk	400–1,954	108.93 82.99	Vektor
2 Mar 1995	Cosmos 2306	Cosmos 3M	Plesetsk	469–517	94.47 65.85	Romb
24 Apr 1996	Cosmos 2332	Cosmos 3M	Plesetsk	295–1,565	103.62 82.96	Yug

RUSSIAN MILITARY SPACE PROGRAMMES IN PERSPECTIVE

Granted the state of the Russian economy in the 1990s, it is remarkable that Russia has been able to maintain a military space programme at all. Unlike Mir and, as we shall see, some of the civilian programmes, military programmes cannot readily attract outside investment or create many commercial opportunities (although, with Kometa, this was tried).

The rate of Russian military launchings in the 1990s was much slower than during the Soviet period in the previous decade. Even if the government had not changed in 1992, there would have been a much reduced rate of launching, as satellite lifetimes were extended and frequent but ultimately uneconomical methods of putting satellites into space were phased out. Several Soviet-period programmes that had reached the end of their production run were also phased out (e.g. Zenit/Oblik, Tselina D).

The following table illustrates the fall in the total number of military launches.

Russian military launches, 1992–2000

	1992	1993	1994	1995	1996	1997	1998	1999	2000	Total
Military	31	27	22	15	9	10	10	4	4	132
Total	54	47	46	32	23	26	24	26	20	298
% military	57%	57%	48%	47%	39%	38%	42%	15%	25%	44%

As may be seen, military launchings fell steadily in both absolute and relative terms, in the latter case from over half of the programme to a fifth. Partly, this reflected the growth of the commercial space programme. There were other reasons for a reduced level of military programmes in the 1990s. First, two programmes proved to be less than successful. These were the Orlets 1 series of photo-reconnaissance satellites, few of which appear to have operated in orbit for any length of time before their deliberate destruction; and the Prognoz series of 24-hour missile early warning satellites, which were also plagued by poor lifetimes.

Second, the reduced level of military tension in the 1990s meant that Russia could afford to operate its military space programme at a lower level. From the mid-1990s, Russian and American missiles were no longer targeted on each other. There was a significant scaling down in the nuclear arsenal of both superpowers and two other countries effectively disarmed their deterrents (Ukraine and Kazakhstan). American intelligence increasingly focused on China as its main potential nuclear enemy and considered that the most likely future threats were likely to come from countries like Iraq, Libya and terrorists on the loose. For the Russian military space programme, round-the-clock systems of global surveillance could be stepped down, as long as some satellites were in orbit which could cover potential targets of interest in a day or two. Thus, four- and six-strong constellations could reasonably be replaced by pairs.

In the light of these considerations, the continuation of a military space programme was a striking feature of the Russian space programme: 132 launches were made in the course of 1992–2000. Within the military effort, four programmes were prioritized: photoreconnaissance, electronic intelligence, ocean surveillance and the Oko early warning system. The continuation of the digital imaging programme, Neman, and the ocean surveillance system, US-P, required the operation of two relay systems, Potok and Parus respectively. Moreover, the operation of the military space programme required the use of two of the most expensive rockets in the fleet: Zenit and Proton. Most remarkably, the Russians made maiden flights of two new photoreconnaissance satellites—Orlets 2 and Arkon. The programme is diminished, but remains focused.

Summary of Russian military launches reaching orbit, 1992–2000 (multiple launches count as one launch)

Photoreconnaissance	36
Electronic intelligence	
Tselina D	3
Tselina 2	10
Ocean surveillance	
US-P	10
Communications	
Strela 1	1
Strela 2	2
Strela 3/Gonetz	9
Potok/Geizer	3
Navigation	
Parus	17
Tsikada	5
Nadezhda	4
GLONASS	10
Early warning	
Oko	13
Prognoz	5
Minor military	
Yug	2
Vektor	1
Romb	1
Total	**132**

The table below demonstrates the priorities within the Russian military space programme.

Clearly, with some programmes, the Russians struggled. Within the four priority programmes, there were two long periods of blindness (several months) and two very short periods of deafness when Russia could not observe or listen in to military

Russian military programmes, by category and year, 1992–2000

	1992	1993	1994	1995	1996	1997	1998	1999	2000	Total
PR	10	6	7	3	2	3	3	1	1	36
Elint	4	3	2	1	1		1		1	13
Ocean		4	1	2	1	1		1		10
Comm	3	3	4	1	1	1	1		1	15
Nav	7	7	5	7	3	2	3	1	1	36
Early	6	3	2	1		3	2	1		18
Minor	1	1	1		1					4
Total	31	27	22	15	9	10	10	4	4	132

targets abroad (several days). The programme was clearly at a low ebb, compared to what the United States and its NATO allies could put over war zones, for example the Balkans in 1999. The Russians clearly struggled to maintain the GLONASS system and the Strela 3 constellations of communications satellites.

However, to imagine that the much contracted service that was left was decrepit or second rate would be a mistake. A minimum service was maintained in key priority areas. By 2000, the Russian military space programme was still very much alive and able to supply strong intelligence information to the GRU in its command and analysis centre in Golitsyno. The Western insight into the Kometa system,

Russia struggled to maintain the GLONASS constellation

afforded as a result of the SPIN experience, suggested that its sister Kobalt spycraft could supply highly-detailed close-look information on any part of the world if required. The Neman system of digital imaging continued to provide round-the-clock photographs of all parts of the Earth throughout the year. The Tselina 2 and US-P land and naval electronic ferrets ensured continuous eavesdropping on the movements and intentions of traditional and new military rivals; the Parus system made sure that the Russian navy could strike back quickly at potential threats; and the Oko system guaranteed that no surprise missile attack could ever arrive without warning.

5

The scientific frontier

Russia's scientific and applications space programme from 1992

Chapter 4 showed how, despite the acute economic pressures on the country, the Russians were able to maintain a viable military space programme during the 1990s. What about the civilian unmanned programme? Here, at least, there were some prospects for attracting interest and revenue from abroad. Russian civilian programmes may be divided into communications, meteorology, Earth resources, biology, materials processing and science.

Russian civilian space programmes, 1992–2000

Weather satellites		*Communications*		
Meteor 1	1969–76 (approx)	Molniya 1	1969–	
Meteor 2	1974–93	Molniya 3	1974–	
Meteor 3	1989–94	Raduga	1975–	
Elektro	1994	Gorizont	1975–2000	
(Meteor 3M	due 2000)	Ekran	1976–92	
		Ekspress	1994–	
Earth resources		Gals	1994–	
Okean	1988–	Luch	1994–5	
Resurs F	1989–			
Resurs O1	1994–	**Scientific**		
		Geodetic		
Materials processing		Geo 1K	1994	
Foton	1988–	Prognoz/Interball	1972–96	
		Koronas	1994	
Biological		Mars 8	1996	
Bion	1972–96			

Space applications had originally been the poor cousin of the Soviet space programme. Throughout the early years, efforts had been concentrated on military programmes, manned missions, the Moon race and the planets. The Soviet Union did not launch its first applications satellites until well into the space age—its first communications satellites in 1965 (Molniya) and its first weather satellites in 1969 (Meteor). The United States had been much faster to appreciate the potential and commercial possibilities of space-based applications, experimenting with satellite communications as early as 1958, with the Tiros weather satellite in 1960 and the first true communications relay satellite, Telstar, in 1962.

WEATHER SATELLITES

Meteor

The first weather satellites had begun during the Soviet period, starting with Meteor 1 in 1969 and Meteor 2 in 1974 (21 of the latter being launched). Soviet weather satellites were a much lower priority than in the space programmes of many other countries and by the end of the Soviet period no geosynchronous meteorological satellites had been launched.

Meteor 2 was an improved version of the original model. Launched by Tsyklon, it doubled the Meteor's lifetime from 6 months to a year and had the capacity to send

Snow cover from orbit (the Alps)

data to automatic electronic tracking centres in addition to the main centres in Moscow, Novosibirsk and Kharbarovsk. Meteor 2s flew in 81.2-degree orbits at 950 km and a three-satellite constellation was able to revisit a site every 12 hours. The last was launched in 1993 and had ceased transmitting several years later, ending the series. It also spun into orbit Temisat, a 30-kg microsatellite built by Kaiser Threde in Germany and Telespazio of Italy to demonstrate commercial data relay services.

The Meteor 3 series of satellites was introduced in 1989 and five were launched during the Soviet period. Meteor 3 metsats weighed 2.15 tonnes (with 700 kg of instruments) and had small electric ion jets for attitude control, based on Glushko designs in GDL dating back to the 1920s. The 5-metre tall spacecraft comprised a hermetically sealed instrument container and two solar panels oriented to the Sun by an electromagnetic solar tracking drive providing 500 W power. Meteor 3s carried scanning telephotometers, infrared radiometers and spectrometers, a multi-channel ultraviolet spectrometer and a radiation measurement device. Its scanning equipment was designed to observe the Earth by night as well as by day. They flew out of Plesetsk on 82-degree orbits at the higher altitude of 1,200 km. Meteor transmitted to over 50 ground stations in Russia, using Automatic Picture Transmission—which means that foreign stations could and did use its data. Information on water moisture in the atmosphere and temperature was transmitted daily down to Moscow and St Petersburg. The hydrological service, with its centres in Moscow, Novosibirsk, and Kharbarovsk, needed its pictures to predict rainfall and snow run-off. Further information was passed on to planes, ships and agricultural institutes. Meteor spacecraft were built at the Research Institute for Electromechanics, VNIIEN. Generally, during the Soviet period, a constellation of three operated at any given time.

Only one was launched after 1991, Meteor 3-6 (January 1994), which also carried French and German experiments, though it had a long lifetime and was still transmitting at the end of the decade. Meteor carried the French experiment SCARAB, or Scanner for Radiation Budget, which was designed to measure the Earth's flow and retention of radiation.

The series was due to be replaced with the introduction of Meteor 3M-1 and 3M-2 in 2000–1. This version would be heavier, 2.5 tonnes, requiring the use of a Zenit launcher, though the intention was also to launch five subsatellites with 3M-1. The Meteor 3M was to orbit at 1,020 km at an inclination of 98 degrees. The subsatellites included 70-kg Badr 2 (Pakistan's second satellite), a Tubsat built in Berlin for Morocco, and a Russian–American geodetic reflector with 32 faces.

Elektro

Despite the fact that America experimented with a 24-hour weather satellite as far back as 1966, a Soviet version was remarkably slow to appear. In 1975 the USSR announced that it would participate in the 1978-9 Global Weather Watch programme. Its contribution would be a 24-hour weather satellite called GOMS, to be located in the first of four geostationary locations for such satellites. The project was repeatedly postponed due to financial difficulties and software problems, with regular promises being given during the 1980s and early 1990s of an imminent launch. A ground station to receive the signals was not built until 1991 in Tashkent in Uzbekistan. However, the following year the station became politically stranded outside Russia: they had to start again and a new one was built in Moscow.

Delays dogged the project to the very end, even the final countdown being held up when the fuel suppliers demanded payment in cash. Eventually, Elektro 1 was launched on 31 October 1994, a full 16 years behind schedule, on a Proton from Baikonour. It reached its geosynchronous destination at $76°E$ several days later. Elektro was a large, 2.5-tonne observation satellite, 6 metres tall with a diameter of 1.4 metres and a solar wing span of 12 metres, equipped with radiometric instruments operating in the visible and infrared bands. White-coated engineers fussing over the final assembly were dwarfed by the size of its main bus, aerials and shaded telescope. Electro was designed to transmit images of Earth 48 times a day to Moscow, Kharbarovsk and local stations. There seem to have been some difficulties with settling Elektro 1 in orbit and the visible channel seems not to have transmitted. A second Elektro was set for 2001, though a more specific launch date was not set. Hopefully, it will not have a wait like its predecessor.

Weather satellites

31 Aug 1993	Meteor 2–21	Tsyklon 3	Plesetsk	938–969	104.12	82.55
25 Jan 1994	Meteor 3–6	Tsyklon 3	Plesetsk	1,186–1,208	109.36	82.56
31 Oct 1994	Elektro 1	Proton	Baikonour	35,851–35,992	1,441	1.3

EARTH RESOURCES MISSIONS: RESURS F1, F2, F1M

From the mid-1960s, scientists began to appreciate the ability of satellites, using different filters and light bands, to photograph the Earth's surface for the purposes of mapping, geology, agriculture, hydrology and mineralogy. The Americans flew the first dedicated Earth resources satellite, Landsat, in 1972—a remarkable spacecraft which became the basis of current space-based applications.

As with many other disciplines, the roots of Russian Earth resources work were buried deep in the Cosmos programme and its Zenit spy satellites. The Russian Earth resources programme has the generic name of Resurs (resources). Between 1975 and 1989, 39 Cosmos satellites carried out remote-sensing work. Subprogrammes carried specific Earth resources work: Hektor–Priroda, 5 missions; Fram, 26 missions; and Oblik, 8 missions.

With *glasnost,* the programme became largely civilianized.[41] However, the terminology used by the Russians is complex and requires some concentration. The Resurs subprogrammes comprise recoverable cabins (the F1, F2 and F1M series); non-recoverable cabins (the O and O1 series); Resurs Okean (which observe the oceans) and its successor Okean O1 and Okean B; and other subprogrammes such as Resurs T (also called Resurs F3) and Resurs R (the Almaz series). To confuse things still further, several variants of these complex nomenclatures are also in use. Some have also been applied retrospectively to the Cosmos programme, so that the first launch in the series was later called the third!

Resurs satellites are built, as were their military counterparts in the Zenit programme, in the Foton KB, part of the TsSKB/Progress plant in Samara on the Volga. Resurs spacecraft have their own control centre at Golitsyno, Krasnoznamensk, Moscow. Resurs pictures are assembled in a huge archive which has more than two million photo images covering most of the world. Some have been sold abroad by the World Map Consortium.

The recoverable Earth resources cabin was called Resurs F, which in turn is subdivided into F1 (short duration: 2–3 weeks) and F2 (long duration: 3–4 weeks). Even after declassification, Russian announcements rarely distinguished between F1 and F2 missions, thereby complicating identification of the category (the table below gives both). These missions are described first.

Resurs F1 was a recoverable Vostok-type cabin equipped with cameras for the study of the Earth's resources. The first Resurs, F1-1, was launched in May 1989 on the R-7 from Plesetsk to be recovered in Kazakhstan. Resurs F1 generally orbited at 250 to 400 km at either 62.8 or 82.6 degrees. Each mission lasted 25 days and involved close-look photography. Resurs F1 carried two SA-34 topographic cameras which provided stereo pictures in three spectral bands; and two SA-20M long-focus wide-frame cameras which took up to 1,800 frames. Pointing was done by use of a star sensor.

Resurs F2 missions lasted longer—between three weeks and 45 days—and used cabins weighing up to 6.3 tonnes. The F2 series concentrated on mapping from higher altitudes. The F1 and F2 spacecraft were similar in appearance, but the F2 carried solar panels. Resurs F2 operated at 210–450 km at either 62.8 or 82.6 degrees. Four Resurs F2 flew.

Small subsatellites called Pions have been released from Resurs satellites. Resurs F16 in August 1993 released Pions 5 and 6 and also carried an American Department of Defence experiment (the first to be carried on a Russian satellite). Generally,

Resurs

29 Apr 1992	Resurs F14	Soyuz U	Plesetsk	231–235	89.22	82.09	F2-5
23 Jun 1992	Resurs F15	Soyuz U	Plesetsk	226–233	89.15	82.32	F1–10
19 Aug 1992	Resurs F16	Soyuz U	Plesetsk	221–238	89.15	82.57	F1–11
	Pion 5,6						Minisatellites
15 Nov 1992	Resurs 500	Soyuz U	Plesetsk	187–303	89.58	89.46	Commem.
21 May 1993	Resurs F17	Soyuz U	Plesetsk	230–237	89.17	82.57	F2–6
25 Jun 1993	Resurs F18	Soyuz U	Plesetsk	223–241	89.14	82.58	F1–12
24 Aug 1993	Resurs F19	Soyuz U	Plesetsk	224–234	89.08	82.59	F1–13
26 Sep 1995	Resurs F20	Soyuz U	Plesetsk	231–235	89.16	82.32	F2–7
18 Nov 1997	Resurs F21	Soyuz U	Plesetsk	180–236	88.66	82.33	F1M
28 Sep 1999	Resurs F22	Soyuz U	Plesetsk	222–230	89.02	82.32	F1M

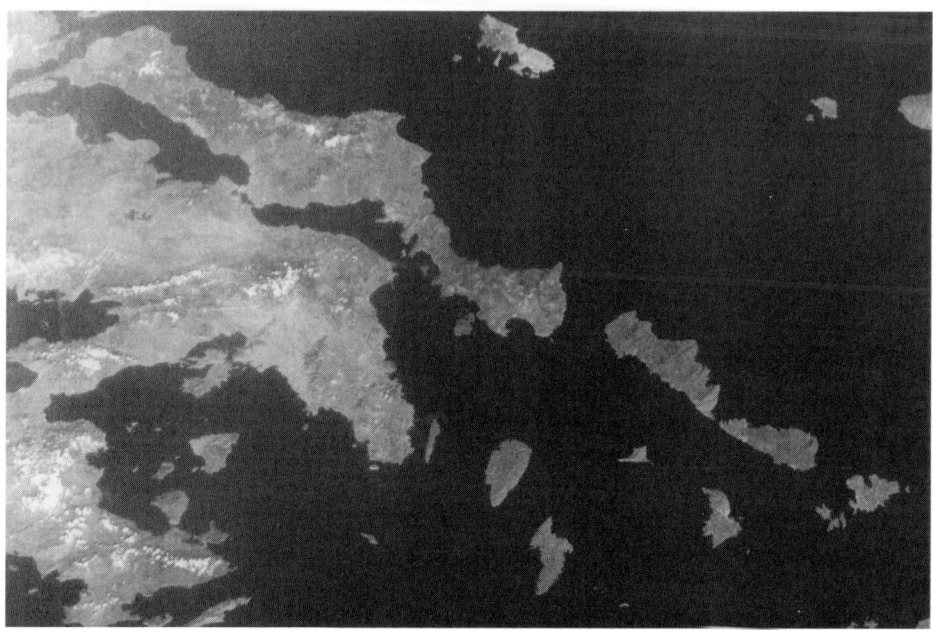

The Greek islands from orbit

Pions orbit for about two months before decay. Pion satellites are small spheres and weigh about 78 kg. They have a diameter of only 70 cm: their only equipment is a transmitter and they are designed to test ground tracking systems. Three Resurs F2 flew.

Thirteen missions were flown during the Soviet period and 10 missions were flown during the Russian period. Originally Resurs F20 was to mark the end of the programme. However, a successor programme appeared: whether this had always been intended, or whether it was an intermediate programme pending the arrival of a wholly new series, is unclear. These were called F1M ('M' for 'modified' or 'modernized'), with the promise of an F2M to follow. The F1M series was designed with improved resolution and wider swaths. The first F1M flew in November 1997 on a mission for the Russian Federal Geodesy and Cartographical Service. The cabin returned to Earth smoothly after 25 days. Two flew altogether.

A special Resurs mission was flown in October 1992 to mark the 500th anniversary of Christopher Columbus's landfall on the islands off the north American coast. Called Resurs 500, it was launched from Plesetsk, Russia, on 15 November and splashed down 320 km southwest of the coast of Seattle on 22 November, conveying anniversary greetings from President Yeltsin to President Bush and a cutglass replica of the Statue of Liberty. The capsule was picked up by the Russian naval vessel the *Marshal Krylov* despite poor sea conditions.

The Russians have proposed a number of spacecraft to succeed the Resurs F programme. These are the Resurs Spektr V, the Resurs Spektr R, Nika K and

Resurs DK. Nika K weighs 7 tonnes with a payload of 1,000 kg in addition to its 2-metre high-resolution cameras. It draws power from 2.5-kW solar panels and can circle the Earth for up to six months, launching subsatellites and sending back a stream of data.[42] Resurs DK, the DK standing for designer Dmitri Kozlov, will transmit information digitally in real time, saving the need to wait for the cabin to return to Earth. Despite having the same name as the earlier series, Resurs DK is quite different. Whereas the main Resurs is based on the design of the Vostok cabin, Resurs DK is a civilianized version of the Yantar spy satellite which has been flying since 1974. Like the Yantar, Resurs DK is a cone-shaped satellite with solar panels at its base, looking downwards at the Earth. Orbiting at 400 km, it will be able to circle the Earth for three years, taking panchromatic, multispectral and infrared pictures down to a resolution of 2.5 metres and, transmitting them back to Earth to a number of ground stations.

Resurs O

Whereas Resurs F capsules are returned to Earth, Resurs O, also called Resurs O1, scan photographs on board and transmit them to the ground digitally in real time, much like the American Landsat series. The comparison echoes the way in which recoverable spy spacecraft gave way to more sophisticated on-board scanning reconnaissance missions. Resurs O is based on the Meteor weather satellite design and was first introduced as Cosmos 1939 in 1988. As this satellite transmitted successfully for six years, much longer than suspected, the first so-named Resurs O did not need to fly until the mid-1990s. Resurs transmits its signals to the Meteor weather centres in Moscow, Kharbarovsk and Novosibirsk and the pictures are sold abroad by a Swedish company, SSC Satellitbild, at prices between €1,100 and €1,800 each. The builders of the Resurs O are VNII Elektromekaniki.

The first Resurs O, also confusingly called Resurs O1-1, O2-1 and O1-3, was flown on 4 November 1994 on the Zenit from Baikonour, a rocket greatly more powerful than what was strictly needed. The 1,900-kg satellite entered an orbit of 661–663 km, 97.9 minutes, 98 degrees. Resurs O1 carried high-resolution optical sensors with a resolution of 15 metres. The second also took a Zenit from Baikonour on 10 July 1998 and released a number of subsatellites. This carried a much improved range of instrumentation and scanners. It received the designation O4 or O1-4 and was accompanied by mini-satellites from Australia, Germany, Thailand, Israel and Chile. Each of the five mini-satellites weighed between 50 and 60 kg. The German Safir 2 relayed telemetric data, the Israeli Techsat and the Thai MYSat mini-satellite carried out remote sensing, the Chilean Fasat-Bravo monitored the ozone layer and provided digital store-and-forward messaging, while the Australian satellite carried passive reflectors.

Okean

The main sea observation resources programme related to Resurs is called Okean (ocean). The series was first tested within the Cosmos programme (Cosmos 1500)

Okean

and was then identified as a civilian programme (Okean 1, 1988). Okean is a tapering cylinder 3 metres tall, and 1.4 metres wide at the base where two solar panels are attached, and weighs 1.9 tonnes.[43]

Okean carries a multizonal scanner, microwave scanner, radiometer, visual and infrared sensors, 1.5-km resolution side-looking radar and a device to receive and transmit information from buoys. The radar is able to measure the height of waves. Designed by the Ukrainian Institute for Radiophysics and Electrics, it has a swath of 450–500 km. Images may be transmitted to up to 570 ground stations. Information on the state and nature of polar ice was transmitted to the State Centre for Remote Sensing and the headquarters of the fishing industry. Okean flies at 670 km. Okean and its predecessors have had an important role in tracking hurricanes, warning of floods in the Gulf of Finland, noting the break-up of the Arctic icepack in the summer and helping ice-breakers through the Arctic and Antarctica. Early Okean satellites helped to rescue several convoys stuck in ice. Generally, Okean satellites transmit for three years, one being launched by Russia in 1994 (Okean 4), probably the last in the series, giving Russia coverage up to 1997–8.

A Ukrainian version of Okean, called Sich, was launched in 1996, with a second one, called Sich 1M, due in 2001 with two microsatellites, Mikron 1 and 2. Granted that all the Okean satellites were built in the Ukraine (in the Yuzhnoye design bureau), it seems odd to call it Ukranian, but the reason was that the mission was wholly controlled from Yevpatoria as part of the Ukrainian national space programme. Even still, it was developed with Russian cooperation and is listed here with the other satellites in the Russian programme. Although the same weight, Sich was quite different from the Okean in design, being rectangular box shaped with a single solar panel and equipped with a radar system and infrared sensors. Sich carried a microsatellite, Fasat Alpha, paid for by the Chilean Air Force, designed to measure the ozone layer and route electronic mail. Unfortunately it failed to separate from Sich.

Okean O, also called Okean 5, was launched by Zenit in July 1999. Good early data were received, but transmissions suddenly ceased and it was some time before they were restored. Although it has the same name and purpose as the older Okean, it is a radically different satellite. Okean O had been some time in preparation and the flight model had waited five years at Baikonour for financial problems to be ironed out. It was built in Ukraine, funded by Russia and operated jointly by both countries. Okean O is three times heavier, weighing 6.3 tonnes. It is box shaped, 12 metres long, 1.8 metres in diameter, with a single detached solar panel able to generate 5 kW of electrical power. It carries a formidable package of nine instruments—a sideways looking radar, two scanning radiometers, three-channel scanner, a visible and infrared scanning radiometer, an eight-channel conical scanner, a spectroradiometer and two microwave radiometers. The sideways-looking radar has a resolution of 2.5 metres. A year later, the Ukrainians announced that the mission had proved to be a considerable success in its first year, studying the oceans, looking for oil and gas deposits, measuring changes in agricultural soils and tracking the ripening of sugar beet. It had repaid the cost of its investment 200 times over, they said. Just as there was a Ukranian version of Okean, so too there was a fully Ukranian version of Okean O, called Sich 2 in preparation. This will carry a double-sideways-looking radar, synthetic aperture radar and three-channel scanner.

Resurs R, Resurs T, Obzor

Other Cosmos Earth resources variants have been reclassified under the Resurs programme, such as the Resurs R, which is also the Almaz radar observatory and Resurs T (also called Resurs F3), which was introduced within the Cosmos programme in 1993 when Cosmos 2260 carried two new KFA-3000 cameras with a resolution of 75 cm. These are enormous and peer out of the recovered spacecraft like giant eyes. The status of the Resurs T system has never been entirely clear. Cosmos 2260 was labelled Resurs T (also Resurs F3) and it was subsequently announced that a number of its Cosmos predecessors from 1986 had also really been Resurs T. It seems reasonable to presume that Resurs T is indeed similar to the main Resurs family but with a different imaging system.[44] Finally, Cosmos 2285 was a remote-sensing satellite developed under what was called the Obzor 'survey' series. It caused some mystery, with various interpretations advanced for its purpose until it was identified as an experimental Earth resources satellite in 1999.[45] Later, the Arsenal Design Bureau announced an operational Obzor series (photo, p. 155) with which it may be connected.

1870 (July 1987) and Almaz 1 (1991). Sales of images from Almaz 1 were slow, certainly insufficient to make a profit, and the station was de-orbited in October 1992. Its back-up, Almaz 1B, has been in the hangar ever since, awaiting launch.

Almaz was a typical example of a large Soviet-style project. Satellite manufacturers have since been moving towards much smaller platforms which can be developed and flown sooner, taking advantage of new microcircuits. Proposals have been put forward by the Khrunichev company for a new system of small visible and radar resource mapping satellites, starting in 2007, called Monitor and Yacht.

MATERIALS PROCESSING: FOTON

Soviet experiments with materials-processing date to the Almaz and Salyut space stations in the 1970s. In the 1980s, starting with Cosmos 1645, the Vostok cabin was adapted as an unmanned, recoverable, materials-processing laboratory called Foton (also written as Photon), able to fly 14–16 consecutive days.

Foton was formally introduced in its own right on 14 April 1988 as a programme for the manufacture in orbit of semiconductors and extrapure materials. Foton was built by the Kozlov TsSKB bureau in Samara; it weighs up to 6.4 tonnes and orbits at 62.8 degrees at 220–400 km from Plesetsk. Chemical batteries provide 400 W of power. The 2.3 metre-diameter cabin has a payload of 700 kg in a volume of $4.5 \, m^3$. Standard equipment includes the Splav-02 furnace, the Zona furnace and the Kashtan electrophoresis unit. In order to protect equipment during the final stages of the descent to Earth, solid rockets fire under the three 27-metre-diameter parachutes to cushion the final fall.

The purpose of the Foton missions is to enable experiments to be conducted in processed alloys and optical materials and in the testing of semiconductors. A feature of the series is that it uses already-flown capsules. Foton has been a modest earner for the Russian space programme, as much as €20m a mission, enabling Western companies and agencies to have access to zero gravity in a way not otherwise possible. There is no equivalent of the Foton programme in the United States or Europe (though there is in China with the FSW programme) and Russia was able to attract several Western companies and agencies, for example the German company Kayser-Threde and the French Space Agency CNES.

On 14 June 1994, Foton 9 carried a French experiment, Gezon, to study the melting of materials and space; and a European biopan experiment to study the behaviour of aminoacids in space. Foton 10 was a joint European Space Agency/ French CNES mission which ended in disaster. Launched on a 15-day mission out of Plesetsk on 16 February 1995, the Foton carried semiconductors and a biobox with shrimps and urchin. The cabin returned to Earth perfectly on 3 March near Orenburg in the Urals but when it was lifted out of the recovery area the next day the helicopter flew into a blizzard. Gusts of wind seized the capsule, causing it to sway alarmingly and the Mil crew dropped the payload from an altitude of several hundred metres. The experiments were smashed.

The Foton 11 mission took place on 9 October 1997. Most of the experiments were reflights from Foton 10 (this was part of the arrangement if something went wrong). Launched from Plesetsk, it carried German experiments devised by Kayser Threde and a small cabin called Mirka (standing for microgravity re-entry capsule). Mirka was a spherical 150-kg, 1-metre-diameter cabin carrying three experiments— on heatshield instrumentation, heat flow, pyrometers and a new type of ablative material. The Mirka capsule was carried on the front of the Vostok-type cabin, separated after retrofire, with its own ablative material, parachutes and beacons. The main cabin carried crystal growth experiments, human cell biology experiments to learn more about cancer, flies (to test for ageing) and beetles (to see how their biological clocks were affected by zero gravity). On the outside of the main cabin was biopan—a container able to expose biological experiments to open space (it closes in time for re-entry). Both the main cabin and the Mirka touched down safely on brown steppe grass near Orsk on 23 October.[46]

Foton 12 flew in September 1999, carrying experiments from Russia, Germany, China, France and Sweden. The French team were given a real treat—being permitted to watch the launch from a protected shelter only 800 metres from the rocket itself. The Soyuz, as usual, performed as advertised, taking off only 0.001 of a second late. Foton 12 came back a day early, because the French Ibis experiment shut down prematurely during one of the plant experiments. The Foton 12 landed in green-brown grasslands 133 km northwest of Orenburg 15 days later, a mere 2,000 metres from the welcoming squad. Foton 12 included a furnace, a machine for measuring microgravity, a biological container and a device to measure electromagnetic emissions. Within minutes, rescue teams were alongside unloading the experiments.

Altogether, Russia made five Foton launches, in 1992, 1994, 1995, 1997 and 1999. It aims to continue the Foton programme, with annual launches to 2005, while replacing it with its successor Nika-T, able to carry a 1,200-kg payload and fly for 120 days at a time, with three large solar panels able to generate 6 kW of power.

Foton series

8 Oct 1992	Foton 8	Soyuz U	Plesetsk	220–359	90.3	62.81
14 Jun 1994	Foton 9	Soyuz U	Plesetsk	221–364	90.4	62.81
16 Feb 1995	Foton 10	Soyuz U	Plesetsk	220–369	90.4	62.81
9 Oct 1997	Foton 11	Soyuz U	Plesetsk	218–375	90.5	62.81
9 Sep 1999	Foton 12	Soyuz U	Plesetsk	217–384	90.5	62.8

COMMUNICATIONS SATELLITES: MOLNIYA

Molniya (the Russian word for 'lightning') was the USSR's first communication satellite, or comsat. Russia was a poor second into the comsat race after the

United States and its comsat technology tended to lag about five years or more behind that of the United States. American comsat development was determined by urgent commercial imperatives that were absent in the USSR. Molniya is cone shaped. Antennae and shiny metallic shapes glisten. At the base are six vane-like panels spread out like a windmill about to turn. Molniya weighs about 816 kg. At 4 metres high and 1.4 metres in diameter, Molniya provides telephone and TV links. Like other Russian comsats, Molniya is built by the Applied Mechanics NPO in Krasnoyarsk, NPO-PM.

Molniya became the basis of the Soviet comsat system. It used a system of orbits not hitherto used by any other type of satellite system, but one easily explicable in terms of the geography of Siberia. In designing a comsat system for Russia, the 24-hour orbit presents difficulties. A stable 24-hour orbit can only be located on the equator and a transmitter from equatorial orbit would simply not be strong enough to reach the far northern arctic latitudes of Siberia for which comsat technology is most useful. At the other extreme, a conventional orbiting satellite in a low orbit covering Siberia would present major tracking problems because it would be overhead only for periods of 10–15 consecutive minutes.

Hence the compromise 12-hour orbit stretching up to the 24-hour altitude of 40,000 km, but with a perigee of 600 km. Apogee was nearly always over Siberia, so that it climbed slowly across that part of the sky over a period of eight hours, before whizzing around its perigee at a much faster velocity. The idea of using such an orbit for communications had first been developed by Bill Hilton in the British Interplanetary Society in 1959–60.[47] He had sketched the idea of such a high eccentricity orbit in 1960, the year before Molniya went into design. This might have been purely coincidental, had not Hilton been immediately accosted by an enthusiastic Aeroflot official(?) wishing to discuss the paper with him and the fact that the paper was at once translated into Russian! One disadvantage of the Molniya orbit is that it takes the satellite through the Van Allen radiation belt and, as a result, each satellite requires some heavy shielding.

Molniya was designed in October 1961 by Korolev's OKB-1. The first attempted launch was in 1964, but it failed and success was not achieved until April 1965. Because the Korolev bureau was preoccupied with the Moon programme, Molniya was then transferred to NPO-PM in Krasnoyarsk. Three satellites were needed for an operational system, though four was optimum and more were helpful to handle rising levels of traffic. A series of 12-metre receiver dishes called Orbita 1 were built throughout the USSR to pick up Molniya signals, and by the 1980s over a hundred such dishes had been built. Molniya 1s operate in a constellation of eight at a time, thus ensuring continuous coverage. Over a hundred have been built in the factory in Krasnoyarsk, indicating an intention to maintain the series for some time yet. A subset received the 'T' designation.

An improved version, Molniya 2, appeared in 1971, but was quickly replaced by Molniya 3 which first appeared in 1974. Molniya 3 was able to use much smaller ground stations and by the early 1970s terminals were as small as 7 metres (the Orbita 2 system). Like Molniya 1, the 3 series operates in constellations of eight. At a general level, Molniya 1 concentrated on domestic, governmental and military

Communications satellites

Molniya 1

4 Mar 1993	Molniya 1–83	Molniya M	Plesetsk	620–39,731	718	62.8
6 Aug 1992	Molniya 1–84	Molniya M	Plesetsk	632–39,721	718	62.8
13 Jan 1993	Molniya 1–85	Molniya M	Plesetsk	606–39,746	718	62.8
26 May 1993	Molniya 1–86T	Molniya M	Plesetsk	401–40,884	736	62.9
22 Dec 1993	Molniya 1–87	Molniya M	Plesetsk	440–39,182	702	62.8
14 Dec 1994	Molniya 1–88T	Molniya M	Plesetsk	441–39,910	717	62.8
14 Aug 1996	Molniya 1–89T	Molniya M	Plesetsk	464–39,887	717	62.8
24 Sep 1997	Molniya 1–90	Molniya M	Plesetsk	449–39,906	717	62.8
29 Sep 1998	Molniya 1–91	Molniya M	Plesetsk	420–40,657	732	62.8

Molniya 3

14 Oct 1992	Molniya 3–42	Molniya M	Plesetsk	477–39,877	718	62.8
2 Dec 1993	Molniya 3–43	Molniya M	Plesetsk	416–39.939	718	62.8
21 Apr 1993	Molniya 3–44	Molniya M	Plesetsk	614–40,618	736	62.8
4 Aug 1993	Molniya 3–45	Molniya M	Plesetsk	412–39,931	718	62.8
23 Aug 1994	Molniya 3–46	Molniya M	Plesetsk	605–39,756	717	62.8
9 Aug 1995	Molniya 3–47	Molniya M	Plesetsk	427–39,969	718	62.8
24 Oct 1996	Molniya 3–48	Molniya M	Plesetsk	610–39,768	718	62.8
1 Jul 1998	Molniya 3–49	Molniya M	Plesetsk	432–39,943	718	62.8
8 Jul 1999	Molniya 3–50	Molniya M	Plesetsk	468–40,811	736	62.8

communications links and Molniya 3 the international and civilian. Russia launched nine Molniya 1 and nine Molniya 3, bringing the series total to 91 and 50 respectively. The venerable series showed no signs of conclusion.

NEW GENERATION: GORIZONT, RADUGA, EKRAN

In 1945, the famous British author Arthur C. Clarke calculated that satellites, orbiting at 36,000 km, circled the Earth every 24 hours. In effect, they appeared to hover over the same point on the Earth. He then calculated how three such satellites, spaced 120° apart, equipped with television and radio relay devices, could in effect cover the globe. Once again, the Americans were first off the mark and their relay satellite, Early Bird 1, introduced global television from 24-hour orbit in 1965. The USSR did not fly a satellite out to 24-hour orbit until 1974 (Cosmos 637/Molniya S) and their first generation of 24-hour comsats did not make their appearance until the following year. Three types emerged: standard TV and communications satellites, Gorizont (the Russian word for 'horizon'), which were civil; Raduga (the Russian word for 'rainbow') which were government and military; and direct broadcast (called Ekran, the Russian for 'screen') which was civilian.

Raduga's channels were used from the beginning to transmit facsimiles of newspapers from Moscow to Irkutsk in the central USSR to Kharbarovsk in the Pacific. Thus, *Pravda* was available in the far east within minutes of leaving the printers'

Gorizont

works in Moscow. No picture of Raduga has been released, though they are thought to have a mass of around 1,960 kg and be similar in appearance to Gorizont. In June 1989, a new version of Raduga was introduced, called Raduga 1 and the series has since been labelled 1-1, 1-2 and so on. Raduga reached 1-4 in February 1999, making 37 launches in the two generations. A new version, Globus, was introduced in August 2000.

Gorizont was the civilian equivalent of Raduga and appeared at around the same time. The first four were put up over 1978–80 in time to broadcast the 1980 Moscow Olympic games. The Gorizont satellite carried the White House–Kremlin hotline. Its main functions, though, are more mundane—the broadcasting of television through-out the Russian federation. Gorizont looked like an automaton out of science fiction, with arrays of shutters, engines, tanks and instruments atop a cylinder base and a dish, pointing Earthwards. A typical Gorizont weighed 2.5 tonnes and carried 11 antennae and eight transponders. This, the first series of geostationary civilian comsats, was coming to its end by the conclusion of the Soviet period. Six were launched in the 1992–5 period and it was announced that Gorizont 31 in 1996 would be the last. In the event, two were launched after Gorizont 31. What was definitively announced as really the last Gorizont was Gorizont 33 in June 2000. The

Gorizont series

2 Apr 1992	Gorizont 25	Proton	Baikonour	35,769–35,796	1,435	1.44
14 Jul 1992	Gorizont 26	Proton	Baikonour	35,792–35,799	1,435	1.48
27 Nov 1992	Gorizont 27	Proton	Baikonour	35,814–36,528	1,435	0.1
28 Oct 1993	Gorizont 28	Proton	Baikonour	35,752–35,789	1,435	1.51
18 Nov 1993	Gorizont 29	Proton	Baikonour	35,039–35,100	1,399	1.48
20 May 1994	Gorizont 30	Proton	Baikonour	35,773–37,787	1,435	1.33
25 Jan 1996	Gorizont 31	Proton	Baikonour	36,763–35,882	1,436	1.52
25 May 1996	Gorizont 32	Proton	Baikonour	35,725–35,852	1,436	1.47
6 Jun 2000	Gorizont 33	Proton	Baikonour			

Raduga series

25 Mar 1993	Raduga 29	Proton	Baikonour	35,981–36,97	1,454	1.42
30 Sep 1993	Raduga 30	Proton	Baikonour	35,849–35,920	1,441	1.52
5 Feb 1994	Raduga 1–3	Proton	Baikonour	24hr		
18 Feb 1994	Raduga 31	Proton	Baikonour	24hr orbit		
28 Dec 1994	Raduga 32	Proton	Baikonour	35,796–35,900	1,439	1.5
19 Feb 1996	Raduga 33	Proton	Baikonour	242–36,502	645	48.6 (fail)
28 Feb 1999	Raduga 1–4	Proton	Baikonour	36,430–36,563	1,472	1.48

Gorizont programme was able to attract some external funding towards the end. Its operator, NPO-PM, was approached by Rimsat, a small company based in Fort Wayne, Indiana, which had the licence to provide telecommunications services for the Pacific Kingdom of Tonga, and in turn for India, Malaysia, Philippines and the southwestern Pacific. Rimsat hired an existing Gorizont comsat—it was already in orbit and had exceeded its planned lifetime—then leased another two already orbiting Gorizonts (31 and 32) from NPO-PM, paying $130m.

Ekran, introduced in October 1976, carried television only and was one of the first national television systems designed to broadcast direct to the homes of isolated communities. The target area was a footprint covering Novosibirsk, Irkutsk and northwest Mongolia. All that people needed on the ground was a small dish rooftop aerial. The government provided about 5,000 initial receivers for the region where they were designed to serve an initial target audience of 20 million people. Ekran continued the tradition of putting more and more power on the satellite and less and less receiving power on the ground, hence smaller dishes. Ekran had a broadcasting power of 200 W, compared to a mere 8–10 W on the Raduga. The service as a whole covered 40% of the USSR, or 9 million square kilometres, with Ekran hovering over Sumatra and Indonesia at 99°E longitude. Ekran weighed 2 tonnes and had a lifetime of over two years. This precision delivery to scattered communities was made possible by the unusual design of the Ekran: a cylinder pointing Earthwards. On top were two solar panels stretching from

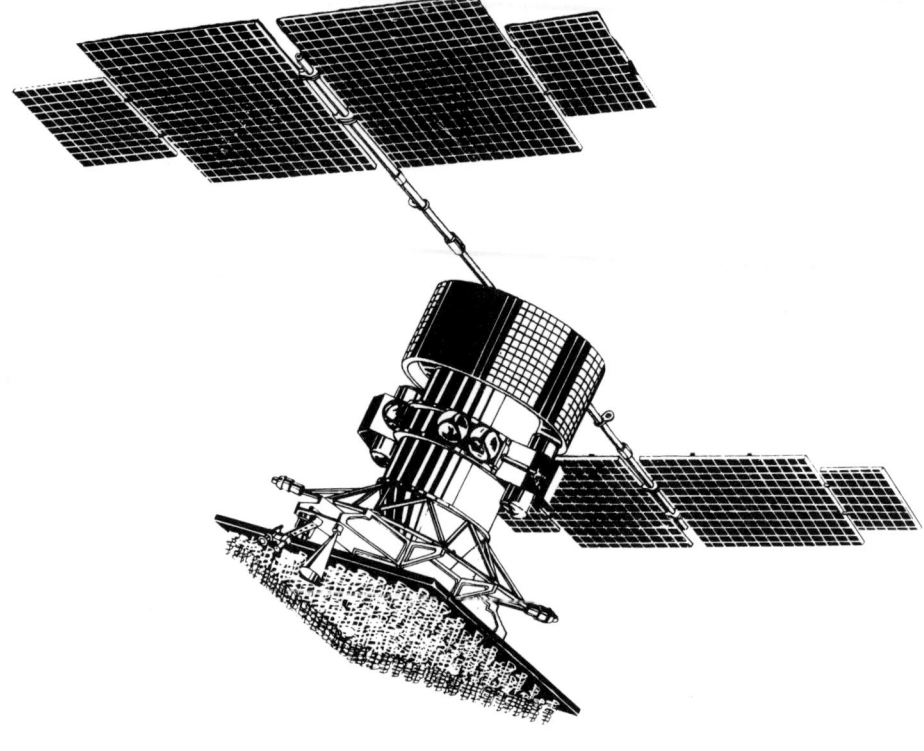

Ekran M

a long beam. Underneath was a flat electrical board, and spread out on it, 200 pencil-shaped antennæ, pinging downwards. An improved version, Ekran M, was introduced with Ekran 16, which was also termed Ekran M1. The last Ekran was Ekran 20, also known as Ekran M5, on 30 October 1992.

Ekran

30 Oct 1992	Ekran 20	Proton	Baikonour	35,575–35,692	1,428	1.58

In the first generation of 24-hour communications satellites, Russia orbited nine Gorizont, seven Raduga and one Ekran. In effect, they brought the Soviet-era comsats to conclusion, before a new generation of Russian comsats was introduced. There were three failures: a Gorizont was lost on the Proton launch failure of 27 May 1993 when the rocket ran out of fuel. Raduga 33 was stranded in a transfer orbit in February 1996. On 5 July 1999, Raduga 1-5, flying into orbit with a new Briz M upper stage, crashed due to an engine fault on the Proton second stage.

NEW GENERATION: EKSPRESS, GALS, KUPON, YAMAL

Starting in 1994, Russia began to replace the old 1970s Gorizont, Raduga and Ekran series with a new, second generation of 24-hour comsats. A real problem with the Gorizont, Raduga and Ekran systems was the length of their design lives. Typical American comsats in the 1980s onwards operated for 7 to 10 years, some even longer. For the USSR, the commercial imperatives were weaker, the competition non-existent and the electronics less advanced. There were few incentives to build longer-lasting comsats and it was cheaper to replace comsats with new ones when they broke down. The average life of a Gorizont was 69 months, Raduga 50 months, and Ekran a miserable 25. Exceptionally, one Gorizont even operated for 10 years (Gorizont 3). The purpose of Ekspress and Gals is to radically upgrade the standard of service, capacity and lifetimes of the Gorizont and Ekran systems.[48] By the 1990s these rates had improved, with 74 months for Gorizont, 75 months for the Raduga 1 series, and a dramatically better 72 months for Ekran M.[49]

Gorizont was replaced by Ekspress, which made its first appearance on 13 October 1994 with the launch of Ekspress A1 by Proton from Baikonour, heading for 24-hour orbit. Weighing 2.5 tonnes, it was stationed on the Greenwich line at 0°E at an altitude of 35,800 km. Ekspress had 17 channels and a lifetime of at least seven years. Russia obtained allocations for six Ekspress satellites at a time: at 35°E, 42°E,

Ekspress

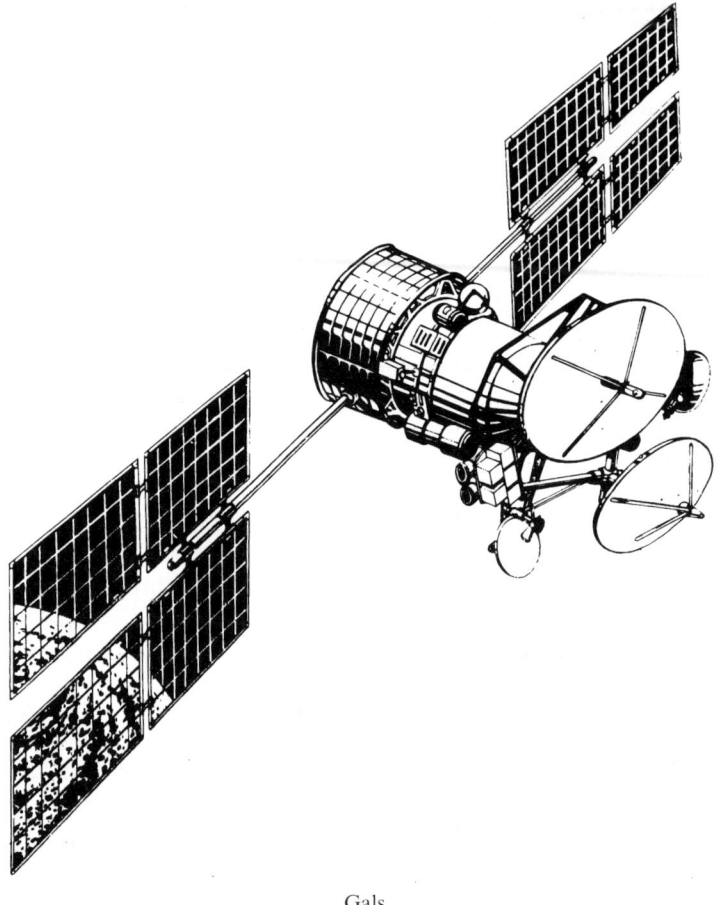

Gals

44°E, 70°E, 96°E and 335°E. The second Ekspress reached orbit in September 1996. The next Ekspress was lost in the Proton failure of 27 October 1999 but was replaced by Ekspress A3 on 12 March 2000 and then a further launching on 24 June, confusingly labelled as 3A. This took place just after the last Gorizont. Successor programmes were planned for later in the decade—Ekspress AM, developed with the Japanese Nippon Electrical, and Ekspress K, in a joint venture between NPO-PM and the French company Alcatel.

Ekran is being phased out by a new series called Gals. The first Gals direct broadcast satellite was lofted on 20 January 1994 by Proton from Baikonour. Weighing 2.5 tonnes, the satellite will transmit to 90-cm dishes. Likewise, Gals is designed to have a much longer operating life—five to seven years—and is the first Russian geosynchronous satellite to use its engines to control its inclination constantly to within an accuracy of 0.1 degree (hitherto, they drifted around the geosynchronous point by up to 1.5 degrees). Gals 2 was launched in November 1995.

A new type of Russian comsat was Kupon. Launched by Proton in November 1997, Kupon 1 was the first of a three-satellite system called Bankir. Built by the Lavotchkin design bureau, the purpose of the system was to provide satellite-based electronic banking systems between London, Russia and the far east for the Russian Federation's Central Bank. Ideally, it would serve up to 40,000 terminals continuously with a satellite-delivered banking service. Unhappily for the bank, the master computer controller crashed as soon as the satellite arrived in geostationary orbit, making it inoperable. This was the first time such a failure had ever been recorded, though that must have been little consolation to the bankers.

Another new type of satellite was Yamal, two of which were orbited together by Proton on 6 September 1999 (called Yamal 100 and 101). These were built for the state gas company, Gasprom. Each box-shaped satellite weighed 1.3 tonnes and was built by RKK Energiya with American Space Systems Loral. Solar panels provided 2.2 kW and 9,000 telephone circuits were installed on the satellite. However, one of the two Yamal seems not to have deployed properly. Although delivered perfectly to its station, it began to spin and would not respond to launch control commands. Despite the setback, proposals were put forward for Yamal 200 (44 transponders) and Yamal 300 (52 transponders) with 12- and 15-year lifetimes respectively.

For Russia, the 1990s were a picture of mixed progress with communications satellites. By 1996, financial shortages had led to a slowdown in the launching of replacement satellites and two-thirds of Russia's comsats were operating beyond their limited natural life. The USSR had gradually built up the system of geosynchronous satellites to the point that 34 were operating in 1991, their peak. By 2000, 24 were in operation, though many of these were operating far beyond their design lifetimes of 6–7 years. Even if their electronics continued to function, and these had visibly improved in the 1990s, many would soon run out of fuel. In summer 2000, Russian Space Agent Yuri Koptev noted that 60% had exceeded their service life. Russia was able to operate only 86 channels in orbit, though it needed at least 180 and probably 600 by 2006. Russia had been allocated 54 satellite locations in the crowded geosynchronous orbit, but was occupying only 18 of them.

Positively, the transition from the first generation of comsats had taken place and Gorizont, Raduga and Ekran were being phased out by a new generation. Two new series had been introduced—Ekspress and Gals—no mean achievement for a

The new Russian comsats

20 Jan 1994	Gals 1	Proton	Baikonour	35,888–35,959	1,443	0.21	Comsat
13 Oct 1994	Ekspress A1	Proton	Baikonour	35,777–35,808	1,436	0.21	Comsat
17 Nov 1995	Gals 2	Proton	Baikonour	35,787–35,949	1,440	0.17	Comsat
26 Sep 1996	Ekspress A2	Proton	Baikonour	35,838–35,908	1,440	0.21	Comsat
12 Nov 1997	Kupon 1	Proton	Baikonour	33,849–38,086	1,445	0.06	Comsat
6 Sep 1999	Yamal 101, 102	Proton	Baikonour	35,512–36,295	1,442	0.01	Comsat
12 Mar 2000	Ekspress A3	Proton	Baikonour	35,779–35,794	1,436	0.19	Comsat
24 Jun 2000	Ekspress 3A	Proton	Baikonour				

programme in such difficulties. Russia had also been successful in attracting commercial investment in satellite programmes, such as Kupon and Yamal, even if results had fallen short of these expectations. In addition to these improvements in satellites in 24-hour orbit, the Russians expressed interest in a large constellation of low Earth-orbiting satellites called Signal. A project for 48 Signal satellites received federal Russian approval in 1999.

LUCH

Luch, the Russian word for 'light', was a series introduced during the late Soviet period to facilitate communication between the cosmonauts on the Mir space station and ground control. It was also called the Altair system The Americans devised a similar system for the shuttle, called TDRS. Luch looks like a flying eagle—a 2.25-tonne bus with two giant 1.8-kW solar wings and three transponders which look like claws.

Until Luch, cosmonauts could only contact ground control when overflying the Soviet Union, or through one of the comships. Keeping comships at sea cost $5m a week, but 24-hour satellites operating as data relays offered a cheaper, more convenient and effective alternative. The logic of Luch was that instead of communicating *downward*, Mir's crews communicated *outward*, to one of three large comsats in geostationary orbit, which then acted as relays with ground control. Each Luch provided 40 minutes of stable communication each orbit, and a constellation of three would provide round-the-clock coverage. Ground listeners in Britain could make out the switch to the Luch system when a series of pips opened the radio relay, with a slight humming sound and echo on the circuit in the background.

The first Luch launches took place in the Cosmos series (Cosmos 1700 in 1985) until 1994 when the formal Luch designation was used. Cosmonauts Leonid Kizim and Vladimir Solovyov were the first to use the Luch system operationally during their stay on Mir in 1986. Two more Luch followed—Cosmos 1897 and 2054. This should have given the Russians round-the-clock relays for Mir. However, the problem of short lifetimes which affected the first generation of 24-hour comsats seems also to have been a problem with Luch. Luch coverage was much less than what the Russians might have wished. Luch satellites seemed to have insufficient fuel to stay on station as long as the American TDRS. From 1992 to 1994, only Cosmos 2054 was available, restricting communications to only short periods each day.

This should have been rectified in the mid-1990s when the dedicated Luch programme entered service. On 15 December 1994, Luch 1 was launched by Proton into 24-hour orbit and located at 95°E. The second Luch, designated Luch 1-1, reached orbit in October 1995. As a result, by early 1996, Mir had round-the-clock global coverage and cosmonauts could communicate with the ground wherever they were.

This positive situation did not last—indeed, full coverage was achieved for only a year. The veteran Cosmos 2054 (December 1989) packed up in January 1997, followed by Luch 1 in August 1998 after less than four years and Luch 1-1 in January 1999 after just over three. As a result, the Mir crew reverted to the

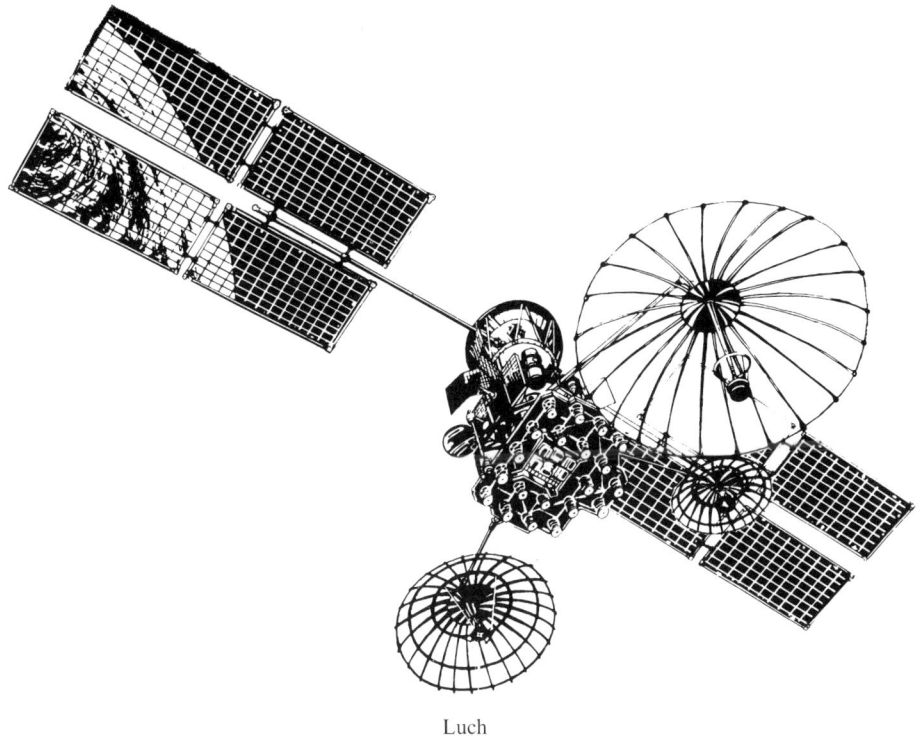

Luch

situation that existed during Salyut and the early manned flights, of communicating only when over ground stations in Russia itself.

Although the full Luch network was achieved for a brief period, the overall series cannot be reckoned successful because of its short duration. By the time the system finally broke down, Mir was nearing the end of its life (or so everyone thought) and as the new International Space Station would use the American TDRS, it was allowed to lapse.

Luch relays

16 Dec 1994	Luch 1	Proton	Baikonour	35,775–35,815	1,436	2.58
11 Oct 1995	Luch 1-1	Proton	Baikonour	35,767–35,810	1,436	3.07

BIOLOGICAL PROGRAMME: BION

Despite its military orientation, the Cosmos programme included an international biological programme. Its roots date to 1962 with the bilateral agreement between NASA and the USSR Academy of Sciences for the compilation of a treatise called

Foundations of Space Biology and Medicine. Although this was never stated publicly at the time, it is likely that cooperation in space medicine was considered to be the area least likely to cause political problems or difficulties between the two sides. As a result, there was structured information exchange between the two countries during the 1960s.

Arrangements were made for cooperation on joint missions under the Nixon–Kosygin agreement on space cooperation signed in Moscow in May 1972, at the same time as the Apollo–Soyuz Test Project was agreed. The first jointly planned biological mission was Cosmos 936 in 1975. Remarkably the series survived the freezing of cooperation in other fields which took place under the Carter administration in 1978–9. The programme provided the only opportunity for space scientists and biologists to have regular access to dedicated medium-duration orbital space biology missions during the cold war period. On the Soviet side, the leader of the programme was Dr Yevgeni Illyin, who had been selected as a medical doctor to fly a subsequently cancelled Voskhod mission in 1965.

The series was called Bion, though this designation was only used formally at a later stage in the series. The objectives of the programme were to study the effects of weightlessness on living organisms on short flights, the effects of radiation on animals, to better understand the adaptation of living bodies to weightlessness and to study means of protecting animals from the effects of weightlessness. The prevention of space sickness, suffered by up to 50% of astronauts and cosmonauts during their first five days in orbit, was an important aim. Nine Bion missions were flown during the Soviet period. One, Bion 8, famously attracted public attention when one of the two monkeys freed himself from his restraint, and began tampering with the controls, sparking the classic headline of: 'Monkey to ground control: "I've taken over!".' Worse was to come, for the capsule was misaligned at re-entry (not the fault of the monkey) and came down 3,000 km off course in Siberia. The monkeys were kept warm until villagers arrived.

Bion used a 2.3-metre-diameter Vostok-derived cabin with a hatbox battery pack located on the front. Bion produced important results on the effects of weightlessness. In December 1992, Cosmos 2229, or Bion 10, flew into a 218–376-km orbit at 62.81 degrees, orbiting the Earth every 90.45 minutes. Monkeys Krosh and Ivasha were on board and they were safely recovered after 13 days. Although suffering some weight loss and dehydration, they were generally in good condition: journalists were soon invited to meet the recuperating macaques who were eating apples. In addition to the monkeys, Bion 10 carried salamanders, flies, plants, seeds and algae. Besides Russia and the United States, Canada, the Czech Republic, Lithuania, Poland and, for the first time, China participated in the programme.

However, what brought this joint Russian–American programme to a premature end was nothing to do with international politics but an unlikely adversary: the American animal rights movement. Animal rights activists were never happy about the programme, nor about American participation in it. In autumn 1996 activists barricaded themselves into NASA administrator Dan Goldin's office in an effort to prevent future missions, objecting to the cruelty involved. They attacked the way in which sensors were implanted in the monkeys before the

Salamanders in orbit

mission and removed afterwards, each exercise requiring an operation under anaesthetic. Samples were also taken of the monkey's tissue.

Goldin permitted American participation in the next mission, Bion 11, to go ahead, but on condition that the monkeys be treated 'humanely'. Flying with them in December 1996 were newts, snails, flies, insects and bacteria. All appeared to go well with Bion 11, macaque monkeys Lapik and Multik greeting their handlers when they were taken out of their cabin on landing 130 km north of Kustanai after 14 days. They were flown to Orenburg and then to the Institute for Medical and Biological Problems in Moscow. Multik died during the post-landing operation, suffering a sudden cardiac arrest. It is possible that the strain of the flight, followed by the operation, proved too much. However, this tragic end made the campaigning point for the activists.

Several months later, in May 1997, NASA announced that it would not participate in flying monkeys again in the future. That was the end of the Bion programme for the time being and the Bion 12 mission was cancelled. TsSKB Progress in Samara, the maker of the Bion cabin, still hoped to see Bion 12 fly and the mission was entrusted to the Institute for Medical and Biological Problems (IMBP) in Moscow to search for new foreign partners for the flight. Ideally, it would like to carry out radiation studies at the proposed altitude of the International Space Station. Ultimately, Bion may be replaced by the 7-tonne Nika-B cabin—but whether it will find any passengers is another matter.

Russian Bion missions

29 Dec 1992	Cosmos 2229/Bion 10	Soyuz U	Plesetsk	218–376	90.45	62.81
24 Dec 1996	Bion 11	Soyuz U	Plesetsk	217–379	90.48	62.8

GEODETIC PROGRAMME: MUSSON, GEO-1K

Turning now to scientific programmes, the Soviet Union began to launch geodetic satellites in 1968, starting with Cosmos 203, the first of 17.[50] Their role was to precisely measure the Earth's gravitational field. In 1981, the Tsyklon launcher was used to launch the first of 12 second-generation geodetic satellites, the Musson series, which carried Doppler transmitters, a flashlight system and laser reflectors. Two Musson launches were made by Russia. The first was given a Cosmos designation (2226) and the series was then declassified satellite as Geo-1K. It was launched on 29 November 1994 on Tsyklon 3 from Plesetsk into an orbit of 1,480–1,527 km, 73.6 degrees, 116 minutes. Geo-1K was a 900-kg cylinder 2 metres in diameter, 2.1 metres long, with a gravity-stabilizing boon, star trackers and 10 vanes of solar panels. Signals were transmitted 12 hours a day and could be heard on 150 MHz and 400 MHz. Geo-1K marked the end of this minor series. It is possible that some of the work undertaken by this series has partly been taken over by Nadezhda navigation satellites.

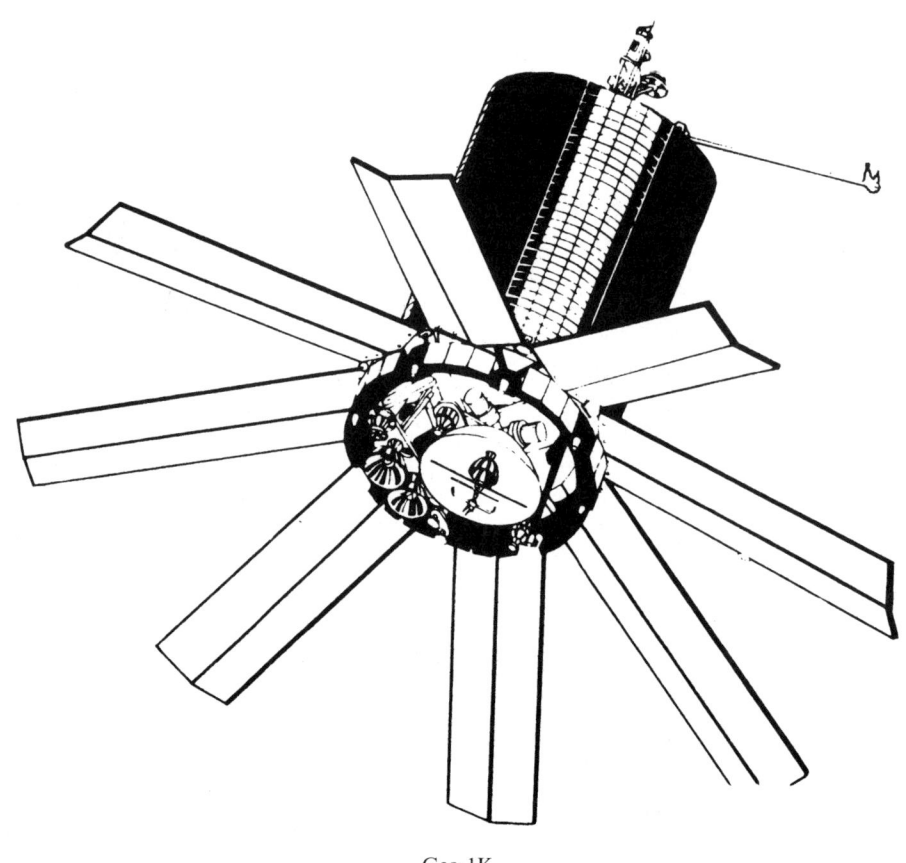

Geo-1K

Musson/Geo-1K series

| 22 Dec 1992 | Cosmos 2226 | Tsyklon 3 | Plesetsk | 1,479–1,526 | 116.03 | 73.63 |
| 29 Nov 1994 | Geo-1K | Tsyklon 3 | Plesetsk | 1,480–1,527 | 116 | 73.61 |

SOLAR OBSERVATION SATELLITES: INTERBALL AND KORONAS

The Soviet Union was late to develop a specialized programme of solar observation satellites. The United States launched the first in its Orbiting Solar Observatory series from 1962. Several early Cosmos satellites carried equipment to study the Sun, for example Cosmos 230 (1968) and Cosmos 481 and 484 (1972). There were several reasons for a series of solar observatories: the Sun, as a typical star, could be studied accurately from space, free of the interferences of Earth's atmosphere; second, the Sun's emissions and electrical disturbances interfered with Earth's physical environment, producing such effects as auroræ, which were poorly understood; and third, solar radiation was a potential threat to cosmonauts in low Earth orbit and early warning of danger was desirable.

The Soviet Union's main programme of solar observatories was called Prognoz (not to be confused with the military early-warning satellites of the same name).

Instrument section on Prognoz

Prognoz had a square base, a dome-shaped top from which an array of instruments emerged and four stretched-out solar sails. The first flew in 1972, taking a highly eccentric orbit to 200,000 km, nearly half the distance of the Moon, giving it a good vantage point for studying conditions on the Sun and perfect to intercept and analyse solar radiation. The larger, second-generation Prognoz followed in 1983, flying as far out as 720,000 km every 27 days.

Interball

The third generation, Prognoz M, also known as Interball, followed and was developed by Russia with the Czech Republic to study the tail end of Earth's magnetosphere. Each satellite weighed 1,270 kg. As radiation from the Sun reached the Earth, it streamed into the Earth's doughnut-shaped magnetic field. Arriving over the polar regions, the reaction of the radiation with the Earth's magnetosphere triggered off auroral displays and electrical storms. As the Earth travelled in its orbit, its magnetic field created a tail trailing behind. For the Prognoz M/Interball project, one probe was injected into the magnetospheric tail with a subsatellite, and the other, with its companion, into the auroral zone. The four satellites, operating simultaneously, would provide a comprehensive picture of solar-electrical activity in the ionosphere.

After very long delays, the first mission, the magnetospheric tail, was finally launched from Plesetsk on 2 August 1995 into a four-day orbit at 193,000 km. Nine and a half hours after launch, the Magion 4 Czech subsatellite separated to begin its independent programme of activities. Interball, made by the Lavotchkin design bureau, carried nine plasma detectors, three detectors for energetic particles and solar rays and a magnetometer. Magion satellites had first been developed by the Ionosphere department of the Geophysical Institute in the Czech Republic and the first had been launched in 1981. Magion 4 was a 59-kg boxed small satellite with X-shaped solar panels carrying a magnetometer, photometer and search coils with a 32-Mbit computer and transmitter. No less than 13 countries contributed to the project, the leaders being, besides Russia and the Czech Republic, Canada, Sweden and France.

The second, auroral spacecraft, Interball 2, weighing 1,250 kg, was launched a year later into an elliptical 5-hour orbit of 19,202 km. Interball 2 likewise released its 58-kg Czech subsatellite, Magion 5, this time equipped with a cold plasma detector and an even smaller 33-kg Argentian microsat which was taking a piggyback ride on the launch. Initially, no signals could be picked up from Magion 5 and it was first thought that it had failed to separate. Later, very weak signals were received, probably due to the failure of one of the solar panels to deploy. Interball 2 carried seven plasma detectors, wave analysers, radio detectors and imagers.

The Interball programme carried six French experiments—Memo, Hyperboloid, Ion, Elektron, Ikarus and Opera. The results, when published, provided detailed charts and information on aurorae, electrons, cold plasma and electromagnetic fields.[51]

Koronas

A new solar observatory was launched on 2 March 1994. Again, this project was developed in the Soviet period by the Lebedev Institute with Poland, Czechoslovakia and the German Democratic Republic. Koronas stood for Comprehensive Orbital Near Earth Observations of the Active Sun. Although other countries participated, it became primarily a Russian–Ukrainian project and was built by NPO Yuzhnoye in Dnepropetrovsk. The spacecraft was in the shape of a multisided cylinder, with four solar panels and four instrument booms.

The general aims of Koronas 1 were to study the Sun's internal structure (helio-seismology), to investigate solar activity, predict the impact and arrival time of solar disturbances in the Earth's atmosphere and discover the reasons for periodic discharges and eruptions from the Sun. Its specific aims were to learn more about solar eruptions—why they took place, their structure, characteristics, cycles and solar plasma emitted. Equipment included a heliometer, coronograph, polarimeter, photometer and X-ray telescope. Koronas 1 was sent up by Tsyklon 3 from Plesetsk into an orbit of 501–541 km and was given an operational lifetime of five to seven years. A companion, Koronas F, was scheduled.

Spektr

Besides Koronas and Interball, Russia's other scientific project in the 1990s was Spektr, not to be confused with the Mir module of the same name. Spektr was a large, 6-tonne X-ray and gamma-ray observatory that had been in planning since the late 1980s, but was repeatedly delayed because of the funding crisis to affect the Russian space programme. It was intended to be the last in a four-part series of astrophysical observatories that included Astron, Kvant and Gamma which flew in the 1980s but that it would in turn lead on to a series of follow-up missions with other Spektr designations (Spektr RG, R and UV). Spektr's aim was to map 30,000 X-ray sources and 3,000 active galactic nuclei. Built by the Lavotchkin design bureau, it is a new rectangular handle-box-shape design with two solar wings and intended to fly in a 500–200,000 km orbit at 51.5 degrees, circling the Earth every four days. Approximately R200bn was injected into the project in 1997 ($34m) in the hope of generating some momentum. India joined the project in early 2000, but by then it was running 11 years behind schedule. The most recent launch date set is 2002. Spektr is believed to be based on the Prognoz series of early warning observatories.

Solar observatories

2 Mar 1994	Koronas 1	Tsyklon 3	Plesetsk	487–528	94.78	82.49
3 Aug 1995	Interball 1	Molniya M	Plesetsk	870–191,752	5,458	62.9
	Magion 4					
29 Aug 1996	Interball 2	Molniya M	Plesetsk	769–19,211	347	62.77
	Magion 5					
	Microsat					

INTERNATIONAL COLLABORATIVE MISSIONS: EKSPRESS

During the period of the Soviet Union, there was a significant level of international collaborative programmes. The Soviet Union ran a programme of launching small scientific satellites called Intercosmos with the socialist block, launching 25 such satellites between October 1969 and December 1991. The Soviet Union launched India's first three small satellites (Aryabhata, Bhaskhara 1 and 2) and its first two applications satellites (IRS 1A and 1B). Generally, these were launched for minimal charge, in exchange for access to the data collected.

Two international collaborative missions of this nature took place during the Russian period. The first was the launch of the last of India's early Earth resources satellites, part of a programme dating to 1962. This saw the launch of the IRS 1C Earth resources satellite on a Molniya M from Plesetsk in December 1995. This time, though, a commercial fee was charged.

The other international scientific collaborative project was unusual in conception, dramatic in execution and bizarre in outcome. In 1991, the German company OHB, in conjunction with the German space agency DARA had the idea of flying some microgravity experiments and returning them to Earth. The completion of the design studies coincided with a period of retrenchment in German public spending, not least in the space industry.

In an effort to cut and spread costs, DARA sought out international partners to provide hardware and equipment at low cost. The Khrunichev company in Russia responded by making available a cone-shaped re-entry capsule originally used for the military Fractional Orbital Bombardment System. This was a project, tested in the 1960s, to send small warheads around the Earth on a three-quarter orbit to descend on the United States from Mexico. Khrunichev called the FOBS capsule the anodyne but more civilian title Ekspress and DARA bought an Ekspress for DM100m.

The 405-kg Ekspress had sufficient volume to carry up to 165 kg of payload. Accordingly, DARA fitted out the cabin's interior with six Japanese and German microgravity experiments, experimental packages and data-handling system. The 365-kg service module contained an attitude control system, retrorocket, transmitters and battery supplies for up to a week. Together, they brought the weight to 762 kg, with an overall size of 2.3 metres in length and 1.4 metres in diameter. A system was set up to link the German and Russian computers to ensure effective control over the spacecraft. Japan offered a free launch in exchange for full access to the research results. Australia offered a desert landing site near Woomera.

Ekspress duly rode into orbit on a Japanese Mu-3SII rocket in January 1995. It ran straight into trouble. An attitude control problem with the second stage 130 seconds into the mission meant that control was lost for 20 seconds and fuel was depleted as the rocket engines swivelled and tried to compensate. The planned orbit of 210–398 km was not achieved. Some signals were received from what seemed to be a very low orbit that was only skimming the edge of the Earth's atmosphere.

Attempting to salvage something from the mission, German ground control in Oberpfaffenhofen commanded the satellite to re-enter, hoping that they might at least recover the capsule rather than wait helplessly for it to burn up. But there was no indication that the message was received.

So the story ended. It was presumed that the Ekspress had burned up somewhere over the Pacific after about three orbits. The failure of the Mu was blamed on the excessively heavy payload—twice that of the Mu's previous heaviest assignment.

There it should have ended, but several months later in Britain, Geoffrey Perry, the science teacher known for his role in observing the early secret Soviet space programme, was alerted to a report in the *Ghanaian Times* of 3 February 1995 which reported that a strange object with Russian markings had descended from the skies with an orange parachute at Kotorigu in west Mamprusi, near the border with Togo. The newspaper recorded the fact that the deputy commissioner of police, Patrick Agboda, had visited the area and noted that the bushes surrounding the fallen object were burned. Kotorigu is very rural, lacking electricity, water or telephone and the road can only be used by four-wheel-drives. The people living in the traditional African villages in the area must have been startled by the sonic boom which preceded the red-hot object descending under its orange and white parachute. The local chief, it was later learned, told people to keep away from the object from space. He ordered his brother to guard it while he went to the district chief, Mr Gumah of Walewale.

A photograph later appeared in the *Ghanaian Chronicle* on 20–23 February, which showed the strange object hanging from a tree and local people posing behind the spread out parachute. It was clear to Perry that it looked very like a FOBS capsule and indeed the article quoted a statement from the commander of armed forces in the region, Group Captain Aryetey, that it was Russian. Perry noted that the ground track of Ekspress brought it over west Africa four hours after launch. Perry contacted the Ghanaian authorities a number of times in the course of the year to ascertain if the strange object was indeed the Ekspress capsule, but fruitlessly. Having made no progress, he then wrote an article for the Western Australian Astronautical Society with his speculation as to the true fate of the Ekspress. Eventually, this found its way to the DARA in Germany who phoned the German embassy in Accra. After a few quick enquiries, it was ascertained that the strange object was almost certainly the Ekspress.

Meantime, what had happened to the Ekspress? District chief Gumah had organized a transport to take the Ekspress away. Ten strong men had been required to put the capsule onto a truck. Hundreds came to look at the object in Walewale, which had become something of a local attraction, where it was handed over to the army. Two weeks later, the Ghanaian air force took the object to the town of Tamale, 50 km south, where it was stored in a huge aircraft hangar. This hangar had been built by the Russians in the 1960s during a period of intense Soviet–Ghana cooperation and the base commander, who had been there at the time, at once recognized the Cyrillic script on the parachute (lettering on the cabin itself had burned off).

In January 1996, officials from the German Space Agency DARA arrived in Ghana, identified the Ekspress and asked: 'Please may we have our satellite back?' They found that the capsule had been moved to Tamale airport where it was lying, untouched and unopened, in a hangar. It was a little dented, not from its brief sojourn in space but from its truck ride along rutted African roads. The Luftwaffe duly sent a Transall plane out to Ghana which brought back the Ekspress to Germany in a wooden container. Subsequent examination of the cabin found that the Ekspress had entered the atmosphere that January day 80 km above the Earth. Barrelling nose first into the atmosphere, the cabin survived the intense heat. Sensing the onrush of air 6 km high, the barometric system commanded the parachutes to open. The experimental packages were in perfect order and the experiment to test ceramic materials on the nose cone was declared a complete, if belated, success! So the Fractional Orbital Bombardment System, devised as a weapon of the cold war, was now vindicated in this strange, multinational adventure.

PIGGYBACKS

Finally, in examining Earth orbiting satellites, mention should be made of piggyback satellites launched. In the course of several launches in the Russian space programme, there was spare space and weight capacity to permit the launching of small satellites for a nominal or low charge. Some were connected to the main

Piggybacks

25 Jan 1994	Meteor 3-6 Tubsat	Tsyklon 3	Plesetsk	1,186–1,208	109.36	82.56
24 Jan 1995	Tsikada 1 Astrid Fasat	Cosmos 3M	Plesetsk	965–1,021	104.97	82.97
28 Mar 1995	Gurwin UNAMSAT EKA 2	Start/Topol	Plesetsk	(Fail)		
31 Aug 1995	Sich Fasat Alpha	Tsyklon 3	Plesetsk	632–669	97.73	82.53
28 Dec 1995	IRS 1C Skipper	Molniya M	Baikonour	816–818	101.24	98.59
10 Jul 1998	Resurs O1-4 Safir 2 Techsat MYSat Fasat Bravo Reflector	Zenit 2	Baikonour	817–818	101.24	98.79
15 Jul 2000	Champ Mita Rubin	Cosmos 3M	Plesetsk	460		87.3

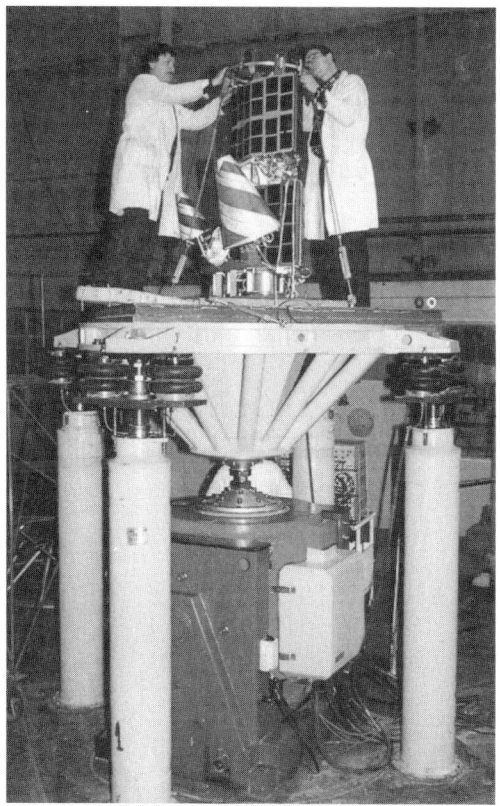

Fasat is prepared for launch

payload, whereas others had no necessary connection. Many of these small satellites came from other countries—the Resurs O1-4 group, for example, coming from Germany, Chile, Australia, Israel and Thailand.

The first, Tubsat, was one of a series of very small satellites made by the Technical University of Berlin (TUB), whence its name (Tubsat), and was followed by two more Tubsats launched on the first Shtil mission. Some have already been described in their appropriate sections. The largest group of piggybacks was that carried by the Resurs O1-4 Zenit launcher of July 1998.

Astrid was a 30-kg Swedish satellite, deployed with a Tsikada navigation satellite on a Cosmos 3M out of Plesetsk on 10 December 1998. Astrid was sprung away from its mother craft just west of the British Isles on its first orbit and transmissions were picked up in Solna, Sweden, soon thereafter. Its task was to study electrons and electric–magnetic polar phenomena.

Skipper was a 230-kg American military piggyback built by Utah State University. It was designed to make aerodynamic studies and to test two scanning spectrometers, but unfortunately the mission was not a success as the battery drained only one day after deployment. Fasat was built by a pioneering American technology company called Final Analysis. It was an octagonal cylindrical-shaped satellite 46 cm in diameter, 89 cm high, with a 3-metre boom designed to test store-and-forward satellite transmission techniques. One of its intended applications is meter-reading for utilities: household electricity and water meters will transmit their readings to overhead satellites, saving personal calls. The last, Mita and Rubin, were launched with the German Champ satellite from Plesetsk. Mita was a 170-kg Italian technology satellite to detect cosmic particles, while 37-kg Rubin was a Russian–German project to develop new tracking systems.

MOON PROGRAMME

In its last plan for space development, published in 1989, the Soviet Union proposed a lunar polar geophysical orbiter to fly in three years' time (Luna 92), to be followed by a farside sample return and rover by the end of the decade, but these projects were swiftly overtaken by the financial crisis. In summer 1997, the Institute for Space Research (IKI) in Moscow announced plans to send a small spacecraft into lunar orbit, using a Molniya rocket from Plesetsk cosmodrome in northern Russia in 2000. The orbiter would deploy three 250-kg penetrators. They would dive into the lunar surface at some speed, burrowing seismic and heat flow instruments under the lunar surface, leaving transmitters just above the surface. With small nuclear isotopes, they would transmit for a year, operating as a three-point network to collect information on moonquakes and heat flow. However, the project disappeared as quickly as it came and no progress was made.[52]

MARS PROGRAMME

Russia made only one deep space mission during the post-1991 period. This was Mars 8 and it failed within an hour of its launch.

The Soviet Union had never been lucky with its Mars probes. None of its two 1960 missions got away from Earth orbit. Of three attempts in 1962, only Mars 1 left Earth orbit, and it failed after a month. In 1964—the next opportunity—only one mission (Zond 2) left Earth orbit and although it passed quite close to Mars, its radio was long since dead. With the Proton rocket available in the late 1960s, the designers of the Babakin bureau were able to prepare large probes up to 5 tonnes in weight. Even then, the two 1969 missions failed, the first due to a Proton third-stage explosion at 438 seconds and the second due to a first-stage engine malfunction.

Three Mars attempts were made in 1971. The first spacecraft failed to leave Earth orbit due to a guidance failure and was renamed Cosmos 419. Mars 2 got away successfully and reached the planet on 27 November. Supposed to soft land, it failed to reach the surface intact but did manage to deposit there a pennant with the Soviet coat-of-arms, making this the first direct human contact with the planet Mars. The next craft, Mars 3, soon after, came down in the southern hemisphere in a region called Electris, and this time a soft landing was achieved. Four petals opened and the

Earlier Soviet Mars probe

domed shape of the capsule rested there on the sands of Mars. Antennae popped out, aerials searched skywards and TV cameras began scanning. However, only a fragment of a picture was received and the radio link broke down 20 seconds later. The mother craft for Mars 2 and 3 entered orbit around the planet and transmitted data for several months.

The Soviet Union made an all-out assault on Mars during the next launching opportunity, 1973. Four spacecraft were launched and all managed to leave Earth orbit. Despite this, three were total failures: Mars 4 failed to fire its engine to enter Mars orbit, contact with Mars 6 was lost during the descent and Mars 7's engine misfired, causing it to miss the planet by 1,300 km. Only Mars 5 achieved any degree of success, entering Mars orbit and sending back pictures and a modest amount of information.

The Americans went on to great success with Mars, their two Viking landers touching down in 1976, sending back superb pictures and sampling the surface with grab arms. A group of scientists, jealous of American achievements, persuaded the government that the Soviet Union should return to Mars and take samples from its surface. A Mars sample return mission, called project 5M, was ordered for launch in 1980.[53] The Lavotchkin bureau threw its best efforts into this excessively ambitious project but it was—probably wisely—cancelled the following year.

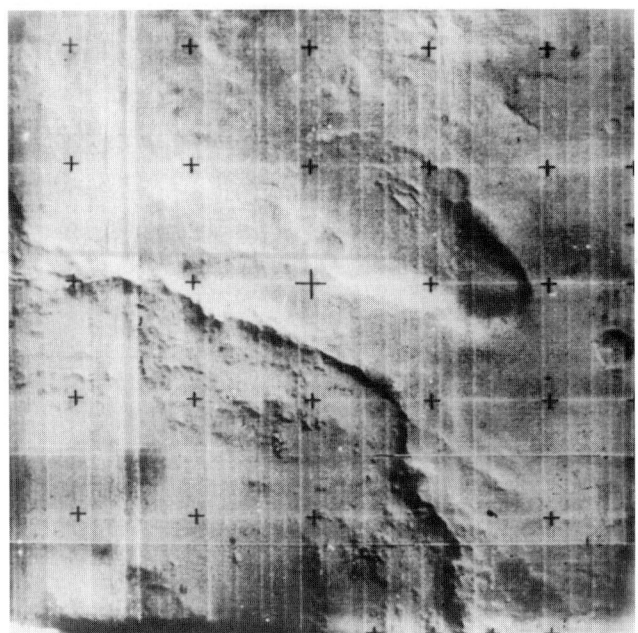

The surface of Mars—taken by Mars 5 in 1974

The Soviet Union resumed its Mars missions in the late 1980s, and, in the spirit of *glasnost*, they were announced in advance. Two new, large, 6-tonne spacecraft were constructed at a total cost of R272m, plus R60m of foreign investment from 12 other nations. This was an ambitious project with a different target: Mars's tiny 27-km-long moon Phobos. The new probes would enter Mars orbit, rendezvous with the little moon Phobos and land experimental packages there.

Phobos 1 and 2 were successfully launched to Mars in July 1988, but they were unable to shake off the ill-luck which dogged their predecessors. On 2 September, ground control in Yevpatoria keyed up its standard command to Phobos 1. The unfortunate technician left one hyphen out of the series of keyed commands and the one which left Earth was the end-of-mission command to close down all systems. It was an embarrassing, expensive mistake—and a foolish one to have designed computer commands in such a flawed manner.

All was now dependent on Phobos 2. Although it entered Mars orbit on schedule in January 1989, there were computer malfunctions and problems with its high-speed transmitter, which reduced dataflow badly. Despite them, Phobos 2 drew ever closer to Phobos, snapped its first pictures of the little moon and by the end of March was preparing to send down the first of its two landers. Then disaster struck. Only 150 km from its target, Phobos 2 lost orientation towards Earth, lost Sun lock and its batteries quickly drained.

The subsequent commission of inquiry set off an acrimonious round of recrimination within the Soviet scientific community, *glasnost* making possible the letting off of years of stored-up feelings and resentments. Scientists blamed engineers and engineers blamed scientists. The Phobos 2 postmortem, announced within a month, found that there was poor computer programming, inadequate batteries (they had only five hours of power) and substandard orientation systems. Lost in the exchanges were the considerable scientific results achieved both en route to Mars and in Martian orbit and the almost concluded rendezvous in Martian orbit. The Moscow Institute of Space Research, IKI, still hoped to send the back-up model, Phobos 3, to Mars in 1992, but the proposal received no political support. Ultimately, the probe was auctioned in the West to raise hard currency.

MARS 8

Phobos 2 had come tantalizingly close to success, reminding scientists of the Martian secrets that could be unlocked by a successful mission. Limited funding for a new mission in 1994, called Mars 94, was made available, though the project was postponed to 1996 (when it was temporarily renamed Mars 96). The purpose of the Mars 96 mission, developed by the Lavotchkin NPO, was to place an orbiter over Mars in September 1997, equipped with cameras, sensors, relay systems and 24 experiments; to send down two small landers on Mars; and to drop two penetrators each carrying a camera and miniaturized sensors which would tunnel into the surface. This was Russia's only interplanetary mission and a lot of hopes were riding on it.

It was an ambitious mission, arguably one of the most adventurous planetary missions ever mounted by any nation, matching the American/European probe Cassini to Saturn launched the following year. Mars 96 weighed a record 6.5 tonnes, including over 3 tonnes of fuel. Twenty countries participated in the experimental packages and the computer software was supplied from France.[54]

Keeping the project going was a major challenge. Only $122m of government money ever arrived for the project, all of it late and taken from other projects, such as Spektr and Luna 92. Foreign countries actually contributed much more—about $180m. The project was repeatedly in an on/off situation with threats of further delay or even cancellation. Final assembly and integration took place during the worst period of power cuts at the Baikonour cosmodrome and engineers completed the job by the light of kerosene lamps.

Mars 96 was to take a long, slow, curving trajectory to the planet. Its powerful and versatile upper stage, called Fregat, already well proved on the Phobos mission, would give the station a final kick on its way for trans-Mars injection and would fire again for a big burn at Mars orbit injection in September 1997. As it travelled to Mars, a suite of instruments would study radiation, solar activity, gamma bursts and the stars. The objective was to enter a polar orbit covering the whole planet (300 by 22,577 km, 106 degrees, 43 hours) and then burn again to adjust the orbit to working altitude. Two 65-kg penetrators, each 1.5 metres long and 120 mm in diameter, would be released a month later, spun up to 95 rev/min and crash-land in Arcadia and Utopia. Inflated gas bags would break their fall, but the penetrators were still expected to impact at some speed, stopping at up to 500 times the force of gravity.

Many tests were run to verify the landing technique, including dropping the package 60 metres down the liftshaft of the Moscow Aviation Institute and later from helicopters. Out of the tail would peep a camera, magnetometer, transmitter, weather detectors and wind meter. The front end of the penetrator would dive into the Martian soil as deep as 6 metres below the permafrost layer. The underground part was equipped with a magnetometer, accelerometer, X-ray spectrometer, alphaparticle spectrometer, thermometer, and neutron detector which would relay their findings to the transmitter about the surface through a cablerope. The penetrators would transmit for a year.

The most exciting part of the mission, however, was the release of the Mars landers before orbital insertion. The two small 40-kg landers, shaped like Luna 9, would separate, aerobrake into the atmosphere, parachute down from 19 km and, encased in inflated air bags, bounce over the surface as high as 70 metres before settling. Once stable, the spacecraft would open up and begin transmitting photographs, to be relayed back by the orbiter on its daily passes. On each orbiter was a compact disk called 'Visions of Mars', a time capsule containing the sum of Earthly knowledge about Mars at that point, including the famous broadcast of H.G. Wells's *War of the Worlds* in 1938 which, when transmitted on American radio at the time, caused panic along the east coast. Each small lander carried a weather meter, thermometer, seismometer, landing and panoramic camera and spectrometer. They would transmit data back to Earth for a Martian year, using the orbiter overhead.

Meantime, the orbiter would aerobrake in its orbit a year after its arrival, gradually lowering its closest point to the surface to 200 km. Thirty instruments were on the nuclear-powered orbiter, designed to make weather studies and scan below the surface for water. The most prominent was the Argus television complex, so named after the many-eyed giant of mythology. Argus, made by the Institute of Precision Mechanics and Optics in St Petersburg, would scan the Martian surface both from altitude with wide-angled lenses and from as close as 300 km with zoom instruments. Argus's resolution was 10 metres and the purpose was to compile a complete mineralogical and topographical map of the planet. At the same time, a videospectrometer would compile an infrared map of the surface. An instrument called thermoscan would make a temperature map of the Martian surface. The orbiter carried an altitude radar to map the mean level of Mars and eight spectrometers to measure everything from cosmic rays to ions, plasma, neutrons and particles.

Despite appalling funding difficulties, with engineers giving their time without pay, the Russians eventually managed to launch Mars 96 on 16 November 1996. The Proton rocket soared eastward into the cold clear night sky and the scientists at the launch site celebrated as the block D fired briefly to place Mars 8 into an initial parking orbit of 51.5 degrees, 139–155 km. But come dawn, it was all over. Passing over Africa an hour later, the block D upper stage should have fired the spacecraft into a highly eccentric orbit, at the end of which the Fregat stage on Mars 8 would kick the spacecraft on its final route to Mars.

It is unclear if the block D then failed to ignite or burned for only a short period and then shut down. It seems that Mars 96 payload, pre-programmed to believe that the trans-Mars manoeuvre had already taken place, separated from the block D and used its engines to make what should have been the kick burn to adjust its course to Mars. Instead, the Fregat's burn put Mars 96 in an erratic, low, unstable Earth orbit as high as 1,500 km, but, more importantly, as low as 87 km. The block D stage crashed into the Pacific Ocean near Easter Island a day later. Mars 96 itself is believed to have crashed into the Andes mountains only hours afterwards and fireballs were spotted. Some of the equipment was designed to survive a tough Mars entry and landing and there was speculation that the probes may have ended up in the Atacama desert, searching for life on Earth. The Chileans were furious with the Russians for not warning them of its hazardous cargo, but lacking tracking ships, the Russians found it difficult to predict the re-entry area.

The Mars 96 disaster came at the worst possible time for the Russian space programme, with finance and morale already at rock bottom. An atmosphere of deep gloom prevailed. The exact cause of the block D failure was never determined, although as many as 20 possible sequences of events were explored which might have explained it. Originally, it had been planned to send the tracking ship *Cosmonaut Viktor Patsayev* to the Gulf of Guinea to follow Mars 8 at the crucial point of Trans Mars Insertion, but lack of money prevented this. Having the *Patsayev* on station would not have saved the mission, but it would have been able to receive the block D telemetry during the crucial moment. The cause was recorded as a random failure. The failure of Mars 8 was the tenth Russian Mars failure in succession—and that only included those that were formally announced. Taking the unannounced missions into consideration, it was nearly the twentieth in succession. The volume of scientific return from a successful mission would have been huge and equalled that of the American probes for the whole decade.

Mars 8 was one of the last great interplanetary explorers. During the 1990s, the trend swung decisively away from large spacecraft of this type. In the United States, NASA administrator Dan Goldin took a new approach. Alarmed by the fact that larger interplanetary spacecraft took so long to design and construct (up to 10 years), their considerable cost (over $1bn) and that a single-point rocket failure could ruin the whole project (something Mars 8 amply demonstrated), he announced the new philosophy of 'faster, cheaper, better'. New planetary probes would be smaller, developed more quickly and launched more frequently in the expectation that, even if there were some losses, more data would ultimately be returned at less cost and risk.

RETURN TO PHOBOS?

Despite the Mars 8 failure, a number of discussions took place with Western countries over the following two years about the possibility of a new Mars mission early in the new century. The most likely candidate for collaboration was France, but the French eventually considered the Russians to be too much of a financial risk and instead developed a cooperative project with the Americans for a sample return mission. Positively, the European Space Agency, which is led by

France and Italy, did select the Soyuz Fregat launcher for Europe's first Mars probe, Mars Express, scheduled for June 2003.

In 1999, the Russian Academy of Sciences began a feasibility study of recovering rock samples from the moon Phobos and invested an initial R9m in the project. Called Phobos-Grunt, the probe was similar to, but smaller than, the failed Phobos missions of 1988–9 and would follow a similar mission profile. The 2,370-kg spacecraft would use five SPT-100V electrical engines originally developed by the Fakel company for communications satellites. The electrical engines, with a thrust of 130 MN, would use 425 kg of xenon to speed the probe on its 475-day trajectory towards Mars in 2005. After entering Mars orbit, the probe would close in on Phobos in the course of a three-week rendezvous. A couple of days after scooping up 170 g of rock samples from Phobos, a small capsule with the samples would be fired towards Earth, to be recovered 280 days later in Russia in May–June 2008.[55]

Although the Americans had their successes with Mars (they landed a miniature roving vehicle called Pathfinder in 1997), they also had their failures. Mars Observer inexplicably fell silent in August 1993. It is possible that a pressurization fault led the engine to explode at the point of Mars orbit insertion. Even more embarrassingly, the Americans lost a Mars orbiter in 1999 because ground control confused its measurements between metric and imperial units. Worse followed later that year when Mars Polar Lander fired its retrorockets to soft land near the south pole. The subsequent postmortem suggested that the jolt of deploying its landing legs wrongly triggered a signal to tell the computer that it had already touched down. This caused the retrorockets to turn off, leaving the craft to fall to destruction from a great height. The great galactic ghoul that allegedly lurked in the vicinity of Mars, snapping up arriving spacecraft, was well fed in the 1990s. The Americans abandoned plans for a lander in 2001 and decided to return to the bouncing-ball approach the next time, which Pathfinder had vindicated and we now know would probably have worked on Mars 8.

Other new projects were also considered during this period. A star-mapping satellite, called Lomonosov, completed its design but never flew. An experimental solar sail project, called Regatta, also concluded design studies but hardware does not seem to have been built. The Jet Propulsion Laboratory in Pasadena, California, with RKK Energiya, made a design study called *Mars Together* of a 150-kg spacecraft with a solar panel of 30 m^2 that would use a sidewards-looking radar to digitally map the surface. NASA and the Russian Space Agency made a paper study of two distinctly different missions: the first, called *Plamya* ('fire'), was to send a Russian probe to within 10 radii of the Sun; the second, called *Lyod* ('ice'), was to send an American probe to reach the only unexplored planet of the solar system, far-distant Pluto. *Plamya* appeared to be the successor of a 1990 design study of a Jupiter–solar probe called Tsiolkovsky. The 85-kg *Lyod* ice probe was expected to pass Pluto at 15,000 km and its moon Charon at 5,000 km, sending down a 10-kg Russian capsule to go even closer and possibly to impact. However, neither country managed to attract support for the fire-and-ice missions and they remained on the mutual wish list.

Interplanetary missions

16 Nov 1996	Mars 8	Proton	Baikonour	139–155	87.43	51.5	Mars, fail

THE UNMANNED CIVILIAN RUSSIAN SPACE PROGRAMME IN RETROSPECT

Between 1992 and 2000, Russia was able to maintain a modest civilian space programme. The table below illustrates the scale and priorities of that programme. It can be seen that Russia launched 78 unmanned civilian spacecraft during this period, an achievement at a time of such economic difficulty. This compares to 132 launches in the military programme in the same period. Although the number of launches declined, the programme was maintained, but its development was uneven.

The meteorological programme was clearly a low priority, even though the two satellites launched, Meteor 2-21 and 3-6, had long operating lifetimes. Ironically, the Elektro 24-hour weather satellite, which had been promised for 16 years during the generously funded Soviet period, did not fly until more difficult financial times in the 1990s.

Using old and tested equipment based on the Zenit/Vostok design, Russia maintained a lively Earth resources programme, Resurs, flying a total of 10 cabins in the Resurs F series. So successful was the F1 model that it was brought out of retirement for an extra two missions. Resurs offered the possibility of commercial sales, both at home and abroad and this undoubtedly contributed to the sustaining of the programme. Not only did Russia fly the Resurs programme to its conclusion, but also introduced a new generation of Earth resources satellites—the advanced, digitally transmitting Resurs O, using the Zenit rocket.

Just as the Resurs O programme replaced the Resurs programme, so too did the Okean O take over in 1999 from the Okean. This also featured a radical upgrade, the new satellite being three times heavier and able to carry a wide range of advanced

Russian scientific and applications programmes, 1992–2000, by year and category

	1992	1993	1994	1995	1996	1997	1998	1999	2000	Total
Weather		1	2							3
Resources	4	5	2	2		1	1	2		17
Materials	1		1	1		1		1		5
Molniya comsats	3	6	2	1	2	1	2	1		18
24-hour comsats	4	4	7	2	4	1		2	3	27
Biology	1			1						2
Scientific	1		2	1	2					6
Total	14	16	16	8	8	4	3	6	3	78

scientific instruments. Despite these advances and the phasing out of old programmes by new ones, one has the impression of a programme falling short of its
full potential. The Resurs DK, for example, clearly an advanced form of Earth
resources satellite, has the potential to be a world leader, but was still seeking
either domestic or foreign customers.

Using old Zenit/Vostok-based equipment, the Foton programme of materials
processing in microgravity maintained a steady pace, with five evenly-spaced
launches by Russia. Here, Russia was able to offer a distinct niche product for
Western scientists and commercial manufacturers, more successfully than the rival
Chinese FSW programme of recoverable micogravity satellites.

Communications were the largest single area of the unmanned civilian Russian
programme, accounting for over half the programme. Here, Russia maintained the
venerable Molniya system, phased out the first generation of 24-hour communications satellites, and introduced a new range of geosynchronous communications
satellites suitable for the early twenty-first century. No less than 18 Molniyas were
launched, maintaining a long-established system first introduced in 1965. Of the first
generation of communications satellites, Ekran was phased out in 1992 and the
others, with reduced rates of launching, between then and 2000, the last Gorizont
flying in 2000. The new system of Gals and Ekspress satellites was introduced in
1994, followed by the first of several commercial satellites for the domestic market
(Yamal, Kupon). The process of transition was hardly smooth, for several launches
were marred by errant Protons and faulty equipment. The Russians overcame the
great weakness of the Soviet-period comsats, namely short operating periods, and
extended the lifetimes of their satellites to over six years, but will continue to need to
extend this further to remain competitive. Longer operating lifetimes were crucial as
launch rates fell, but by definition were insufficient to maintain pace with the
exploding global communications demand. Technically, the transition from first-
generation to second-generation comsats has now taken place, which is no mean
achievement, but whether this will prove to be enduring will depend on the commercial development of the system and the installation of the appropriate financial,
commercial, regulatory and ground structures.

Scientific programmes were a minor part of the Russian civilian space
programme. Although the space programmes of the two superpowers were always
portrayed publicly as being about science and pushing back the frontiers of
knowledge, in reality purely scientific work has always taken a back seat to
manned spaceflight, military programmes and applications which have gobbled up
the bulk of the resources. At a time of retrenchment, scientific programmes were
likely to suffer—and they did. Russia inherited a stack of scientific programmes from
the final days of the Soviet Union. However, although many of these were already
being delayed or cancelled, Russia was able to carry out the two Interball missions
and launch the Koronas solar probe. The Spektr programme remained on the books
for 10 years and made only minuscule progress towards launching. Another set of
programmes in the pipeline—Lomonosov, Luna 92, Tsiolkovsky—never got beyond
the drawing board. By the end of the 1990s, several paper studies were made in the
certainty that they would never get any further, like fire-and-ice. Only two biology

Astronomical missions suffered badly: this is the telescope from Astron

missions were flown, although their ending was the outcome of political rather than financial factors.

And then there was Mars 8. Almost all the resources for unmanned Russian scientific programmes had been transferred to Mars 96 in a desperate attempt to sustain this mission. It was probably a well-justified gamble, for Mars 8 was, with Cassini, the last of the great interplanetary mega-missions, one which could have, at a stroke, opened up huge new areas of understanding of the planet Mars. The disaster which befell this unfortunate project was the cruellest single blow inflicted on the programme at one of its lowest points. After so many years of delay, the devotion of its scientists and the overcoming of one obstacle after another, it was tragic to lose the mission to a random block D failure. Had it worked, and the Phobos 2 mission intimated some of what the Lavotchkin probes could achieve, it might have provided an infusion of new confidence into the programme. Instead, it drew a line in Russia's Mars programme—one to which it is unlikely to return until the financial climate is much improved. For the time being, Mars will be left to the Americans (far from invulnerable to self-inflicted blunders), the Europeans (Mars Express) and the Japanese, who launched their first probe to Mars in 1998 (Nozomi). Mars 8 was a true expression of the old Russian proverb: better a terrible ending than terribleness without ending.

In summary, Russia maintained an unmanned civilian space programme during the years 1992–2000. The big loser was science, with only four spacecraft launched, one being a heart-breaking failure. Meteorological satellites also remained a low priority. Steady progress was maintained in the materials processing programme, Foton, with a regular series of flights in the Earth resources programme and the continuation of the biological programme, Bion. Russia replaced the old set of Earth resources spacecraft with new, larger and much more capable satellites in the Resurs O and Okean O series. It was in the area of communications, the largest single area of unmanned civilian investment, that progress was most marked. The fleet of first-generation geosynchronous satellites was phased out and replaced by a new generation, no mean achievement at a time of economic hardship, and Russia successfully confronted the problem of short satellite lifetimes with new and much improved equipment.

6

The ground frontier
Cosmodromes and space facilities from 1992

Chapters 3, 4 and 5 examined the development of the space aspect of the Russian space programme. This required a huge support structure on the ground—cosmodromes, rockets, engines, design bureaux and organization. This chapter examines the ground aspect of the Russian space programme from 1992 to 2000.

The dominant consideration for the Russian space programme from 1992 was financial: how could the space programme modernize, adapt, or even survive in the face of declining state revenues, national economic contraction, falling exchange rates, altered political priorities and the transition to capitalism? However, one of the most urgent problems facing the new space programme was a political one that no one had anticipated: its main cosmodrome was in another country, Kazakhstan, and some of its main production centres were in another, Ukraine. Both countries had taken advantage of the coup against Gorbachev to declare their independence and by the end of 1991 it was clear that they intended to exercise their new self-determination.

THE COSMODROMES

The first problem facing the new Russia was that its main cosmodrome, Baikonour, was no longer part of Russia! Instead, Baikonour was firmly in the middle of the Republic of Kazakhstan.

Russia had three cosmodromes:

- Baikonour, in Kazakhstan, its largest and busiest, used for manned and interplanetary missions, flights to 24-hour orbit and most commercial missions;
- Kapustin Yar, also called Volgograd station, which once fired the V-2, now disused; and
- Plesetsk, in the Arctic, used mainly for military missions.

Russia's cosmodromes

Map 1: Russia's cosmodromes

BAIKONOUR

Baikonour was the centre most publicly associated with Soviet cosmonautics. The cosmodrome was located on the endless, flat and arid desert of Kazakhstan, on the rail line along the Syr Darya between Moscow and Tashkent. In fact, the real Baikonour was a sleepy railhead far to the north, but the USSR called the cosmodrome Baikonour in the hope that the duped Americans, should they ever attack, would target the railhead named Baikonour in error. Most of the workers at Baikonour live in the city of Leninsk, between the river and the railway and adjacent to Krainy airfield to the southwest of the cosmodrome. In the post-communist renaming of formerly Soviet cities, Leninsk was formally renamed Baikonour in December 1995, creating a new problem which went unconsidered at the time: two Baikonours!

It is probably the largest cosmodrome in the world, 90 km from east to west, 75 km from north to south, and almost as large as the new state of Moldova. Baikonour employs as many as 35,000 people with 95,000 dependants, with an area of 6,717 km^2 and a downrange fall zone of 104,305 km^2. There is a long commuting distance from Leninsk to the launch pads and processing areas, at least 20 km.

It was not only its location that made it startlingly different from its American rival at Cape Canaveral. The first was that rail was the principal mode of transport. The cosmodrome had 470 km of track, although there is a longer but poorly maintained road network (1,281 km). Rockets were normally carried flat on their back on railcars from the assembly hangars to the pad where they are then raised to a vertical position. It was a system made easier by the fact that the gauge of the Russian railway is the widest in the world. By contrast, the Americans brought their rockets to the pad on giant road crawlers. Rocket launchings were handled by uniformed military officers and personnel, members of the 18,000-strong Strategic Rocket Forces (RVSN), rather than contracted civilian companies, as is American practice. The other difference to Cape Canaveral was that launches could take place in extreme cold, with temperatures as low as $-35\,°$C. In summer it can be at least as hot as Cape Canaveral and even reach $50\,°$C. Baikonour also had extremely fast turnaround times: it was not unknown for a Zenit to be brought down to the pad, fuelled up and fired in a space of less than five hours.

Baikonour cosmodrome was first settled in 1955, when construction workers were sent there to build the rocket pad that launched Sputnik. Unlike the old American launch pads at Cape Canaveral, which were allowed to rust as new gantries were built further up the Cape, this original pad is still in use. From pad 1 rose Sputnik, the first Lunas, Vostok, Voskhod, most Soyuz and Progress spacecraft. It was renamed 'the Gagarin pad' and crews for the International Space Station will leave Earth from there.

A launch from the Gagarin pad is quite unlike a mission from Cape Canaveral. The Soyuz rocket, 50 metres long, clunks and trundles its way down to the pad on a railway flatcar at walking speed, accompanied along its overgrown verges by scores of rocket workers and sight seekers. The transport reaches the pad, which is a concrete platform on heavy cement legs. Around it is a giant flame trench, looking like a reservoir empty of water. Once at the pad the transporter arm lifts the rocket up to the vertical. The booster is held up to the vertical while clamps rise up like a bear trap to grasp it so that engineers may inspect it at all levels. An hour before lift-off the high gantries are lowered. Any cosmonaut on board will now feel the Soyuz gently rocking in the wind. New fuel still has to be pumped on board until the very end. Liquid oxygen boils at $-190\,°$C and wisps of it always surround a rocket's stages. The fuel hoses are pulled away at 60 seconds before lift-off. With 20 seconds to go, the electrical lines are removed, and the rocket's electrical systems must now use their own batteries. The ignition command is sent and flames roar out into the trenches. When the thrust exceeds the rocket's own weight, the four lower arms still restraining it fall back like petals on a flower and let the rocket go free.

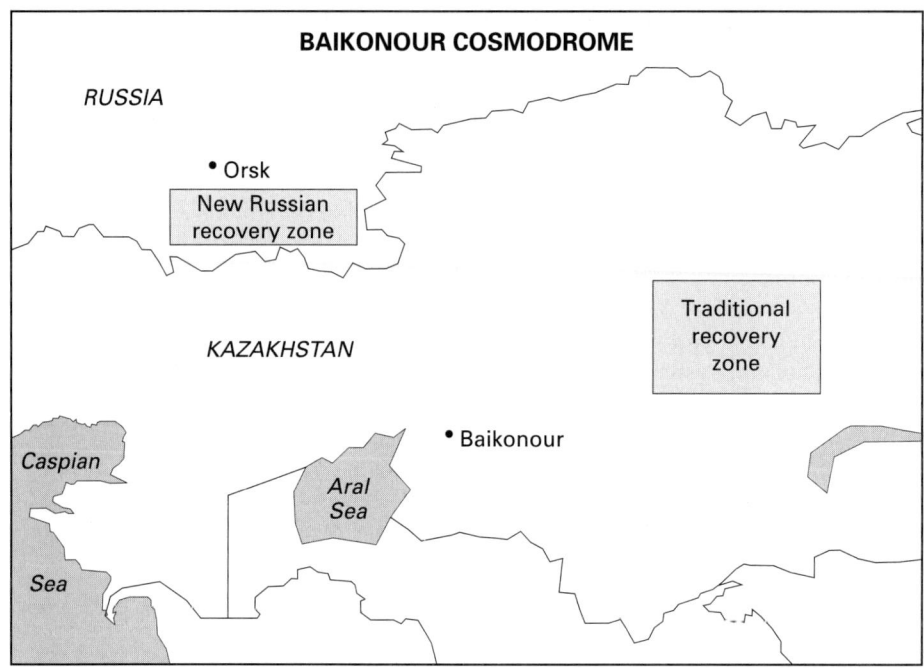

Map 2: Baikonour cosmodrome

The Gagarin pad became so busy that in 1964, work began on a second R-7 pad, called pad 31, a distant 30 km to the northeast and this subsequently became the base for the launching of many Soyuz, Progress, Zenit and Yantar missions.

At around same time, about 35 km northwest of the original Sputnik pad, construction began of two pads to support the UR-500K Proton booster built for the man-around-the Moon programme. Each pad was flanked by two 110-metre high towers which combine the functions of lightning conductor, TV camera point and floodlamp location. There are now four Proton pads in two neighbouring complexes. The first pads, launch complex 81, were built in the 1960s and served until 1988 when a refurbishment programme began. A second set of pads, launch complex 200, was built in the 1970s only 600 metres away. These double Proton pads are all some distance from Baikonour town—a full 1½ hours on the train. An adjacent assembly area was laid out like an industrial estate.

Protons are assembled in a large horizontal assembly and integration facility which can hold up to six full Protons at a time. The assembled rocket is moved by diesel railcar to the pad about five days before lift-off. Because the Proton uses storable fuels, there are no tell-tale signs of valving cool fuels to herald an imminent launch. There is a single, dull thud, Proton lifting off in 2.2 seconds and clearing the tower in 6 seconds. Riding a pillar of blue flame, Proton pitches over in 18 seconds. A sonic boom is heard a minute into flight. On a clear day, Proton may be followed

5½ minutes into the flight to second- and even third-stage ignition, leaving a wispy contrail behind in the sky.

In the next wave of expansion, in the mid to late 1960s, construction began of the N pads, set up to support the man-on-the-Moon effort. They are located a mere 3.5 km from the first R-7 pad. The completed N complex was impressive: the two enormous pads, matched by 183-metre-tall towers, were fed rockets from a 250-metre-long assembly hangar, about the size as the old Zeppelin airship sheds. After cancellation of the N-1 they lay idle for several years until, in the 1980s, engineers began reconstructing them for the N booster's replacement—Energiya. Huge, white-painted facilities were built to support Energiya and the Buran space shuttle it was designed to fly: a Buran integration hall, an Energiya integration hall, test facilities, two parallel launch pads for Energiya–Buran (built on the exact site of the N-1 pads), one of which was used to launch Buran in 1988. A third, quite different, adjacent pad was used to fly Energiya on its first, Polyus SK1F-DM mission in 1987. At one stage, 4,000 technicians worked here, but the pads fell into disuse when the Buran programme was cancelled in 1993. They are now staffed by a small core of patrolling security personnel. The real second Buran is still there, called Baikal, in storage, as are three complete ready-to-go Energiya rockets.

In summer 1995, the decision was taken to modify the Energiya–Buran facilities for use as integration halls for Western commercial payloads and equipment for the

Assembly halls, launch pads in winter

International Space Station.[56] Soyuz and Progress spacecraft en route to Mir are now kitted out there and it is also the core of modernized Baikonour. To the north of these pads and east of the Proton pads was built the runway for the space shuttle Buran. Called Anniversary airfield (Yubeleniye), it is 4,500 metres long and 84 metres wide and runs from SW to NE. In the early 1990s it cracked and deteriorated. In 1995, repair work began on the runway which was designated the principal airfield receiving components of the International Space Station from Europe and north America and was repaved. Boeing 747 and Airbus airliners now fly in there.

Finally, two pads were built in the 1980s for the Zenit launcher, to the south of the second R-7 launch site, pad B. By 2000, Baikonour had nine active launch pads, of which two are double pads.

Baikonour's launch pads

1, 31	Soyuz	90	Tsyklon 2/M
45	Zenit (double)	109	Dnepr
81, 200	Proton (double)	131, 175	Rockot

In addition, there are pads 41 (Cosmos 3M) and 110 (Energiya) which are disused.

Tug-of-war over Baikonour

Neither Russia nor Kazakhstan was sure how best to manage Baikonour when the Soviet Union broke up in late 1991. Kazakhstan had declared its independence during the coup in the Soviet Union, and in September 1991 new Kazakh President, Nasultan Nasurbayev, announced that it was taking over the cosmodrome, except for the military facilities there.

In the course of 1992–3, Baikonour was a kind of no-man's land. Technically, it was run by an interstate commission. Workers on the ground made jokes about who owned the table in the room: Russia? Kazakhstan? Both? Half one side and half the other? The Russians had already gone to some effort to keep Kazakhstan on side. In 1991, they had invited Kazakhstan (then a state inside the Soviet Union) to fly a cosmonaut to Mir. Two cosmonauts—Toktar Aubakirov and Talgat Musabayev—had been sent to join the squad and Aubakirov had flown on Soyuz TM-13 (he later became Deputy Minister of Defence and member of the Baikonour interstate commission). Musabayev had stayed within the cosmonaut corps with the aspiration of making a mission himself. The Russians probably hoped that with these kind of arrangements it would be possible to maintain the operation of Baikonour for the foreseeable future. The Russians took the view that they had paid for, built and continued to maintain Baikonour and they did not feel that they owed anything to the Kazakhs. In early 1992, there was some expectation that Baikonour would be managed by the Commonwealth of Independent States, the federation that was expected to take over the former role of the Soviet Union in regard to its 15 member states.

In fact, joint operation did not work as easily as the Russians had hoped. There

Soyuz winter recovery

was much awkwardness as to who actually gave the orders there. Whenever Kazakhs tried to assert their authority, Russians responded by saying that they were part of a military unit, answerable to Russian military law (technically they may have been right). The Kazakhs established a customs post in Baikonour to decide what could move in and out of the cosmodrome, to considerable Russian annoyance.

In July 1993, Russian Defence Minister Pavel Grachev flew to Baikonour for talks with his Kazakh counterpart, Sagadat Nurmagambetov, in what turned out to be a fruitless effort to resolve the issue. The question of the legal authority in the site was indeed a crucial issue, but not the most decisive one. Money was. Later in 1993, Kazakhstan made it plain that it intended to charge Russia for the use of the cosmodrome and for recovering crews from orbit. Things came to a head in January 1994 when Kazakhstan began to charge Russia prohibitive rates for basing its recovery helicopters in Kazakhstan during the return of the Soyuz TM-17 crew of Vasili Tsibliev and Alexander Serebrov. The capsule was even impounded by customs on landing.

As a result, Russia based its helicopters in Chelyabinsk on Russian territory, flying them into Kazakhstan only for the immediate period of the recovery itself. Even then, they had to file flight plans in advance and carry parachute bags full of cash for fuelling stops. Even recovery Mil helicopters would be boarded, the commanders being required to pay $400 cash on the spot in landing fees. The Kazakh government also introduced a new element into the equation: pollution. It was certainly true that the desert downrange of Baikonour was littered with the débris of impacting rocket stages. In 1993–4, the Kazakh government carried out an ecological survey, instancing toxic fuels in the ground soil, rusting rocket bodies polluting the land and sewerage discharges from Leninsk. However, Russia regarded the Kazakh exercise as less to do with environmental concerns than the subsequent large compensation claims that followed in their wake.

Russia considers options

The Russians responded to the Kazakh threat in several ways. First, they considered how to transfer as many launches as they could to the Plesetsk cosmodrome in northern Russia. They also began to recover capsules in Russia, in preference to Kazakhstan, when they could. The first spacecraft to be landed in Russia was the Raduga capsule of Progress M-18, which came down in the Russian steppe on 4 July 1993, using a landing area between the southern Ural mountains and Kazakhstan. Second, they began to cast around for an alternative launch site to Baikonour, but within Russian territory. Third, they began to work out a more permanent arrangement for the use of the Baikonour cosmodrome.

It was not feasible to move all space operations to Plesetsk. The manned space station and flights to geosynchronous orbit required the more southerly latitude offered by Baikonour. Almost all the lucrative commercial flights were of comsats to 24-hour equatorial orbit and these simply could not be reached from Plesetsk. The cosmodrome in Baikonour was the only one with Proton and Zenit rocket pads and Russia needed the use of these pads for the foreseeable future. In trying to move launches to Plesetsk, Russia was bucking the trend. Once the world's busiest space port, Plesetsk's launch rate was actually falling as the military programme contracted. The new commercial business was going to Baikonour.

The deal

Negotiations between Kazakhstan and Russia rumbled on. Eventually, in March 1994, agreement was reached whereby Russia would take a lease on the cosmodrome till 2024 (to be more precise, until 2014, with a 10-year option on extension). The area of the cosmodrome would be sovereign Russian territory, under the command of Russian troops. Russia had hoped to strike a bargain for the use of the cosmodrome, offering Kazakhstan access to its space programme, sweetened by seats on missions to Mir (Soyuz TM-13, 19). In the end, the Kazakh fee was $115m a year, backdated to 1991, payable in cash in hard currency and they took up an outstanding offer of a flight to Mir in any case (Talgat Musabayev was quickly promoted and given an assignment). No sooner was the ink dry than the Kazakhs explained that this was just the basic rental and that there would have to be fees on top to actually use the site!

In no time, Russian parliamentarians were complaining that the rental was using up half the manned space programme budget. The rental was a running sore for Russia and granted that its space programme was virtually running on empty anyway, this was money it could ill afford. On the positive side, Russia could now operate the cosmodrome without interference, even if landings were another matter. The Kazakhs, for their part, argued that some missions flying out of Baikonour were extremely profitable, namely the Proton commercial missions and that they derived no direct benefit from these profits. Few Kazakhs actually worked there—the staff were mainly Russians.

Russian hopes that this deal marked the end of the matter for the time being were not realized. Relationships between Russia and Kazakhstan worsened in summer 1999. On 5 July, a Proton rocket carrying the first new Briz-M upper stage for a Raduga comsat failed. Only 280 seconds into the mission, controllers began to notice deviations from the expected performance. By 390 seconds, the rocket had veered 14 km off its planned route and was destroyed. Débris fell over a wide area, some individual chunks of 200 kg coming down. Some dramatic television pictures showed that a 200-kg fragment of what was indisputably a twisted rocket body had fallen into a Karaganda back garden, having narrowly missed the resident 39-year-old woman and her 5-year-old child.

Testing the deal

Kazakhstan protested and expressed its fears of an ecological disaster, warning that poisonous heptil fuel might have fallen to the ground. As an indication of its seriousness, Kazakhstan banned the launch of the waiting Progress M-42 freighter to Mir. The ban was not lifted until Russia checked for heptil, cleaned up the area affected, paid compensation to those in the débris field, made a cash advance of $50m to Kazakhstan and agreed to come up with a further $65m early in the new year to settle the year's rental.

Worse followed: on 27 October another Proton exploded at 277 seconds. Once again, Kazakhstan insisted on a clean-up and compensation, Russia insisting that no heptil was found and that it would have evaporated in the upper atmosphere. In the event, Russia paid €400 000 compensation. Kazakhstan suspended Proton launches until the cause was determined (a bad batch of engines due to poor quality control) and Russia came forward with a plan to safely requalify the Proton.

Modern Baikonour

By mid-2000, issues around the operation of Baikonour appear to have calmed down. At the April 2000 meeting of the intergovernmental commission governing the operation of the cosmodrome, the principal issues raised were pensions for Kazakhs working in the cosmodrome and whether phone calls from the cosmodrome to surrounding Kazakhstan should be charged at the international or the local rate. Although it probably galled the Russians, rent payments were now being made generally on time, the 1999 norm being $50m in cash and $65m in goods.

Leaving aside the argument that was raging over its ownership and future, the 1990s were not happy years for the cosmodrome. As early as 1992, Moscow Television gave a report to the effect that Baikonour was 'in the grip of cold and confusion'. The cancellation of the Buran project led to a big exodus of personnel. There was a vacuum of power, authority and organization. There was a high level of pilferage and theft, with copper cables and sheet metal disappearing. Rubbish accumulated and was not cleared away. Air conditioning did not work in the summer and heating failed in the winter. Food was scarce and there were weeks when basic commodities just did not arrive. As people left, schools, kindergartens and cultural facilities closed. Crime rose and people were afraid to go out at night. The rocket soldiers were mutinous and there were even accounts of riots (though these were later denied). Energiya rockets indoors were guarded by soldiers in felt boots—because there was no heating in the building.

By 2000, Baikonour presented a picture of contrasts. Many parts of the cosmodrome had rusted and fallen into disuse. There was a small core of modern buildings where work was under way on the International Space Station, the resupply of the Mir space station, military programmes and the Proton launchers. Here, the assembly rooms were to world standard and work proceeded methodically. Jumbo jets flew in and out of Yubeleniye airfield with satellites and components. By contrast, there were serious problems in the city of Baikonour, with water and fuel supplies being turned off for long periods. Some parts of the city had been boarded up and were no longer occupied, launch workers being flown in from Moscow only for the periods immediately before and after the flights for which they were responsible.[57] Inevitably, these wider problems impacted on operations. At the launch centres themselves, there were intermittent power blackouts. The international areas of the cosmodrome brought in a German company to provide supplies, which it was more than able to do when it was paid on time.

VOLGOGRAD STATION

The problems associated with Baikonour forced Russia to consider ways of better utilizing its other two cosmodromes, the Volgograd station and Plesetsk, or even to consider a new location altogether.

Volgograd station (also known as Kapustin Yar) had been used for the German V-2s from 1947 and for the small Cosmos launcher from 1962, as well as for other minor or experimental missions (e.g. BOR spaceplane). A total of 139 orbital and suborbital missions were fired from Volgograd before 1987 when it fell into disuse.

Indeed, during the 1990s, it was used, under international supervision, as a decommissioning ground to take old missiles out of use. Arms inspectors would arrive there from time to time to check that missiles were either blown up or put beyond further use. The only space-related activity was in January–February 1997, when two MR-12 sounding rockets were fired under a joint programme between the Institute for Dynamics of the Geosphere of the Russian Academy of Sciences, Johns Hopkins University in Maryland and the respective military forces of Russia and the United States. The sounding rockets released artificial plasma clouds to test for their effects on radio communication. Later that year, President Yeltsin visited the cosmodrome to mark the 50th anniversary of its opening when it first fired captured German V-2s across the south Russian desert. He inspected the rocket troops there, awarded medals, and promised to pay the cosmodrome's debts.

Cosmos 3M instrumentation stage: the rocket which reopened Volgograd station

Signs that Russia might start to move some of its missions there were confirmed when, in April 1999, the cosmodrome was reopened for the launching, by a Cosmos 3M rocket, of a German X-ray scientific satellite called Abrixas which stood for broadband imaging X-ray all-skies survey. The Cosmos 3M put the 470-kg Abrixas into the planned orbit, accompanied piggyback by a 35-kg Italian minisatellite called Megsat, built in Brescia to test high rates of data transmission from space. Unfortunately for the German scientists, the Abrixas thermal protection system broke down, the satellite overheated, and contact was lost three hours later. At the turn of the century, Kapustin Yar had two airfields (the old Kapustin Yar, Vladimirovka), three old R-12 pads, two Cosmos 3M pads, a launch site for the Start rocket, tracking facilities and some assembly facilities.

NEW COSMODROME: SVOBODNY–BLAGOVESHENSK

The development of an entirely new cosmodrome was the radical alternative to Baikonour. Although the economic circumstances could not have been more unfavourable, this is exactly what the Russians did.

The idea of a fourth cosmodrome emerged as the difficulties over using Baikonour grew. In November 1993, the Council of Ministers of the Russian Federation issued a decree for a feasibility study of a new cosmodrome should Baikonour no longer be available. Three new sites were considered, all in the far east: Vladivostok, Kharbarovsk and Svobodny. The choice was for Svobodny, an intercontinental ballistic missile base called Svobodny 18 built in 1968 with 30 underground rocket silos, although all but five of these had been decommissioned in 1993. It is 51°N, 400 km west of the Sea of Japan and 96 km from the border with China. The latitude of 50°N is significant, for that is the traditional inclination for the manned space programme. The nearest town is Blagoveshensk.

Svobodny offered a number of advantages. It was close to the trans-Siberian railway—important, granted that all Russia's rockets were transported from Moscow by rail—and nearer to the Pacific rim economies, whence some satellite launching business might be hoped for. It was not much further north than Baikonour, which meant that it was still a relatively economic location for

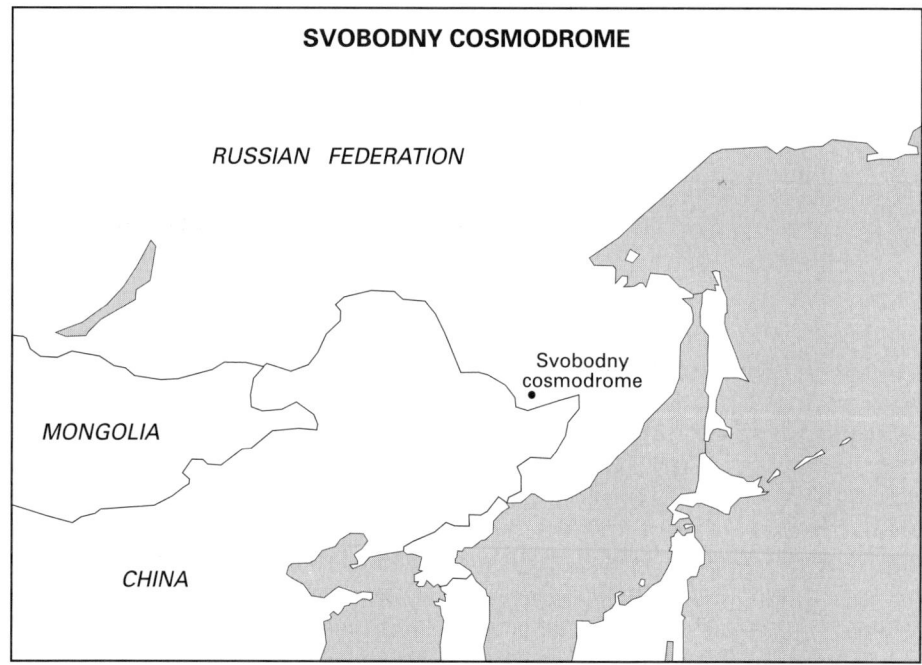

Map 3: Svobodny cosmodrome

reaching 24-hour orbit. Downrange tracking facilities must be exceptionally good, since rockets leaving Svobodny arch over Sakhalin island, one of the most radar-intensive zones on our planet and site of the notorious 1983 incident in which a Korean passenger jet was shot down. There was some local opposition from green activists, worried about the environmental consequences of spilt rocket fuel and falling upper stages to the region. Political interests lobbied strongly for the idea of a cosmodrome as a means to regenerate the economy of the region.

Official authorization for the conversion of Svobodny 18 into a cosmodrome was given by the Russian government in March 1994. When agreement was reached over Baikonour later that same month, there were reports that the Svobodny project would be abandoned, but in the event, it went ahead anyway. President Yeltsin visited Blagoveshensk in summer 1994 and the modification of the launch pads in Svobodny began soon after, principally the updating of the old Rockot pads. New power supplies and a command centre were installed that autumn. Although the authorization for the cosmodrome provided for the construction of Soyuz and Angara launch pads, the first launch from Svobodny was more modest, using a former military Start 1 rocket. The inaugural launch duly took place on 4 March 1997, placing in 400-km orbit a small, 87-kg Strela-class military communications satellite called Zeya (named after the local river). The apprehensions of the environmentalists were borne out when the second stage impacted on Keptin, Yakutia, 35 km downrange, sparking local protests. Later that year, a small American imaging satellite called Early Bird was launched from Svobodny.

By the new century, Svobodny comprised a Rockot launching area, Start launch pad, an industrial area, a fuelling plant, an airport and prospective sites for the new Angara launchers. A development plan for the cosmodrome envisaged a technical staff of 30,000 and an eventual total population of 100,000.

PLESETSK

Meanwhile, what of the world's once busiest cosmodrome, Plesetsk? How did it fare in the 1990s? Was Russia able to transfer launches from Baikonour to Plesetsk?

Plesetsk had been the Soviet Union's premier missile base, with four R-7s targeted on the United States from 1960 onwards. Use of Baikonour in the manned and lunar programmes was so intensive by the mid-1960s that from 1963 Plesetsk was developed as the USSR's main base for military missions and for placing satellites into polar orbits. An area of 200 km^2 was cleared around the town of Mirny, although the site was named Plesetsk, which was actually a village 4 km away. Like Baikonour, names were never what they seemed.

Cosmos 112 was the first orbital launch from Plesetsk in 1966, an event noticed first by the boys of Kettering Grammar School in England who claimed the credit for the identification of the base (even though the Americans almost certainly knew before that but kept it to themselves). Most casual observers would consider either Cape Canaveral or Baikonour to be the busiest spaceports in the world, but in fact neither is: the honour for launching the most satellites ever belongs to Plesetsk. The

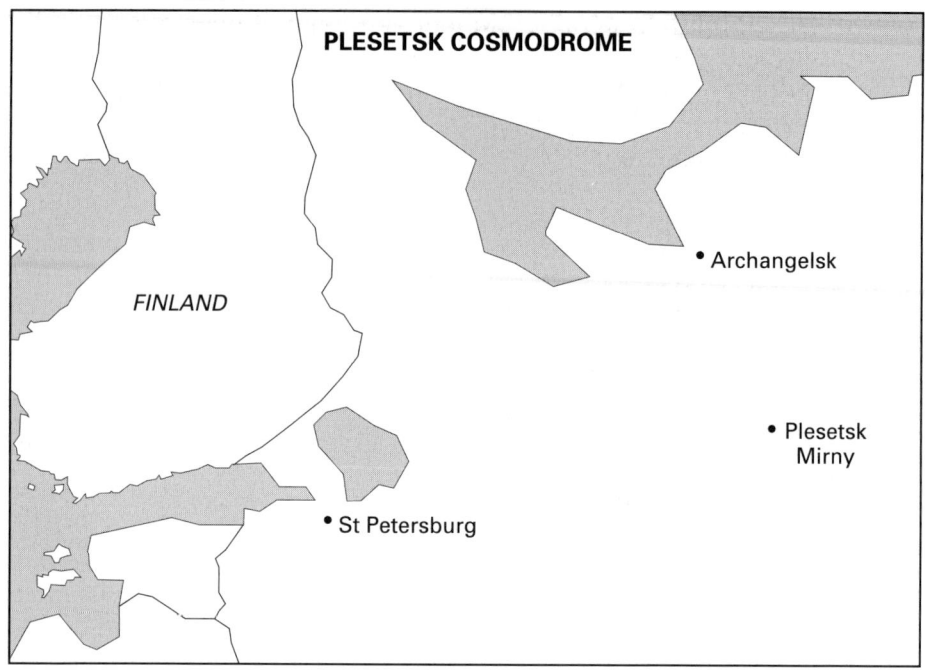

Map 4: Plesetsk cosmodrome

Soviet press was not permitted to acknowledge its existence until 1983. Officially it was just a 'military test site' until it was renamed a cosmodrome in the late 1990s.

The cosmodrome is located 15 km to the northeast of the towns of Mirny and Kochmas on the banks of the Emtsa river and is surrounded by forest. Most of the cosmodrome's workers live in Mirny, which houses up to 40,000 people in 9-floor apartments. Beside a lily-covered lake in Mirny lies the memorial to 51 rocket workers who died in a launch explosion there on 18 March 1980, a day always commemorated there and a day on which, by custom, no launches ever take place. The nearest big city is Archangel, 180 km far to the north. The area of the cosmodrome is 1,762 m^2, or 46 km from north to south and 82 km from east to west. Plesetsk is at 63°N, near the Arctic circle. The summer nights are long and it never really gets dark. In winter there are only a few hours of greyness at midday amidst remorseless night. The temperatures are even more extreme than Baikonour, reaching down as low as −46 °C, not that this has ever affected launchings. Plesetsk is near enough to Sweden for observers there to see the occasional launch in the far eastern sky arcing into the distance.

Although it covers a large land area, the core of Plesetsk is actually much more compact than either Baikonour or Kapustin Yar. Because it is built in forestry and in ravines, there is less space to spread the facilities. People go to work on a morning diesel train which leaves Mirny for all the different sites, although some staff are sufficiently close that they can cycle to their stations. Because it is military it is

Cosmos pad at Plesetsk

heavily protected by surface-to-air missiles. The present cosmodrome comprises 10 pads: two Tsyklon 3, two Cosmos 3M, four Soyuz–Molniya, a Rockot and a new Angara pad in construction. Plesetsk has the largest oxygen and nitrogen plant in Europe and has seven assembly shops and integration halls. The road network was in such an atrocious condition by 1992 that it nearly broke the chassis of Boris Yeltsin's limousine when he visited the centre. A presidential decree soon led to improvements.

As the military space programme contracted, Plesetsk's launch rate declined from one every two weeks to one every two months. In 1994, Baikonour overtook Plesetsk as Russia's busiest launch centre, even though Plesetsk had launched so many satellites in the 1970s and 1980s that it would still head the list for some time to come (a total of 1,500 launches by 2000, compared to Baikonour's 1,100). The fall in launch rate in Plesetsk was gradual and the northern cosmodrome experienced no sudden exodus paralleling the collapse of the Buran/Energiya project in 1993.

Conditions deteriorated in Plesetsk in the mid-1990s. Less information is available about Plesetsk than about Baikonour, since it receives fewer Western visitors. Living quarters for some of the rocket troops were very poor, which must be no joke in the Arctic. One of the city's five schools was made of prefabs. Over two-thirds of the soldiers had to supplement their food by growing potatoes in the grounds of the general hospital, right beside President Yeltsin's quarters during his visit. For the conscripts working there, conditions were harsh, with run-down buildings and food limited largely to bread, gruel, soup and eggs.

Because it is a military area, there is limited access for foreign traders able to supplement supplies. As an economy measure, the city's bakery, although it produced high-quality bread and cakes, was closed and thereon supplied only with military bread. During the mid-1990s, there were frequent blackouts—a distinct problem when one is preparing a rocket for launch—due to the local electricity utility not having been paid for some time. There was evidence of corruption, soldiers being court martialled in 1996 for reselling military petrol to civilians. Their defence, a plausible if not commendable one, was that they were only selling the balance of the backpay they were owed by the government.

Plesetsk's launch pads

16, 41, 43 (2)	Soyuz–Molniya M
32 (2)	Tsyklon 3
35	Angara
131, 132	Cosmos
133	Rockot

The pads are grouped closely together, amidst assembly and processing areas. Leaving aside the road improvements resulting from their damage to Boris Yeltsin's limousine, Plesetsk began to see the first signs of renewal in the mid-1990s. With investment money supplied by the German partners of its owners, a new launch tower was built for the Rockot launcher (hitherto, Rockot was launched from silos, but this was unsuitable for civilian payloads). In 1995, construction began of the first new pad in Plesetsk for some time, designed to take the Zenit. This was significant, for it was the first clear evidence that Russia intended to move at least some Zenit operations out of Baikonour, where the Zenit had, until then, exclusively been based. Even more significantly, in 1999 the new pad 35 was then reconverted for the Angara launcher and the Zenit idea was dropped. The Angara is a family of new launchers which can replace not only the Zenit, but the Proton as well. This would enable Russia, over the course of a number of years, to move all Zenit and Proton missions north to the Arctic. The Kazakhs may have overplayed their Baikonour rental hand and the desert may yet reclaim the rocket base that launched the Soviet Union to the stars.

TESTING ROCKETS

Finally, one should mention the main centre for the test firing of rocket engines. Niikimash rocket test centre was set up in Novostroyka in Sergeev Posad in the late 1940s (then known as Zagorsk), although it operated under the cover of a machine-building centre for the chemical industry. Surrounded by 17 km of barbed wire fences, the test centre was built on ravines overlooking the river Kunya. All rocket engines for the Soviet and Russian space programmes have been tested there. Sergeev Posad continued in operation throughout the 1990s.

OTHER GROUND FACILITIES

The cosmodromes are the most visible and the largest of the ground facilities sustaining the Russian space programme. Equally important are other key facilities such as Star Town, mission control and the tracking services. Was Russia able to maintain them?

Star Town

Star Town, also called Star City, or Zvezchny Gorodok, dated to 1960, when facilities had to be found to train and house the newly formed cosmonaut squad. Some 310 hectares of land were cleared in a birch forest 40 km northeast of Moscow. It was a closed city until the mid-1990s. Now, visitors can get off at the Tsiolkovsky stop 17 stations from Moscow city centre on the line to Monino. The central point of Star Town is a man-made lake originally built by conscripts, around which are a series of 15-floor blocks where the cosmonauts, their families and Star Town workers live. The lake freezes over in winter and it is possible to walk, sledge and ski on the lake. Further away lie health and sports facilities, a museum, post office, shops, nursery and hotel.

The main working facility in Star Town is the Yuri Gagarin Cosmonaut Training Centre, or the TsPK. This is a series of 12 blocks comprising full-scale space station training replicas, simulators, centrifuge, running track, administration, swimming pool and the hydrolab where cosmonauts practise spacewalking. The hydrolab is 23 metres across and 12 metres deep, holding 5 million litres of water. The centrifuge was built by Swedish engineers in 1980. Weighing 300 tonnes, it is 18 metres long and can fling two unfortunate trainees around at a time at up to 68 revolutions a minute, treating them to up to 30g forces.[58]

During the 1990s, Star Town became a much more open and international place. In addition to Russian cosmonauts training and living there, there were contingents of Americans, Japanese, Europeans and Chinese. Star Town did not suffer the same deterioration as did the cosmodromes, even if there was little new building and if maintenance was problematic (a general problem in Russia in any case). The most visually striking addition was special houses built for the NASA astronauts and, being built in New England style, they seemed incongruous and out of place amidst the Soviet functionalism of the surrounding architecture.

MISSION CONTROL KALININGRAD

The Apollo–Soyuz Test Project was the first occasion on which Westerners penetrated the inner sanctuary of Star Town—indeed the facilities built for American visitors then were converted into the health centre. Similarly, the project permitted the first American access to the flight control centre in Kaliningrad. The existence of a major control facility in the area had not been acknowledged until then—indeed, maps of the area were not available until the late 1990s.

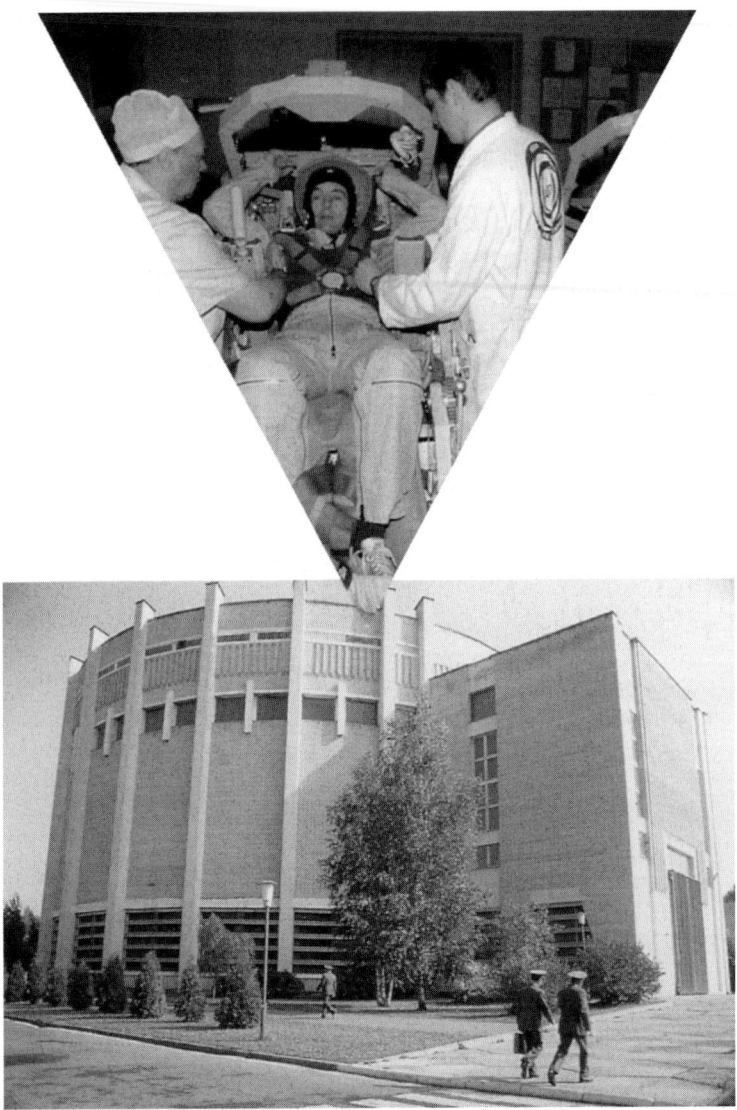

Centrifuge in Star Town

Until 1975, mission control for Soviet space missions had been Yevpatoria in the Crimea, a location whose function had always been acknowledged. Korolev had been a frequent visitor there for the interplanetary missions. Yevpatoria had been supplemented by and linked to other tracking facilities in the 1960s, enabling cosmonauts to talk to the ground over Soviet territory: Dzhusaly, Kolpashevo, Tbilisi, Ulan Ude, Ussurisk and Petropavlovsk. They got a few minutes talking time as they flew over, what the Americans call a 'communication pass' or 'comm pass'.

Mission control, the TsUP

For the Apollo–Soyuz Test Project, the USSR built an entirely new flight control centre in Kaliningrad, now called Korolev in honour of the great designer. In fact, the flight control centre is part of a much larger complex of space centres, facilities and complexes.[59]

Kaliningrad was, before the revolution, a forest north of Moscow where wealthy Muscovites had country houses—there were about 50 in the area and Lenin came to live in one of them from January to March 1922 before his final decline. The centre of the district was the government's forestry institute, located there in 1890. At that time it was called Podlipki and a railway station opened there in 1914. The railway station alone retained its name in the face of subsequent revolution and counter-revolution. Podlipki was renamed Kalininsky district in 1928 and then Kaliningrad in 1938 after the Soviet Union's first president Mikhail Kalinin (president, 1919–46). The township was renamed Korolev on 8 July 1996 by President Yeltsin, its third name this century.

Kaliningrad was viewed by the government as a centre for the development of rocketry as far back as 1940 when rocket gliders were tested there. Arms factories were moved there in autumn 1941 when the Germans moved in on Leningrad, though they were in turn evacuated to the Urals when the Germans reached Moscow later that year. The factories were rebuilt there in 1946, the principal one

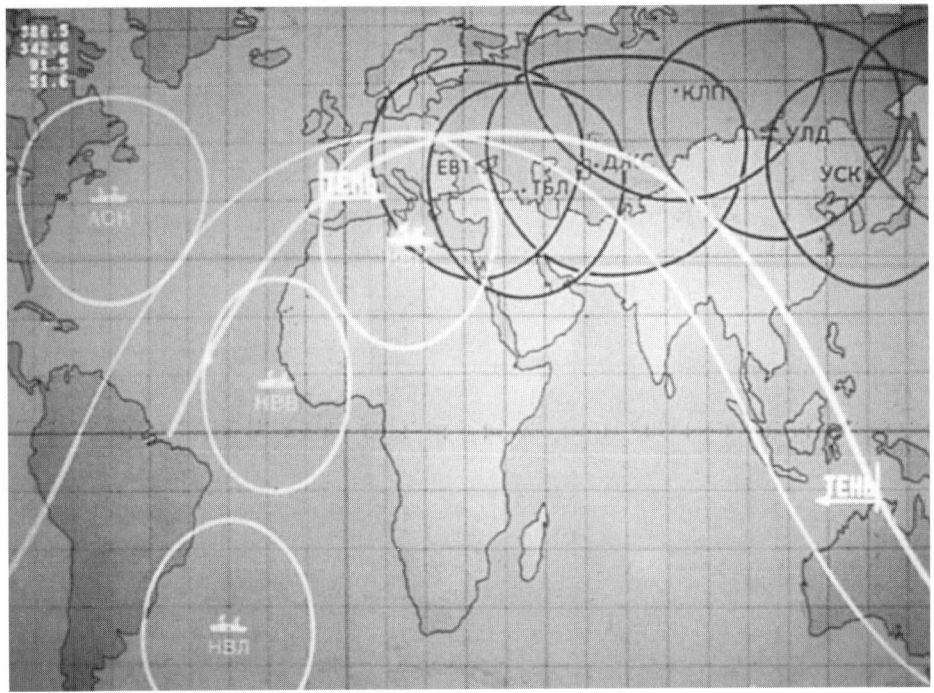

Tracking map

being OKB-1, or the Korolev design bureau. The V-2 Germans lived there briefly, before being moved to Seliger lake and the security-conscious nature of the area was emphasized by the building then of kilometres of brick walls. By 1960, there were eight design bureaux and rocket factories in Podlipki employing 200,000 people. Now, many of the main design bureaux may be found there, such as Energiya, Strela and Zvezda. Indeed, one of the ironies of the great rivalries between the design bureaux in the 1960s that partly cost the USSR the Moon race was that some of them were in close physical proximity to each other.

Like Star Town, Kaliningrad became a much more open place in the 1990s, though some sections are still closed off. Those factories involved in commercial alliances with Western companies have been able to improve their facilities. American and European engineers are involved in a range of projects in the area. Mission control is where most change is in evidence. This is called TsUP (*Tsentr Upravleniye Polyotami*), pronounced Tsoop, the present building dating to 1973 when it opened to handle the Soyuz 12 mission. It is not unlike Western mission control centres, with banks of consoles, central wall display, a television link to the Mir crew and maps of operational tracking stations. There are two very similar large control rooms: the main mission control, which has traditionally handled the Mir space station; and Buran control, which masterminded its 1988 mission. During the 1990s, the Buran control room was converted to the mission control centre for the

International Space Station and this may, in the course of time, become the part of Korolev best known to Western viewers. There are two smaller rooms: one to handle link-ups between the cosmonauts in orbit and their families; and another to control Progress and Soyuz missions when flying independently. Two thousand people work in TsUP, of whom some 300 are mission controllers.

TRACKING AND CONTROL

The scale and scope of the Soviet space programme required a national and international network for the tracking, control and recovery of spacecraft. The original Soviet space surveillance system, called SKKP (System for Monitoring Space) was set up in 1962, with radio, radar and optical devices and, later, lasers.

Radar tracking stations were set up in Irkutsk, Murmansk, Pechora, Sevastopol, Uzhgorod, Balkash, Mingechaiur and Riga, but only the first three, on Russian territory, may now be available. Optical trackers were installed in Dushambe, Byurakan (Armenia), Maidanak (Uzbekistan), Sanglok (Tajikistan) and Zvenigorod (Moscow). The largest optical tracker, in Zvenigorod, weighs 25 tonnes, has a focal length of 75 cm and can track satellites in geostationary orbit. In addition, the Zelenshuk observatory high in the north Caucasus mountains is used to follow the orbits of geostationary satellites. The north Caucasus is the home of the Krona complex, which has a battery of lasers, reflectors and phased radars to locate orbiting satellites to an accuracy of 40 metres. Sanglok, which is 2,300 metres high, has 1.1-metre telescopes to follow satellites in Molniya orbits. Like the Americans, the Russians now keep a full catalogue of objects circling the Earth.[60] The number of objects tracked has risen from 250 a year in 1969 to 1,000 a year in 1975 and 7,500 in 1994.[61]

Before the opening of Kaliningrad mission control, the main control centre was in Yevpatoria, western Crimea, chosen in 1957 by Korolev himself, offering a southerly latitude. It was called the TsDUC, or Centre for Long Range Space Communications. The eight original tracking antennae were built on converted battleship turrets. Since then, these 16-mm towers were joined by 32-metre and 70-metre dishes. During the period of the USSR, all lunar and interplanetary missions were controlled from Yevpatoria and it was the main point of communications for the early manned flights. After Ukrainian independence, the Yevpatoria system briefly withdrew from the network, but has since returned, although subsequent services are not seamless and are still hit by occasional strikes by ground controllers and by Russian–Ukranian disputes.

The main military tracking centre is in Golitsyno. Like many such facilities, it went through many changes of nomenclature, starting as facility 413, then Golitsyno 2 and now Krasnoznamensk, its current name. Officially, of course, it did not exist during the Soviet period (it was never marked on maps) and formal information on Golitsyno was not published until 2000. Golitsyno is 41 km west of Moscow on the highway to Minsk. The centre is full of small consoles, with none of the big screens to be found at Korolev. Golitsyno can control up to 120 satellites at a time and can handle huge quantities of data. In August 2000, the centre was equipped with mobile command posts, meaning that its functions could be dispersed in the event of conflict. Although not formally involved in the civilian space programme, it tracked the Zarya space station during its first few minutes in orbit and may have saved it from some untimely problems.

In addition to Korolev and Yevpatoria, there are other important centres for the control of space flights. These are the Priroda Scientific Centre (Earth observation satellites), Bear's Lake (civilian unmanned missions) and Noginsk (the military ocean surveillance programme). These remained in operation throughout the 1990s.

An important feature of the Soviet tracking system was the comships, or communications ships. Comships were used to track lunar and deep space missions and maintain contact with cosmonauts when their orbits took them far from the USSR. Large tracking ships were constructed for the Moon effort—the *Cosmonaut Vladimir Komarov* (17,500 tonnes), the *Cosmonaut Yuri Gagarin* (45,000 tonnes), which became the flagship, and the *Academician Sergei Korolev* (21,250 tonnes). Another large ship, the *Academician Nikolai Pilyugin*, was laid down in Leningrad in April 1988. They were large, impressive, streamlined white ships, and with giant aerials and huge telescope-like domes they looked futuristic. A series of smaller tracking comships was commissioned in 1974: the *Pavel Belyayev* (1978), the *Georgi Dobrovolski* (1978), the *Viktor Patsayev* and the *Vladislav Volkov* (1977). They were accompanied by a fleet of smaller, converted comships—*Borovichi*,

Comships: scrapped

Kegostrov, Morzhovets and *Nevel.* The Soviet navy also commissioned its own comships, presumably for use in association with its naval reconnaissance satellites. These were the *Marshal Nedelin,* assigned to the Pacific fleet and the *Marshal Krylov.*

In 1992, as the economic crisis began to bite in Russia, all the tracking ships were recalled, even though this meant that the then-orbiting Mir cosmonauts were now out of touch with ground control for up to nine hours at a time. The *Borovichi, Kegostrov, Morzhovets* and *Nevel* were sold and the *Vladimir Komarov* became, briefly, an environmental monitoring ship in the Gulf of Finland. In 1994, the *Yuri Gagarin* and *Sergei Korolev,* after lying idle in Odessa for some time, came under the control of the Ukrainian Space Forces who at once tried to sell them— but no buyers appeared. Eventually, in 1996, the *Yuri Gagarin,* the *Sergei Korolev* and the *Vladimir Komarov* were scrapped at a price of $170 a tonne. Russia could not afford to put a tracking ship on station for even a week for the Mars 8 launch in 1996. Some of the smaller ships, like the *Viktor Patsayev,* may still be seen, idle, in St Petersburg.

With the end of the tracking fleet and the demise of the short-lived Luch network of relays, mission control became reliant on ground stations alone to communicate with its cosmonauts. However, not all are in the Russian Federation and, the tracking network was in danger of shrinking even further. As a result, Russia's ability to communicate with its orbiting spacemen and spacewomen was even more restricted than it had been when Yuri Gagarin first circled the Earth.

COSMODROMES AND GROUND FACILITIES IN PERSPECTIVE

The table below summarizes launches from the four cosmodromes.

Russian launches by cosmodrome, 1992–2000

	1992	1993	1994	1995	1996	1997	1998	1999	2000
Plesetsk	33	25	17	13	10	9	5	6	3
Baikonour	21	22	29	19	13	15	16	19	16
Svobodny						2			
Kapustin Yar								1	

These are successful launches, in which a payload reached orbit (even if not the correct orbit). In addition, one demonstration launch was made by submarine from the Barents Sea. To 14 August 2000. For project Sea Launch, see pp. 227–231 below.

During the period, Plesetsk had 121 launches, Baikonour had 170 and the two other land-based cosmodromes 3. During the Soviet period, Plesetsk had accounted for 55% of launches, Baikonour 40% and Kapustin Yar 5%. Now, of the launches during the Russian space programme, Baikonour was the clear leader with 58%,

Mission in progress

Plesetsk 41% and others 1%. This difference may be largely attributable to the fall in the launch rate of military satellites, which were concentrated on Plesetsk, and the growth of commercial satellite launches, almost all of which took place at Baikonour and which had scarcely been a feature of the Soviet programme at all.

Like the rest of Russia's space programme, the cosmodromes went through difficulty and hardship in the 1990s. Overall, the physical conditions of Baikonour and Plesetsk declined sharply during the decade. However, parts of Baikonour were modernized and are now a busy, international spaceport that will service commercial operators and the International Space Station for at least another 10 years. The period was dominated by the protracted set-to with Kazakhstan over the future of Baikonour. Although the 20-year rental agreement brought some stability and legal clarity to the situation, for the Russians to pay out $115m a year to Kazakstan for rental was a considerable sum of lost money at a time when they could least afford it. Rather than bemoan their situation, the Russians cast around for alternatives which would at least enable them to keep their options open when the agreement came up for renewal after 2024. Russia reopened the Volgograd station, began construction of two new pads at Plesetsk and opened an entirely new cosmodrome in the far east.

Russia was able to maintain its principal ground facilities in the 1990s: Star Town, Kaliningrad and mission control. The mission control centre was adapted to serve the new International Space Station. The one feature of its ground infrastructure which suffered most was the tracking system, which almost collapsed. Russia lost the automatic use of tracking stations outside the Russian Federation. Its inability to maintain the Luch network was a handicap in the operation of Mir. Most importantly, Russia lost its entire, once proud, tracking fleet and its many fine ships. They

Soyuz U heads to the pad for Baikonour launch

were recalled early in the 1990s and none ever set sail again. By dramatic contrast, the Chinese had three new big tracking ships in their fleet—and they would play an important part in China's new manned space programme. But by 2000, the Russian tracking network was threadbare. Russia then faced the problem of how to adapt its rocket fleet.

7

The rocket frontier
Development of rockets in Russia from 1992

Russia had managed to maintain its manned, military and civilian space programmes, but would it be able to develop and modernize its launching potential? How did Russia develop its rocket fleet in the 1990s?

The old Soviet space programme was remarkably economical and used only six rockets: the R-7 (and its many versions), the R-12 and R-14 Cosmos, the R-36 Tsyklon, the UR-500 Proton, the unsuccessful N-1 and the Zenit–Energiya system.

In the Russian space programme, Energiya was abandoned and a range of small rockets was introduced, generally using old hardware left over from the cold war. Existing rockets were modernized and new upper stages were introduced. An entirely new rocket was designed and prepared for the twenty-first century, the Angara. The table below lists the Soviet-era rockets which continued in use under the Russian space programme.

OLD RELIABLE

More R-7 rockets have been built than any other make in history. By 2000, an astonishing 1,630 had been launched. One Western expert calculated that the factory which built its engines must have turned out a new R-7 nozzle every 12 minutes of the working day since the start of the space age! With its core and four strap-ons and 20 nozzles roaring at lift-off, it is unique in the rocket world. As the years went by, the engineers simply added more and more to the top, till the rocket reached to nearly 50 metres with the Soyuz escape tower, compared to the more modest 29 metres when it started as Russia's first intercontinental ballistic missile.

There have been no less than 16 versions of the R-7. These were given project codes, starting with the first test before Sputnik (8K71) to the last Soyuz version (11A511U2). Some have been fairly minor variants, with only one or two launches. The most used versions have been the Voskhod model (11A57, 306 launches) which was used for all but the very last phase of the Zenit photoreconnaissance programme, Molniya M (8K78M) (with Molniya, over 300 launches) and the Soyuz U (11A511U (680 launches by 2000)). The Soyuz U was a consolidation of a number of 1960s variants and the U design has remained unchanged since.[62]

Russia's rockets from the Soviet period

Name	Designer	Intro.	Length (m)	Weight (kg)	Payload (tonnes)
R-7 Soyuz U	Korolev	1966	49.5	309,000	7.5
R-7 Molniya M	Korolev	1960	45.2	305,000	1.6 to high LEO
R-14 Cosmos 3M	Yangel	1964	31.4	109,000	1.78
UR-500 Proton	Chelomei	1965	59.5	690,000	20.6 LEO, 2.3 GEO
R-36 Tsyklon 2M	Yangel	1969	35	180,000	4
R-36 Tsyklon 3	Yangel	1977	39.3	193,000	5.5
Zenit	Yangel	1985	57	459,000	13.7

During the Russian period, three, then two versions of the R-7 were in use: the Soyuz U, the Soyuz U2 and the Molniya M. The most powerful, the Soyuz U2, was retired after 70 launches (not a single failure) due to the non-availability of its main fuel source, sintine, leaving the Soyuz U and Molniya M continuing to operate. The Soyuz was used principally for Soyuz, Progress, Bion, Foton and Yantar missions, while the Molniya M was used for Molniya comsats and the Oko early warning system.

There is absolutely no sign of the R-7 being retired. In the early 1980s, the Progress factory was making R-7s at the rate of 50 a year, but this was scaled dramatically back to less than 12 a year. With the investment of the Western company Starsem and its success in winning Western contracts (Globalstar, Skybridge), the rate of construction was increased to over 20 a year by the new century. The introduction of a number of new, successful upper stages to the Soyuz is likely, if anything, to lead to further production runs.

The U2 version was introduced in 1982, but retired from service in 1996. It was essentially the same as the U, but it carried a fuel additive, a synthetic hydrocarbon called sintine. This gave an added kick compare to the normal kerosene. However, the process of making sintine was expensive and could no longer be afforded. In 1996, the factory which made sintine in Ufa, Bachkiria, closed. For manned flights, this created problems: a reduced kick in getting into orbit meant that Soyuz spacecraft must rendezvous with Mir at lower altitudes, or carry a reduced payload, or that the cosmonauts had to diet! The back-up Kurs rendezvous system was removed, meaning that Soyuz had only one chance to dock, and personal effects were reduced from 1 kg to 500 g.[63] Soyuz TM-22 was the last flight of the sintine-fuelled U2 model.

In a programme under financial pressure, the R-7 offered huge advantages, for the development costs and problems had been paid for and sorted out 30 years earlier.

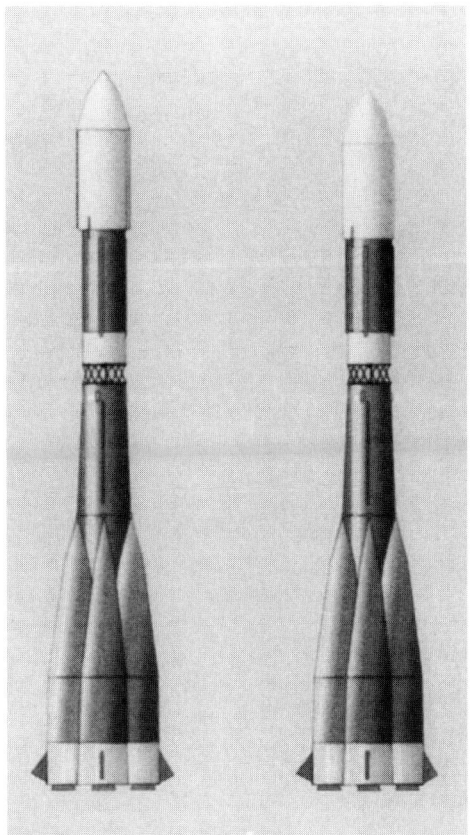

Molniya (left) and Soyuz U

For example, the original Molniya version was introduced for the first Mars probes in 1960 and had its first success with the Venera 1 Venus probe in 1961. Mastery of the block L upper stage was an exasperating process and right up until the early 1970s it continued to give trouble. Molniya M, which was introduced with Luna 14 in 1968, used a number of different fourth stages (block L, ML, 2BL) but these now proved to be extremely reliable. A critical feature of the Molniya was the successful development of a special instrument unit called the BOZ (*Blok Obespechniya Zapushka*, or Ignition Unit) to ensure that the unit was pointed in the correct direction, that the engine tanks were properly pressurized in zero gravity and that ignition actually took place.

There were two R-7 failures during the Russian period, taking place a month apart. On 14 May 1996, a Soyuz U carrying a photoreconnaissance satellite broke up suddenly 49 seconds into launch. The investigation must have put the accident down to a random failure, to its cost, for when the next Soyuz U took off a month later, on 21 June, exactly the same thing happened and at almost the same time into the mission! It turned out that the manufacturing process for the Soyuz shroud had

R-7 rocket in its cradle

been modified and the glue of the glass-reinforced cone had been applied unevenly. The subcontractors, the Plastik company of Syzran, had not informed the TsSKB in Samara which had been unaware of the change. The welds and bolts attaching the shroud to the rocket were insufficiently strong to hold the shroud when the rocket went through maximum dynamic pressure. They were strengthened accordingly and the old gluing process restored. Even at this stage of launching rockets, problems regarding shrouds had been underestimated. In 1996, two Chinese rockets carrying Western satellites (and their shrouds) broke apart in similar circumstances.

Key features: Soyuz U

Stage	Engines	Maker	Fuel	Thrust
Strap-ons	4 × RD-107	Energomash	LOX/kerosene	82,918 kg each
1	RD-108	Energomash	LOX/kerosene	81,130 kg
2	RD-461	KhimAutomatiki	LOX/kerosene	30,239 kg

Key features: Molniya M

Stage	Engines	Maker	Fuel	Thrust
Strap-ons	4 × RD-107	Energomash	LOX/kerosene	82,918 kg each
1	RD-108	Energomash	LOX/kerosene	81,130 kg
2 (Block I)	RD-461	KhimAutomatiki	LOX/kerosene	30,391 kg
3 (Block L)	RD-110	Energiya	LOX/kerosene	6,804 kg

COSMOS 3M

The smallest Soviet-period rocket which continued in operation was the Cosmos. This started life as the Cosmos 1 (1964–5), then the Cosmos 3 (1966–8) and ever since then its improved version, the 3M. This rocket was developed by the Mikhail Yangel bureau in Dnepropetrovsk in 1961 (project name: 11K65) to launch satellites in between the 500-kg capacity of the small Cosmos (R-12) and the 7.5-tonne limit of the Soyuz. Its first mission was the three Strela military communications satellites of August 1964. The R-14 was 30 metres tall, had two stages and could orbit payloads of about 1,000 kg. After nine test flights from Baikonour, all subsequent missions were shifted to Plesetsk, with just a few from Kapustin Yar and it took over the functions of the R-12 when it retired in 1977.

The R-14 was originally produced in Krasnoyarsk (the Cosmos 3) but since then the only production line appears to be in Omsk (the Cosmos 3M) where it is made by Polyot enterprises and was the one used during the Russian period. The Cosmos 3M rocket had been launched on orbital missions over 400 times by 2000 and had a reliability rate of 94%, almost all accidents being in the early stage of its development, including a launch disaster at Plesetsk on 26 June 1973 causing nine fatalities. The Cosmos 3M rocket made 30 missions between 1992 and 2000, being principally used for Parus, Strela 1 and 2, Tsikada, Nadezhda and geodetic missions. The Cosmos suffered one failure in the Russian period, on 10 October 1995 when it placed a Parus satellite in far too low an orbit. An obstruction in an oxidizer pipe was blamed. Despite this, Cosmos 3M was an otherwise entirely reliable rocket, and despite reports that its production line had closed was still orbiting satellites in the summer of 2000 (Nadezhda 6).

Some attempts have been made to market the Cosmos 3M in the West, the agency being Assured Space Access Inc, the rate being $10m for a full launch and less than $10,000 for small piggyback payloads. A modernized version of the Cosmos, called Vzliot, is in design.

Key features: Cosmos 3M

Stage	Engines	Maker	Fuel	Thrust
1	RD-216	Energomash	N_2O_4/UDMH	154,842 kg
2	11D49	KhimMash	N_2O_4/UDMH	15,966 kg

UR-500 PROTON

Vladimir Chelomei's UR-500K Proton had dual origins in the early 1960s: as a city-buster nuclear-carrying rocket and as a powerful rocket able to send a spacecraft to circumnavigate the Moon.[64]

The 44-metre tall Proton became in the course of time the workhorse of the deep space programme, geostationary satellites and the orbiter of heavy payloads into

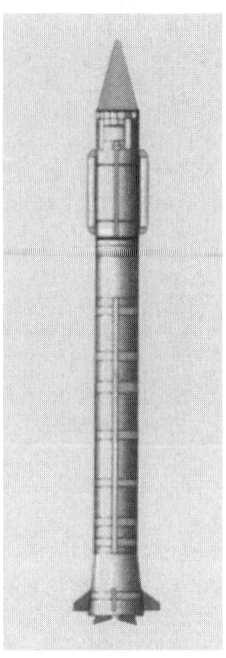

Cosmos 3M

Earth orbit. Despite its unreliability in its early years—and problems which cost the Russians the Moon race—it became very solid and reliable, rarely malfunctioning after 1970. Proton was able to lift up to 21 tonnes into low Earth orbit (e.g. Mir, its modules, Zarya, Zvezda) or 6 tonnes to the Moon or Mars (e.g. Mars 8) or 2.5 tonnes to geosynchronous orbit (e.g Luch, Ekspress) or the global positioning system (GLONASS). Proton was also used for a number of domestic military programmes (e.g. Raduga, Potok, Prognoz). Altogether, Proton made 67 launches between 1992 and summer 2000. The Proton rocket is now manufactured in the Khrunichev plant in Moscow, now part of American–Russian company Lockheed Khrunichev.

Proton exists in two versions: the three-stage and the four-stage. The three-stage is used for large payloads in low Earth orbit, like space stations and their modules. The four-stage, the most common, uses an upper stage. Originally, Chelomei designed his own upper stage, but in December 1965, shortly before his death, Korolev wrested control of the upper stage from Chelomei and used his own fourth stage, the block D, which is now used on all Protons.

So reliable did Proton become that in 1983 the USSR offered it to Western commercial companies on an economic basis. Proton won its first commercial contract in 1992 with a deal to launch an Inmarsat communication satellite. Since then, a stream of commercial contracts followed to put comsats into 24-hour orbit, mainly for Western communications companies. Proton is able to offer reliability, a proven track record and costs below the American rate, enough to make a difference

Proton: flying since 1965

in the competitive global world of the 1990s. By the late 1990s, the annual Proton launch rate had, despite the contraction of the Russian economy, actually increased.

Proton launchings have two unusual features. First, because the Proton uses storable fuels, there are none of the tell-tale wisps of liquid oxygen burning off during the final stages of the countdown. There is no means of telling that the rocket is fuelled and ready to fly—it just goes. Second, the bottom part looks as if it holds six strap-on rockets, like the four of R-7. In fact, they are not boosters, but fuel tanks positioned on the side feeding into the main propulsion area. The explanation for their 4.1-metre diameter lies in the width of the carrying cars of the Russian railway system, which is not wide enough for the base of the Proton, so the tanks had to be transported separately.

The year 1999 was not kind to the Proton. In July, a Proton carrying the new Briz M upper stage and a Raduga comsat failed at 120 km, 277 seconds into the mission, débris raining down over a wide area. One such occasional failure was serious enough, but on 27 October the Proton failed again, this time carrying the Ekspress A1 comsat. The second stage engines seemed to explode. This was the 269th Proton launch and the 37th failure (though most of these had been in the 1960s). There was plenty of forensic evidence to examine and within two days investigators had found two engines. The investigators found a pattern of poor workmanship in a batch of engines produced in the Voronezh plant in 1993 when there had been a break in production due to a shortage of orders. In the words of the investigation, 'a large number of particles had found their way into parts of the engines. Ducts and welds had not been properly cleaned.' Rigorous checks were run of all engines to ensure that none of the bad batch was installed on a Proton awaiting launch. It was decided that future Protons would be fitted with filters to the inlets of the gas generators, the number of rivets and welds would be reduced and the turbopumps would be made of tougher alloys able to withstand higher temperatures. The Proton was not cleared to fly again until February, when it put the Garuda comsat into orbit. The string of Proton successes that followed suggested that, barring random failure, these measures had been successful. Proton flew again with Sesat on 18 April, the last Gorizont on 6 June, a Potok on 5 July and, supremely, the Zvezda space station on 12 July.

By August 2000, the Proton had flown 277 times. In the Russian period there were six failures: Gorizont (May 1993), Raduga 33 (February 1996), Mars 8 (November 1996), Asiasat 3 (December 1997) and then the two disasters in 1999. Little was understandably said of the May 1993 Gorizont failure once the reasons were discovered. Officially, the reason given was contaminated fuel, but local information suggested that not enough fuel had been put into the second stage and the rocket had fallen short of Earth orbit. Whether this was due to carelessness or, as some suspected, theft, no one was quite sure. In the case of Asiasat, a gas turbine generator broke down. With the Raduga mission, a broken wire caused a joint to leak, the block DM-2 failed to ignite and there was an explosion soon thereafter.

Throughout the 1990s, there were repeated reports that the Proton would be upgraded to a new version, the Proton M (M for modernized or modified). The modified Proton would have an improved computerized control system taken from Zenit, a larger shroud, stronger interstage joints, a new upper stage and be made of lighter materials. As a result, it would be able to lift 22 tonnes into Earth orbit (some figures quoted up to 26 tonnes) and up to 4.2 tonnes to geosynchronous orbit. The upper stage cited was the KVD-1 developed by the Isayev bureau in the 1960s. However, the promised maiden flight kept receding. Although Proton M still remained on the books in 2000, there was increasing reference to it being phased out

The Proton M version

in favour of the new Angara. By inference, the prospects for the Proton M appeared to be dimming.

Whatever its long-term future, the medium-term outlook for the Proton is promising, with a range of commercial and scientific cargoes manifested. A highlight will be the 2001 launch of the European Space Agency's Integral gamma-ray observatory.

Key features: Proton

Stage	Engines	Makers	Fuel	Thrust
1	6 × RD-253	Khrunichev	N_2O_4/UDMH	150,142 kg each
2	4 × RD-210	Khrunichev	N_2O_4/UDMH	59,422 kg each
3	RD-210	Khrunichev	N_2O_4/UDMH	62,143 kg
4	Block D	Energiya	LOX/kerosene	8,528 kg

TSYKLON

The Tsyklon rocket had its roots in the R-16 military booster dating to 1960 and in the UR-200K design of Vladimir Chelomei. Go-ahead for the Tsyklon programme was given in 1965 as an intermediate booster able to lift military payloads larger than the Cosmos 3M rocket but smaller than Vostok (Tsyklon was called project 11L67). The R-36 Tsyklon 2 was introduced in 1969 to fly the US-P EORSATs, the Fractional Orbital Bombardment System and the hunter–killer satellite series. The current Tsyklon 3, with a more powerful upper stage, flew in 1977 (there is no Tsyklon 1). To add to the confusion, Soviet sources would refer to the Tsyklon 2 as the M and the Tsyklon 3 as the plain 'Tsyklon'. Essentially, the difference is between the 2, with two stages used for US-P from Baikonour, and the 3, with three stages, used for other missions, such as Meteor, Okean, Gonetz, Strela 3, Geo, Koronas and Sich.

Tsyklon has two main stages, both burning UDMH fuel with either nitric acid or nitrogen tetroxide as oxidizer. Tsyklon 3 is 39 metres tall and can put payloads into 150-km to 10,000-km orbits from Plesetsk. There are two automated pads at Plesetsk which can launch Tsyklons rapidly in all weathers, from +50 °C to −45 °C.

Tsyklons are a declining part of the Russia's rocket fleet. It is reported that the production line has closed, but that old military Tsyklons have been decommissioned for continuing missions. By summer 2000, Tsyklon 2 and M had flown 118 times, the Tsyklon 3 as many as 215 times and launch rates were down to a handful a year. Its success rate was 98%.

Key features: Tsyklon 3

Stage	Engines	Makers	Fuel	Thrust
1	4 × RD-261	Yuzhnoye/Energomash	N_2O_4/UDMH	269,710 kg
2	RD-262	Yuzhnoye/Energomash	N_2O_4/UDMH	95,981 kg
3	RD-861	Yuzhnoye/Energomash	N_2O_4/UDMH	7,983 kg

ZENIT

Zenit also originates from Mikhail Yangel's bureau in Ukraine, NPO Yuzhnoye.

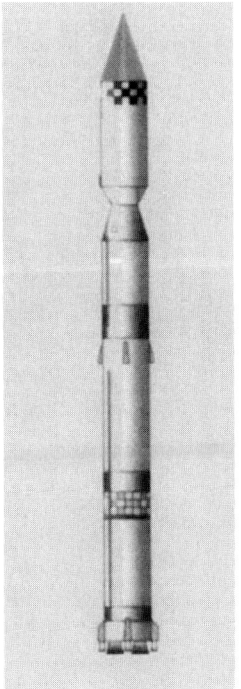

Tsyklon 3

Zenit was ordered in December 1974, first flying in April 1985 in a suborbital test. Its first successful mission was Cosmos 1697 in October 1985 and it became operational with Cosmos 1833 in March 1987.[65] Zenit serves as the strap-on booster for the Energiya rocket, but is a powerful rocket in its own right, standing almost as tall as Chelomei's famous Proton. Zenit exists in three versions: Zenit 1 is the first stage of the Energiya; Zenit 2 is the main model; and Zenit 3 is the sea-launched version used to reach 24-hour orbit (also called Zenit 3SL).

Zenit 2 has a capacity of 13.7 tonnes to low Earth orbit. Its height is 57 metres, weight 445 tonnes, diameter 3.7 metres and its main RD-171 engine burns for 133 seconds, the second stage for 303 seconds. The RD-171 engines and the second-stage RD-120 were developed by the Energomash NPO in Khimki, descended from the Leningrad Gas Dynamics Laboratory (GDL). The Zenit complex was completed at Baikonour in 1985: it comprises supply and service sheds and six-floor multistorey buildings capable of processing several Zenit rockets at the same time. Zenit has a unique, automated system whereby the payload is mated, the Zenit transported on a railcar to its pad and put into position for firing. Once the railcar arrives, automatic systems connect the Zenit to 25 fuel lines and 3,500 electrical circuits. Zenit is clamped to the pad by vices that are sufficient to hold the rocket down at full thrust and it is then raised into a firing position. Once secured, it can be fuelled within minutes by pumps drawing off silos which are located underground close to

Automatic assembly of Zenit on launch pad

the pad. The entire fuelling and launch sequence is conducted automatically. The hoses and pumps are withdrawn 12 minutes before launch. At 4 minutes before launch, the railcar makes its way back to the integration building. A system of sprinklers is available to extinguish fires. At 15 seconds before launch, the cooling system is activated, drowning the base of the pad in water to cool it and dampen vibration. The pad can be ready for another launch in five hours.

A total of 33 launch attempts had been made by summer 2000. Zenit's early and primary payload was the heavy electronic intelligence (elint) satellite Tselina 2 sent into 850 km orbits at 71 degrees from Baikonour. By 2000, Zenit had successfully put 10 Tselina 2 elints into orbit and was indispensable to the Russian electronic intelligence system. Although Zenit fired perfectly on both its Energiya flights, the rocket had a checkered history. Overall, its reliability has been in the order of 75%, one of the lowest of modern rockets, where 95% is the norm. Worryingly, there have been as many explanations as there have been Zenit failures and it has been difficult to discern a common pattern or trend. The Russians have tended to blame a fall in quality control in Ukraine, where the rockets are made, and have increasingly looked to an indigenous Russian rocket for a replacement, despite the Zenit's impressive power.

On 28 December 1985, Zenit suffered a failure when its second-stage engine failed to ignite. On 4 October 1990, a Zenit exploded 70 metres high, right over its launch pad at Baikonour, destroying the Tselina 2 elint payload and ruining the launch tower. There was another launch failure in August 1991 when the second stage exploded during the third minute of flight. Then there was a pad explosion on 5 February 1992, completely destroying the Tselina 2 electronic intelligence satellite on board and the launch pad itself. The next loss was on 20 May 1997 when a Zenit blew up 48 seconds into a mission to put up another Tselina. The investigation reported two weeks later, attributing the failure to a fault in one of the engines when Energomash stressed it beyond specification during ground tests. On 12 September 1998, Zenit suffered its worst accident when the second-stage shut down 4 minutes 32 seconds into the launch of 12 Globalstar satellites. Fragments of the comsats were later found 2,300 km downrange in southern Siberia. Never before had so many satellites been lost on one rocket. The fault was later blamed on a Russian software failure. The value of the lost satellites was $250m and the Globalstar company at once pulled all its remaining bookings on Zenit. It was not only Zenit rockets that exploded, for at the time the United States was also going through a difficult period, losing a Titan IVB, its brand new Delta 3 that autumn and the following April two more Titan IVs, an Athena and another Delta 3. Zenit did not return to flight until the following summer.

Key features: Zenit 2

Stage	Engines	Makers	Fuel	Thrust
1	RD-171	Yuzhnoye/Energomash	LOX/kerosene	740,030 kg
2	RD-120	Yuzhnoye/Energomash	LOX/kerosene	85,005 kg

UKRANIAN ROCKETS TO CHRISTMAS ISLAND: ZENIT 3SL, THE SEA LAUNCH

However, Zenit was to have an exciting new lease of life in an entirely new and unexpected venture—the $500m Sea Launch project. This was a joint venture between the American Boeing company (40%), Russia's Energiya (25%), Norway's Kvaerner (20%) and Ukraine's Yuzhnoye (15%), to put communications satellites and similar payloads into orbit from a mobile equatorial launch site in the middle of the Pacific near Christmas Island using a maritime version of Zenit.

An equatorial site cuts down considerably on the costs of reaching geosynchronous equatorial orbit, obviating the need for any energy-expensive dogleg manoeuvres from higher latitudes. This was not the first use of a sea platform; Scout rockets had been fired from the American—Italian San Marco platform off Kenya from 1967 to 1988. However, this was the first project to use an ocean-based equatorial launch platform for commercial operations to 24-hour orbit and was on a scale much larger than anything previously contemplated.

Hitherto, Zenit's capability to 24-hour orbit was very limited (only 600 kg). However, if fitted with the Proton DM upper stage and launched from near the equator, it could place 5.25 tonnes into geosynchronous orbit, making it a highly competitive launcher. Moreover, Zenit's automated countdown procedures made it ideally suited for operations in confined and difficult spaces, such as on board ship. Accordingly, it was given the new name of Zenit 3SL (3 for three stages, SL for Sea Launch).

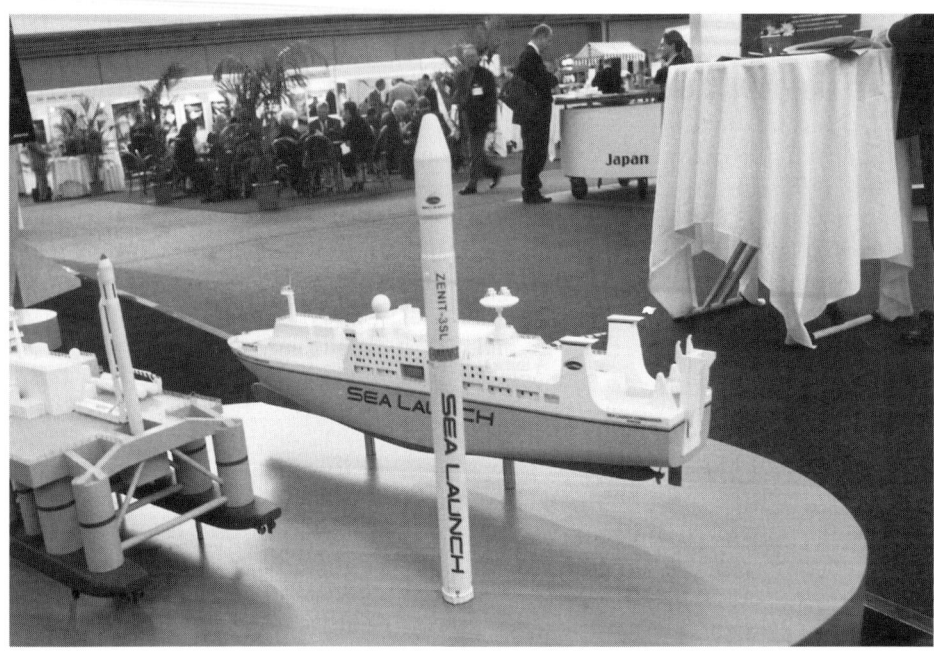

Key elements of Sea Launch: platform, rocket and command ship

The Zenit project involved the construction of a 34,000-tonne rocket transport, assembly and command ship built in Govan, Scotland, in the course of 1996, called the *Sea Commander*, later adapted with mission control and space-tracking gear in the Kanonersky works in St Petersburg. This is a big ship, which rides high in the water like a modern cruise liner. It was a project especially welcome for the work-starved yards in Glasgow. The ship is 203 metres long and 32 metres wide, just narrow enough to fit through the Panama Canal. Inside, it comprises a huge rocket assembly hall, mission control and facilities for handling the hazardous chemicals used on the launches. In the body of the ship are placed the Zenit rockets, up to three at a time, in the space that would be occupied by cars in a sea ferry. The three Zenits fill the centre of the ship in a hangar 70 metres long, 12 metres deep in three bays, as in a railway marshalling yard. For its first journey in 1998, the *Sea Commander* travelled from St Petersburg to Long Beach with two Zenits on board and its 240 crew.

Home port is Long Beach, California, where Boeing took over an old naval causeway which had been used as a depot from 1944, and later a waste dumping ground, and was in a terrible state. The area was cleared and rebuilt in 1997 with a hangar designed to take three Zenits and an integration building. The role of the *Sea Commander* was to load the Zenits, bring them 1,600 km across the Pacific to the launch site, transfer them to the launch platform, back away to a distance of 5 km and act as the launch control and tracking ship.

The floating launch platform was a decommissioned 131-metre-long, 78-metre-wide North Sea oil rig called the *Odyssey*, bought in Norway. It was self-propelled, semi-submersible built in Norway in the 1980s but taken out of service when it was damaged by fire in 1988. It lay idle in Vyborg for a number of years. Bought by the Sea Launch project in 1995, it was refitted in Stavanger and Vyborg where its legs were strengthened, new power systems and a crane were installed, and it was adapted to take rockets. There was a considerable amount of plumbing involved in fitting the ship to take oxidizer and fuel tanks, as well as the associated loading and pressurization systems. The helipad and crew quarters for 68 were modernized and a flame trench was fitted to take away the exhausts of the RD-171 engines. The *Odyssey* was refitted in the Baltic and then towed out to Kiribati, otherwise known as Christmas Island.

Getting the launch platform there was no easy undertaking. Averaging 12 knots, aided by its four 3-metre propellers, it was towed all the way from Vyborg, near St Petersburg, Russia, to Christiansand, Norway. It barely got under the new Oresund bridge, then being constructed to link Denmark and Sweden across the Baltic. The *Odyssey* then proceeded down the English Channel, through the Straits of Gibraltar and the Suez Canal, across the Indian Ocean to Singapore and then across the full Pacific before reaching its home port in Long Beach. The journey took 107 days. Once it arrived there, it took on 15,000 tonnes of water ballast to enable it to semi-submerge and stabilize itself in the ocean. The journey from Long Beach to Kiribati took a final 12 days.

As a general role, rockets are transported from the Ukraine by ship through the Black Sea, the Mediterranean, the mid-Atlantic and the Panama Canal and then up the west coast. The fuel is shipped from Russia (due to concerns about American kerosene). When the *Sea Commander* arrives alongside in mid-Pacific, it hoists the 465-tonne Zenit onto the platform, the *Odyssey*, where it can be raised to the vertical for launching. For lift-off, the roof of the hangar is rolled back and the Zenit is raised hydraulically into position (the hangar then closes again). Four hours before lift-off, when fuelling begins, the crew of the platform evacuate and board the *Sea Commander*. The entire process from that point is entirely automated, as at Baikonour. The rocket stands the equivalent of a four-storey house. Launches can take place in conditions of up to 12 metres per second windspeed and wave heights up to 3 metres. During the launch, telemetry is monitored from the *Sea Commander*'s mission control rooms staffed by Energiya and Yuzhnoye personnel.[66]

Following the loss of the Globalstar Zenit, the Sea Launch project became caught up in a technology transfer controversy, and the United States government suspended the project. Progress was only resumed when Boeing paid a $10m fine. Apparently, Boeing had failed to take sufficient precautions against its personnel, some of whom worked in the secret skunk works (builders of the U-2 spyplane) inadvertently giving away rocket secrets to the Ukrainians. Granted that American engineers were working on an advanced Ukrainian project, one might have thought that any technology transfer was flowing the other way, but this was not how the government saw things.

Odyssey

After the Globalstar loss, the Energiya–Boeing–Yuzhnoye team did not feel confident enough to fly a commercial payload on the very first mission (neither probably did their clients). The first Sea Launch mission was therefore set to be a demonstration mission with a mock payload. They need not have worried. Sea Launch soared aloft from its Christmas Island sea platform. Arcing over the blue equatorial Pacific, the block DM stage duly fired 34 minutes later and the payload arrived on station, over the equator, three hours later.

The first operational mission was duly made on 10 October 1999. There was almost no damage to the launch platform *Odyssey* as the Zenit soared on a pillar of flame skyward. At 2 minutes 50 seconds, the first stage dropped off. The second stage then burned until 8 minutes into the mission. There was a nerve-wracking moment: telemetry was lost just at the point of block DM ignition. In the event, the DM lit automatically, burned for 8 minutes and restarted twice again, each time on schedule. Its cargo, the 3,800-kg Direct TV1R comsat, was at its geosynchronous destination 62 minutes later. Despite the novelty of the project, it attracted little media attention, though, what was probably more important for the promoters, the order book lengthened.

Despite this promising start to the venture, there was an unexpected setback on the second commercial launch. Carrying a 2.7-tonne ICO F-1 comsat, the Zenit second stage shut down at 461 seconds into the mission and the payload was lost, the Russians and Ukrainians blaming each other. A Russian–Ukranian interstate investigation eventually settled the matter, attributing the cause to the software system. This was yet another example of computer software programmes being increasingly used to programme complete launch operations, but where a minor error had led to the complete loss of the mission. The classic original example was the maiden flight of the European Ariane 5 where the computer reset itself, thought it was off course, even though it was not, and destroyed the vehicle. On the Sea Launch project, the computer programme neglected to close a valve on the helium pressurization system on the second stage. As a result, pressure seeped away and the second stage lost thrust, closed down early and the Zenit failed to reach orbit. The real failure here was not so much the command that had not been included, but that checks of the software system had failed to spot the omission. This was yet another embarrassing failure for the jinxed Zenit, which still seemed unable to shake off a record of intermittent failures. In 36 launches since 1985, it had now failed nine times.

Despite this, Sea Launch did not lose any customers and was back in business within months. The Zenit 3 soared into a dark blue summer sky on 28 July, delivering the PanAmSat-9 on station a mere 1 hour 45 minutes later. The transfer orbit was so accurate that the perigee was 1,900.5 km, compared to the 1,900-km target. Sea Launch had 17 customers on its books in autumn 2000 and the company was planning how to reduce the launch interval to 50 days in order to clear the waiting list. Its fifth mission succeeded on 21 October 2000.

Key features: Zenit 3SL

Stage	Engines	Makers	Fuel	Thrust
1	RD-171	Yuzhnoye/Energomash	LOX/kerosene	740,030 kg
2	RD-120	Yuzhnoye/Energomash	LOX/kerosene	85,005 kg
3	Block DM	Yuzhnoye/Energomash	LOX/kerosene	7,992 kg

Sea Launch series

28 Mar 1999	Sea Launch	Kiribati	639–36,064	645	1.21
10 Oct 1999	Direct TV1R	Kiribati	35,783–35,786	1,435	0.06
12 Mar 2000	ICO F-1	Kiribati	Second-stage failure		
28 Jul 2000	PAS-9	Kiribati	1,900–35,814 (transfer orbit)		1.2

So much for Russia's rocket fleet inherited from the Soviet Union. The ultra-powerful but customerless Energiya was closed own and the Tsyklon and Cosmos brought towards the end of their production runs. With new missions, the R-7 Soyuz and Molniya M continued to thrive and the Proton attracted a share of the world launcher market. The troublesome Zenit was adapted, improbably but successfully, to launching satellites from Christmas Island. What about new rockets?

RUSSIA'S NEW ROCKETS

Russia realized at an early stage that it would need to innovate both to maintain its tradition in rocket building and to develop commercial income from the world-wide launcher market. Four rockets were introduced during the post-communist period, all derived from surplus cold war missiles. A new rocket was proposed which would replace most of the Soviet family, the Angara.

Russia's new rockets

Name	Designer/bureau	Service	Length	Payload
Rockot	Vladimir Chelomei	1994–	29 m	1.85 tonnes
Topol, Pioner (Start 1,2)	Alexander Nadirazhde	1993–	29 m	700 kg
Dnepr	Mikhail Yangel	1999–	34.3 m	4,000 kg
Shtil	Makeev	1999–	30 m	Up to 700 kg
Angara	Khrunichev	Due 2002	52 m	Up to 28 tonnes

KHRUNICHEV'S ROCKOT

A surprise new entry in 1994 to the rocket fleet was Rockot, a small, 29-metre-tall missile. Rockot, originally project 15A35, was based on Vladimir Chelomei's UR-100 missile and during the cold war received the NATO designation of RS-19. A total of 360 Rockots were deployed in silos around Russia and Ukraine at the height of the cold war. The Rockot was fired 144 times in long-range tests during the period. Now, their manufacturers were searching for ways of disposing of their hardware peacefully and entering the commercial launcher market. Rockot was the first of the cold war missiles to be adapted for civilian missions. Suborbital tests were made at Baikonour in 1990 and 1991 before the first orbital mission in 1994. The word 'Rockot' means 'roar of sound', not 'rocket' ('raket' in Russian).

For commercial applications, a new third stage, called Briz K, was fitted to enable it to put 2-tonne payloads into Earth orbit. Briz K weighed 3 tonnes, twice that when fuelled. This was a versatile stage, which could fly for 7 hours and restart six times, dispensing different satellites into different orbits. Rockot can place up to 1.85 tonnes in low Earth orbit.

Although there were several Rockot pads in northwest Baikonour, they were all underground silos and this presented a number of problems. The noise circulating in the silo during the first two or three seconds of lift-off is tremendous and there were fears that it would damage the satellite payload. At the same time as the Russian Space Agency was considering this problem, the Kazakh authorities got wind of the proposal and demanded a share of the profits of every Rockot mission out of Baikonour, with the result that the Russians threatened to move all Rockot launches to Plesetsk.

Accordingly, a non-silo, open pad was built at Plesetsk. This involved the construction of a large service structure, unusual in the Russian programme. The Rockot is wheeled up to the structure in a vertical position, whereupon it is embraced by the service tower. Here, the payload is lifted by crane and placed on top of the bottom two stages. This was a procedure familiar to most space

Rockot

programmes the world over (e.g. the United States, Japan, India, China), but was new to the rail-based Russians.

Rockot's first stage uses a RD-233 motor, the second stage the RD-235. The original third stage was the Briz K, but from 2000 the Briz KM, which is relightable, was introduced, along with a larger fairing. Fuels used are UDMH and nitrogen tetroxide. The diameter is 2.5 metres, height 29 metres and launch weight 107 tonnes.

Rockot was fired for the first time from Baikonour on 26 December 1994. Rockot placed a Rosto radio satellite into orbit, but the test was somewhat marred by the explosion of the third stage 3½ hours after lift-off, scattering débris in a 2,000-km orbit at 64.6 degrees. A second demonstration launch, this time entirely successful, was made in 1999 from the new pad at Plesetsk.

In 1995, Khrunichev formed a company with German Daimler-Benz Aerospace to market Rockot, offering it at €7m a launch. Within five years, it had built up an order book of 12 launches for €200m. The company bought 45 old Rockots from the Russian strategic missile forces in order to build its inventory. In 2000, Eurockot was part-bought in turn by the German company Astrium GMBH, a shareholder of Arianespace (51%) and Khrunichev (49%). The next launch is the US–German GRACE project in 2001. In late 1999, it won an order for 48 small satellites in the Leo One Worldwide messaging service and is considered to have a bright future.

Key features: Rockot

Stage	Engines	Makers	Fuel
1	RD-233	Khrunichev	N_2O_4/UDMH
2	RD-235	Khrunichev	N_2O_4/UDMH
3	Briz K/KM	Khrunichev	N_2O_4/UDMH

Rockot launches

26 Dec 1994	Rosto	Rockot	Baikonour	1,875–2,253	128.66	64.8
16 May 2000	Simsat	Rockot	Plesetsk	544–558	95.7	86.4

KOMPLEKS' TOPOL AND PIONER (START 1 AND START 2)

Two other converted military launchers were the Pioner or the SS-20 (NATO designation) or Start 1 and its relative the Topol or SS-25, called the Start 2. The term 'Start' was derived from the strategic arms reduction talks. Both were conceived in the 1960s by missile designer Alexander Nadirazhde (1914–87). The SS-25 was an important part of the Soviet Union's nuclear strike force, with 288 missiles deployed in silos at nine locations.

Once the cold war ended, the Moscow Kompleks Technical Centre converted them into modern, solid-fuelled rockets which could place 700 kg class satellites into orbit, with a commercial price around $10m. Start 1, Pioner, is four-stage and Start 2, Topol, five-stage. In a break with tradition for Russian satellite launchers, both were solid-fuel rockets. Although solid fuels are an important part of the space programmes of some countries (e.g. India, Japan), they had played little part in rocket development in the Soviet Union, which had relied on liquid fuels being fed through valves under pressure into combustion chambers for burning.

Solid-fuel rockets were based on a sludge-like grey substance poured into the rocket tube: they were powerful but less accurate and, once ignited, could not be turned off, burning to depletion.

The first Start 1 launch took place on 25 March 1993 from Plesetsk, placing a small test satellite in orbit and marking the first orbital solid rocket launch by Russia. Start 1 was later the launcher used in the inaugural flight from Svobodny cosmodrome in 1997 of the small Zeya payload. Start 1 was then used to launch an American Early Bird high-resolution earth observation satellite into orbit on 24 December 1997 from Svobodny cosmodrome. Unhappily, contact was lost with Early Bird, designed to provide 3-metre resolution maps, after four days, but this was not attributable to any fault of the launcher.

Start 2 was less successful. At first, all appeared to go perfectly at Plesetsk on 28 March 1995. The modified rocket carried a Russian satellite called EKA2, a Mexican meteorite detection system and a private Israeli satellite called Gurvin. The fourth stage shut down 12 seconds prematurely and the fifth stage did not ignite, the payload impacting into the Lama river near the Laptev Sea. The loss of Gurwin was especially unfortunate as it was a prototype of a new, versatile, low-power, low-cost platform. Gurwin was a 52-kg box with a miniaturized set of instruments able to measure ozone depletion in general and, something of particular interest to a middle-east country, the long-term stratospheric effect of the burning of the Kuwaiti oil fields set alight by Saddam Hussein's retreating army. Gurwin also had a CCD Earth observing camera and a store-and-forward communications system for radio amateurs.

Start 1/Pioner launches

25 Mar 1993	Start 1	Start 1	Plesetsk	684–970	101.44	75.76
4 Mar 1997	Zeya	Start 1	Svobodny	467–480	94.06	97.28
24 Dec 1997	Early Bird 1	Start 1	Svobodny	461–528	97.3	93.3

Start 2/Topol launches

28 Mar 1995	Gurwin UNAMSAT EKA 2	Start 2	Plesetsk	Failed to reach orbit

DNEPR

Dnepr was a converted SS-18 Satan cold war rocket. Built by Yuzhnoye NPO, it was deployed from 1974 and production ceased in 1991. Like the Rockot, Dnepr was designed to be fired in rapid succession from silos. As a civilianized missile, Dnepr was able to put much larger 4,000-kg payloads into Earth orbit and the price suggested by Yuzhnoye was $20m to $40m.

Almost 150 of these missiles were left over at the end of the cold war, with strategic arms limitation agreements and probably safety requiring their elimination by 2007. The Dnepr had five silos at Baikonour. Dnepr was ready for a test mission by 1999. Its first payload was a small satellite made by the great British experts in small satellite design, the University of Surrey. UOSAT 12 entered an orbit of 660 km, 64.5 degrees, 97 minutes and carried a radio experiment, GPS receiver and Earth-observing cameras. A second Dnepr that September launched a 55-kg Malaysian satellite called Tiunsat 1.

Several other rockets, derived from cold war missiles, have been offered to Western customers from the late 1990s. These have been given a range of names: Kvant and Strela being just two examples. Kvant is a small two-stage launcher proposed by NPO Energiya, the first being RD-120M powered and the second using the block DM to put five tonnes into low Earth orbit. Strela is a small launcher based on the SS-19 and similar to the Rockot, able to put 1.8 tonnes into low Earth orbit. Their exact status is uncertain.

Key features: Dnepr

Stage	Engines	Maker	Fuel	Thrust
1	4 × RD-264	Yuzhnoye/Energomash	N_2O_4/UDMH	461,163 kg
2	RD-288	Yuzhnoye/Energomash	N_2O_4/UDMH	77,541 kg
3	RD-864	Yuzhnoye/Energomash	N_2O_4/UDMH	8,101 kg

Dnepr launches

21 Apr 1999	UOSAT 12	Dnepr	Baikonour	649–652	97.75	64.56	Test

SHTIL AND RELATIVES

Rockot, Start and Dnepr were land-based missiles. The end of the cold war left the Russian Federation with a surfeit not just of land-based missiles but of sea-based rockets, the equivalent of the American Polaris and Poseidon missiles. Several hundred were available in a number of different categories—for example, 112 in the Shtil class alone.

Consideration was given to the use of submarine-launched missiles as a means of putting satellites into orbit. This had a number of advantages: it was a useful means of disposing of old missiles, provided work for the grounded Russian navy and launch pads were already in existence. Moreover, a submarine-launched satellite could be launched from any (watery) latitude on Earth, facilitating direct access to the intended orbit, especially equatorial orbit.

The Makeev design bureau, based in Chelyabinsk, was the principal maker of these solid-fuelled submarine-launched ballistic missiles. Makeev offered them to

Western companies and scientific bodies for orbital and suborbital flights. Several versions of the submarine-launched ballistic missiles were put forward: Vysota, Shtil, Skif, Surf, Priboi, Volna and Riksha.[67] Vysota used the 30-tonne RSM-40 cold war missile and could put a 130-km payload into circular orbit of 250 km. The Volna performance was much the same, but Surf was a different concept: the rocket was much bigger (104 tonnes) and could launch twice the payload (up to 2.4 tonnes). It was too large for a submarine tube and instead was towed into position in a container which floated vertically in the sea.

The first test launching from a submarine took place on 7 June 1995. This was the Volna, based on the RSM-54, 14 metres long, 40 tonnes in weight and 1.9 metres in diameter. This first test was a suborbital mission, carrying a German science package and it came down in the far east 5,600 km away 30 minutes after lift off. A Kalmar-class submarine, commanded by Captain Vladimir Bashenov, was used to fire the rocket from 50 metres below the Barents Sea near Yagelnaya Bay. The Volna carried a small 700-kg capsule with a microgravity payload developed by the Bremen University Space Technology and Gravitation Centre. The Germans were not allowed on board the submarine but helped to fit it at the naval base of Severomorsk. While arcing 1,270 km over Siberia, the Volna's capsule was able to provide almost 20 minutes of zero-gravity for experiments into electrical fields and the behaviour of fluids.[68]

The first orbital test came on 7 July 1998. A Delta 4 class nuclear submarine, the *Novomoskovsk*, under the command of Capt. Alexander Moiseyev and his 130 crew, dived off the coast of Murmansk and lay below the Barents Sea as the crew worked through the launching drills. The submarine then launched a Shtil missile towards orbit from below the Barents Sea. The rocket curved into the northern sky, placing in 400–773 km orbit, 78.9 degrees, 96.4 minutes two tiny satellites from the Technical University of Berlin. Tubsat 1 was 8.5 kg and Tubsat N1 even smaller, 3 kg. Not many details are available of Shtil or its performance—though it was amply clear that the concept was viable.

Shtil launches

7 Jul 1998	Tubsat N, N1	Shtil 1	Barents Sea	401–777	96.45	78.92	Demonstration

ANGARA

Development of a new large rocket began in 1995, the futuristic Angara. Approved by the Russian government in 1994, first funding began to flow the following year. At a time when ever-newer designs of prospective Russian rockets kept appearing at international air and space shows, it was difficult to know whether Angara should be treated any more seriously than the others—and it was not. However, the approval announcement had two key phrases: the first was that this new launcher would be the primary Russian rocket until 2030 and that pads would *not* be built at Baikonour.

Clearly, Angara was directed towards long-term Russian launcher independence, in particular from Kazakhstan. The seriousness of the Angara project was apparent in 2000 when the first hardware appeared on the floor of the Khrunichev factory in Moscow. Angara will be operational in 2003–2005 and will be the mainstay of the Russian rocket fleet for many years.

Angara builds on the Zenit system and uses the successful liquid oxygen and kerosene RD-171 Zenit engines for the first stage, though the engine is renamed the RD-174 and the RD-191. The second stage uses a high-performance liquid oxygen and hydrogen RD-120M engine. The upper stage will be either the Briz M or the ultra-powerful KVD-1 motor developed in the course of the Moon race in the 1960s.

One of the problems in reporting on the Angara rocket is the many design evolutions through which it had passed over the past four years (a record 18). Currently, four versions of the Angara are under construction, immemorably called versions 1, 1.3, 3 and 4. Angara will replace a number of rockets in the Russian launcher fleet: the Angara 1.1 will take the place of the Cosmos, Angara 1.2 the Tsyklon, Angara 3 the Zenit and Angara 4 the Proton. Angara will mark the most comprehensive programme of rocket replacement in the space programme, Soviet or Russian. It will reduce dependence on components made in the Ukraine and in the case of the Proton discontinue the use of toxic fuels.

In the mid-1990s, it seemed that Angara would be one of many paper projects from the period and that the high-performance Proton would continue to be modified and remain Russia's prime heavy launcher. Plans for the 10% more powerful Proton M confirmed this theory. However, this is not what happened, and Angara won instead. There seem to have been two reasons. First, there was growing political concern about the dangers posed from the fallout of toxic fuels from Proton, a fear underlined when Proton suffered its two accidents in 1999 and Kazakhstan made large clean-up compensation claims. Second, Proton was based on Baikonour. In the event of Baikonour continuing to prove ruinously expensive to Russia, new Proton pads would have to be built in Plesetsk and Svobodny. It was easier to convert the already building Zenit pad in Plesetsk to fire the Angara, with which it had many features in common, than to build fresh Proton pads there.

The table below summarizes the capacities of the Angara versions on offer. Although they present as one launcher family, the capacity of the four versions varies enormously. Angara 1.1 is a lightweight launcher while the Angara 4 is a good 25% more powerful than the Proton. The table illustrates the current versions under construction (the names and figures have undergone numerous changes and they may continue to do so).

Although lighter than the Proton, the Angara has better performance and can lift much heavier payloads. Working from the new pad 35 at Plesetsk, the two-stage version is to place between 26 and 28 tonnes into 63-degree orbits; and the three-stage version, 4.5 tonnes into geostationary orbit.[69]

The Angara takes substantial advantage of existing technology. The engines, taken from the Zenit/Energiya system, have already been proved. Angara 1.1 uses

Angara (basic version)

Angara launchers

	Angara 1.1	Angara 1.2	Angara 3	Angara 4
Lift-off mass (tonnes)	145	179	464	752
Payload	1.7 to LEO	3.4 to LEO	15.2 to LEO 4.5 to GEO	28 to LEO

Angara reusable stage

the Rockot nose fairing while Angara 1.2 uses the Soyuz ST fairing, which is in turn based on the Ariane 4.

The 1.2 version is reusable. After staging, the first stage develops wings and becomes a glider, coming back to Pero airfield where it is made safe, any remaining propellants removed, and preparations begin for the next mission. Also borrowing systems from elsewhere, Angara 1.2 uses the undercarriage of the Sukhoi 17 jet fighter and a manoeuvring jet engine taken from the Yakovlev 130 jet.[70]

Thus by 2000, Russia had maintained its rocket fleet, introduced new types based on cold war missiles (Rockot, Dnepr, Start and Shtil) and began construction of a new rocket which would roll over most of the old rocket force, the Angara. Russia also took a number of steps to get extra performance from its existing rockets. The improvement of rocket upper stages is not the most glamorous form of rocketry, but for a programme under severe financial pressure, it offered the promise of low-cost enhanced performance.

NEW UPPER STAGES: IKAR, FREGAT, ST

Centre of attention for upper-stage improvement was the venerable R-7 Soyuz rocket. Originally, it had been proposed to phase out the R-7 and replace it with a completely re-engineered model, variously called the Soyuz 2 and the Rus. In the event, it proved almost equally effective to re-engineer upper stages from earlier programmes and use them to provide the extra thrust necessary to push heavier payloads into orbit. The three upgrades proved entirely successful and it is doubtful now if the Rus or Soyuz 2 project will proceed.

The first upgrade was called Ikar (in Russian), Ikare (in French) or Icarus (in English). Ikar already existed: a throwaway remark by one of the designers suggested that it had already been used in the Yantar programme. Ikar, as a civilianized version, offered a perfect method for getting groups of communications satellites into 1,400-km orbit. Soyuz–Ikar won the competition for part of the American Globalstar network of low Earth-orbiting comsats, partly because of the R-7's known reliability and the price on offer, less than $40m, compared to the American equivalent, the Delta 2, which would have cost nearly $60m. Winning the competition provided the additional resources and investment necessary to redevelop the Ikar as an operational upper stage for the Soyuz.

Ikar was 2.9 metres long, 2.72 metres in diameter, weighing 3.29 tonnes, with 900 kg of UDMH and nitrogen tetroxide fuel and able to get 3,300 kg into high Earth orbit. The motor was a 17D61 able to generate 2,943 kN and was equipped with 16 steering thrusters. Ikar was adapted with a 390-kg dispenser to spring each of four satellites into their appropriate orbits. All six Ikar Globalstar launches went off with perfect precision.

Ikar launches

9 Feb 1999	Globalstar (4)	Soyuz/Ikar	Baikonour	1,337–1,356	112.61	52
15 Mar 1999	Globalstar (4)	Soyuz/Ikar	Baikonour	1,409–1,417	114.07	52
15 Apr 1999	Globalstar (4)	Soyuz/Ikar	Baikonour	902–944	103.48	51.94
22 Sep 1999	Globalstar (4)	Soyuz/Ikar	Baikonour	901–957	104.36	51.98
18 Oct 1999	Globalstar (4)	Soyuz/Ikar	Baikonour	1,394–1,419	113.93	51.96
22 Nov 1999	Globalstar (4)	Soyuz/Ikar	Baikonour	897–942	103.41	51.97

Fregat was a new upper stage introduced in 2000 to facilitate the placing in orbit of the European Space Agency's Cluster satellites to study the Sun. Like Ikar, Fregat had historical antecedents—back in the Phobos programme in 1988 where it had first been developed. Fregat had eight tanks carrying 3,000 kg of fuel, 28 attitude control thrusters, and a rocket motor which could be used for up to 20 course corrections. Made by Lavotchkin, it was 1.5 metres tall, 3.35 metres in diameter and could burn for up to 877 seconds. It offered the perfect solution to the European Space Agency's problem of how to orbit its Cluster series of satellites. Again, the ESA contact provided the extra resources that enabled Fregat to be redeveloped as an operational system.

The original Cluster series had been lost when the first Ariane 5 exploded in a giant fireball over Korou, Guyana, on its maiden flight in 1998. Now the European Space Agency turned to Russia for help to put a set of back-up models into the appropriate orbits. The agency required an upper stage with considerable thrust and versatility. Fregat offered the possibility of several restarts and could put 5-tonne payloads into precise orbits as high as 450 km.

Cluster: winning test of Fregat

In preparation for the Cluster launch, two demonstration tests were carried out. The first Soyuz–Fregat was duly launched on 9 February 2000 and carried out its two demonstration burns, one at 200 km, the other at 600 km. A month later, on 20 March, Fregat went through its paces again, with repeated engine restarts and a mock separation of the payload.

In a special experiment, the first Fregat carried a novel re-entry system. After five hours, both the Fregat itself and a smaller cargo were de-orbited. As they came in through re-entry, each deployed a rubber inflatable heat shield at 150 km designed both to protect the cargo during re-entry and to cushion its impact on the ground. Fregat's inflatable cone, 14 metres in diameter, was based on devices built to soft-land the ill-fated Mars 8. The empty Fregat itself weighed 1,800 kg so a substantial amount of protection was clearly required. The other payload was the Inflatable and Re-entry and Descent Technology (IRDT) heatshield which had been developed by Daimler Chrysler, Khrunichev and the European Space Agency at a cost of €1.8m. This was a much smaller test, using a 4-metre-wide cone and, inside it, a 110-kg payload. This carried different types of rock—basalt, dolomite and artificial cement/carbonate—to see how meteorites reacted to the heat of re-entry. The cones were made of silicone-based shields and stiff ablative material.

The effect of inflating the cones was to reduce the speed of re-entry from 5,500 to 200 metres per second. At 30 km, the cones fully inflated in order to further reduce the speed of descent, now in the atmosphere. The actual touchdown was cushioned by an inert gas-filled air bag. The smaller cargo was quickly found in the snow near Orenburg and thermometers suggested that the highest temperature it experienced was 25 °C. The larger cargo, the Fregat motor itself, was never found, although the telemetry suggested that it did make it through re-entry.[71]

Following the second test in March, Cluster was eventually launched from Baikonour on 16 July. By this stage, the Cluster satellites had received names, the British winner proposing the names Rumba, Salsa, Samba and Tango to reflect the way the satellites would dance in formation around the heavens. Fregat fired 90 minutes into the mission, putting the first two Cluster satellites into a parking orbit of 240–18,000 km, the first of six firings to send them to a final operational altitude of 19,000–119,000 km.

Fregat launches

8 Feb 2000	Dumsat/Fregat	Soyuz/Fregat	Baikonour	581–607	96.56	64.86	Demo
20 Mar 2000	Dumsat/Fregat	Soyuz/Fregat	Baikonour	243–18,021	320	64.64	Demo
16 Jul 2000	Cluster 1	Soyuz/Fregat	Baikonour	240–18,000		64.8	Commercial
9 Aug 2000	Cluster 2	Soyuz/Fregat	Baikonour				

The next step in the upper-stage development was called the Soyuz ST. This was the Fregat but with the shroud from the Ariane 4. Its lightness and shape gave an improved performance going up through the atmosphere, increasing the payload to 450 km orbit from 5 tonnes to 5.5 tonnes.

Further plans still exist to upgrade the R-7. These involve the replacement of the core stage and strap-ons by RD-120M engines and the replacement of the second stage by a new engine, RD-124 from KhimAutomatiki, topped by the Fregat upper stage. The RD-124 engine is 2.1 metres tall weighing 360 kg, able to operate at 165 atmospheres. Yamal is a proposal from the TsSKB Design Bureau in Samara. Although the R-7 would look much the same, the Yamal would widen the main stage from 2.05 to 2.66 metres and the upper stage from 2.95 to 3.44 metres. Instead of the proposed RD-120M engines, it would use the NK-33 engines from the N-1 1960s Moon rocket. Both developments are still some time ahead, if they take place at all.

INTRODUCING THE BRIZ UPPER STAGE

Equally important to the upgrading of the Soyuz was the introduction of new and more capable upper stages for satellites going to 24-hour synchronous orbit on the Proton rocket. More powerful upper stages offered the promise of heavier payloads (or more distant destinations) without the cost of introducing entirely new rockets.

The standard upper stage was the block D, originally designed as the final stage of the N-1 rocket to bring the first Soviet cosmonauts to the Moon. A squat cylinder, block D weighed 33 tonnes, including casing, fuel tanks, fuel, motors and instrument unit. It was transferred to the Proton programme (still called block D) with a new version introduced from 1975 for satellites going to geosynchronous orbit (block DM, then DM-2). Subsequent minor variations, reflecting control systems rather than changes in the engine design, were introduced from 1982 (block D-1, D-2, D-2M).[72] Block DM-2M was used for Gals, Ekspress, Kupon and Prognoz; block DM-3 and 4 have been used for international commercial missions; and block DM-5 was for Arkon. For the Iridium missions, the block DM-2 was modified with a wider payload adaptor, new pneumatic and hydraulic subsystems and changes to the oxidizer tank.

For the Russian space programme, the 1990s were not happy years for the block D and it encountered more than its fair share of upper stage failures, a problem which most experts thought had been left behind in the 1960s. A block DM failed on the Raduga launch of February 1996 and, notoriously, failed on Mars 8 that November. On 25 December 1997, a block DM failed 1 second into a 110-seconds burn on a commercial Proton launch, stranding Asiasat 3 in a useless orbit of 36,000 by 203 km and forcing a $200m insurance payout (in the end Hughes took over the satellite and, in an ingenious set of manoeuvres never done before, looped it around the Moon with its small motor and eventually settled it in a usable orbit).

These mishaps re-emphasized the desirability of introducing a new, more powerful and reliable upper stage. Here, the new upper stage called Briz M made its appearance.

Briz is a new upper stage, made by Khrunichev, which can be restarted up to 20 times. Briz M was introduced in 1999, to provide additional power for the Proton rocket and with a view to future use on the Proton M and Angara launch vehicles. It is the shape of a flattened doughnut, with the engine, the central unit and instrumentation surrounded by a toroidal fuel tank. The engine provides a slow-burning capacity to put up to 6.6 tonnes into geostationary transfer orbit. What had been intended as the first flight of the Briz M took place on 5 July 1999. However, whether it would have worked was never ascertained, for the Proton crashed before the engine could be lit. This was an inauspicious start to the attempt to replace a problem engine!

Despite this setback, Briz KM, designed for the smaller Rockot launcher, made a successful debut the following year. The small Rockot launched out of Plesetsk on 16 May 2000 and used the Briz KM to put two small payloads, called Simsat, into a 540-km orbit. The Briz M followed three weeks later without a hitch, bringing the last Gorizont to its destination.

Briz launches

5 Jul 1999	Raduga	Proton	Baikonour	Second-stage failure (first Briz M)
16 May 2000	Simsat	Rockot	Plesetsk	Entered orbit—first Briz KM
6 Jun 2000	Gorizont 33	Proton	Baikonour	First successful Briz M

Thus, by 2000 Russia had not only introduced new upper-stage systems for the Soyuz (Ikar, Fregat and ST forthcoming) but was in the process of the replacing the block D with a new, more powerful upper stage, the Briz. What other projects were scheduled? Even if they have not materialized, they give an indication of future Russian space interests and capabilities.

NEW PROJECTS: SKYLIFTER

Several new launching systems have been in design since the early 1990s, but many of these interesting projects remain stranded due to lack of funding. At the 1995 Paris Air Show, for example, the Tupolev 160 Blackjack strategic bomber turned heads when it appeared with a new air-launched rocket, the Burlak, designed to be flown to altitude by the Blackjack where it would be released to carry on to orbit by itself. The Blackjack/Burlak combination could put either 800 kg into polar orbit or 1,100 kg into conventional orbit. Burlak was a two-stage, 32-tonne, 25-metre-long rocket. The concept was similar to Pegasus, a private American air-launched rocket, which since 1990 had been placing small satellite payloads into orbit from an airfield in California, brought to altitude by a Lockheed Tristar civil airliner. Four years later, the Ukranian government sold its cold war Tupolev 160 Blackjacks stationed at Priluki airfield to a California company, Platforms International, for the further development of this and similar concepts.

A successor project to Blackjack/Burlak is Skylifter. This is a project devised by the Yakovlev aircraft company to convert the Tu 160 to the carrier of satellite-launching demilitarized missiles to be dropped at altitude. It is a much larger idea than the Burlak, involving the Tu 160 fuselage, the wings of the Antonov 124, the undercarriage of the even bigger Antonov 225 and other parts taken from the Yakovlev 40. There would therefore be a large, central carrying area 24 metres wide to hold the rocket, so large that ignition could take place while still attached to the aircraft. The Skylifter would bring the payloads up to 10,000 metres at a speed of mach 0.75 before the drop. Yakovlev estimates the investment cost of the project to be $100m and that Skylifter could put 1,200-kg payloads into orbit for $5m.[73] The Molniya company has put forward an even larger proposal, the 450-tonne Heracles.

MAKS AND AIR LAUNCH

The cancellation of the Buran space shuttle, while heart-breaking for those involved, forced some of the participant companies to explore whether they could salvage anything either materially or conceptually from the project. The most promising proposal was put forward by the Molniya design bureau: called MAKS (*Mnogotse-levaya Aviationno Kosmicheskaya Sistema*, or multipurpose aerospace system), the proposal envisages a small space shuttle being launched from the upper atmosphere by the plane that carried Buran, the Antonov 225.

MAKS comprises a small orbiter and teardrop-shaped tank brought to altitude by the huge Ukranian-built Antonov. The MAKS stack of orbiter and tank is 36.3 metres long and weighs 275 tonnes, of which 237 tonnes is propellant. The orbiter, based on the Spiral spaceplane design, is 19.3 metres long, has a wing span of 12.5 metres and weighs 27 tonnes. The system runs on liquid oxygen, kerosene and liquid hydrogen which are held in the aluminium lithium alloy tank. Designed by Boris Katorgin at NPO Energomash, MAKS uses a RD-701 engine which is a closed-cycle, twin-chambered tripropellant with a thrust of 4,000 kN and a chamber pressure of 300 atmospheres. The two-person shuttle could carry a payload of 8.5 tonnes into a 200-km, 51.6-degree orbit.

The MAKS orbiter was similar in size and shape to the European Space Agency's small shuttle, the cancelled Hermes. The mission profile for MAKS is for the orbiter and tank to be ferried to an altitude of 9,000 metres, possibly some distance from the airfield (the Antonov's range is no less than 10,000 km). MAKS is then released for its upward flight to orbit. During the later stages of the ascent, the MAKS switches from tripropellant to liquid hydrogen and oxidizer only so as to get the most efficiency out of the two different fuels.

MAKS was available in three versions: a manned orbiter (MAKS-OS), a reusable unmanned cargo carrier (MAKS-M) and an expendable cargo carrier able to ferry 18 tonnes into orbit. In the course of 1993–4, under a project funded by the European Space Agency, a simplified version of MAKS-OS was designed, using liquid oxygen and kerosene fuel rather than tripropellant. Despite its promising design and several years in preparation, the MAKS project had found no customers and little money by

MAKS on Antonov transport

2000—although Molniya had by no means given up and was confident that it would fly one day.

By 2000, another project had re-emerged in a new form as 'Project Air Launch', presumably as a counterpoint to the already successful Sea Launch. An Air Launch company was formed by RKK Energiya, KB Polyot and the Ukranian Antonov design bureau to launch satellites up to 2.5 tonnes by the smaller (but still enormous by any standard) Antonov 124-100 Ruslan. They claimed that, using a new launcher, they could bring the cost of a launch down to $20m or $5,000 to $6,000 per kg, two-thirds the cost of Sea Launch. They could launch from anywhere in the world within range of an An-124 airfield. Unlike MAKS, which was based on the top of the Antonov, Air Launch carried a much larger rocket inside the hull of the aircraft. The Antonov would fly to an altitude of 11,000 metres, where it would perform a zoom manoeuvre and the 100-tonne two-stage launcher, called Polet, would be dropped backwards out of the cargo hold, stabilize vertically with a parachute and, once the Antonov was safely out of range, fire its engines upwards. Staging would take place at 72 km.

By 2000, the Air Launch project had carried out computer simulations of the launching. The new rocket's first stage would be powered by NK-43 engines, built by Progress in Samara, and the second stage by the 11D58MFD engine made by Motorostroitel. Air Launch would have a capacity of 400 kg to Earth escape, 800 kg to geosynchronous transfer obit, 2.8 tonnes to polar orbit and 3.6 tonnes to 24-hour orbit.

RUSSIA'S ROCKETS IN PERSPECTIVE

During the 1990s, Russia maintained and modernized its rocket fleet. Two rocket production lines were run towards conclusion: Tsyklon and Cosmos 3M. One subtype was withdrawn, due to a fuel problem (Soyuz U2). The old R-7 rocket, flying in its two versions of Soyuz U and Molniya M, received a new lease of life with Ikar, Fregat and ST upper stages. Proton won significant commercial contracts and by century's end was accelerating its production line. The Zenit, despite a series of apparently unconnected failures, continued to fly the Russian Tselina fleet and began a promising career as a seaborne launcher. Four new military-based rockets were introduced: Rockot, Pioner/Topol, Dnepr and Shtil. A project for a new universal rocket, the Angara, drew close to its first launch.

The table below, which shows the changing rates of launch of the different Russian rockets of the period, confirms the dominance of the Proton (70) and the Soyuz (107) rockets as the mainstay of the Russian rocket programme, with minor and niche roles played by Molniya M, Cosmos 3M, Tsyklon, Zenit and others. But how reliable were they in global perspective?

Launch rates of Russia's rockets

	1992	1993	1994	1995	1996	1997	1998	1999	2000	Total
Soyuz U	24	17	15	12	4	10	8	10	7	107
Molniya M	8	8	3	4	3	3	3	2		34
Cosmos 3M	7	6	5	5	4	2	2	2	2	35
Tsyklon 2/M, 3	5	4	7	1	1	2	1	1		22
Proton	8	6	13	6	9	7	7	6	8	70
Zenit 2	3	2	4	1	1		2	1	1	15
Zenit Sea Launch								2	1	3
Others		1	1	1			1	1	1	6

To 14 August 2000.

Soviet rockets experienced many difficulties during their early years and there was a high failure rate on the lunar and interplanetary missions in the 1960s. This was not a uniquely Soviet phenomenon, for similar problems afflicted early American and European efforts—the Americans, for example, took several years to tame their powerful Centaur rocket. Similarly, Russia took almost 10 years to tame the upper stage of the Molniya (block L). The Proton block D upper stage was very troublesome, some problems returning in the 1990s. Even years after a rocket is introduced, there will be random failures, as the Americans found to their cost in the late 1990s.

To make rockets commercially marketable, or to attract scientific payloads, high standards of reliability are now required, generally in the order of 95% if insurers are to be satisfied. Russian rockets have now achieved very high standards of reliability—the commercial claims are 98% for the Soyuz, 96% for

The new Angara rocket

the Molniya, 96% for the Tsyklon 2, 100% for the Tsyklon 3, 97% for the Cosmos, 86% for Proton (this includes its development phase, for recent rates have been higher) and 71% for the Zenit. Top of the class was the retired Soyuz U2, with 100%.

The table of launch failures shows that there were 11 Russian failures over 1992–2000. Of these 11 failures, four were Zenit, two Soyuz U, one Tsyklon M, one Start and three Proton. There were no Cosmos failures, although there was one poor orbital insertion.

Launch failures 1992–2000

Data	Payload	Rocket	Launch site	Circumstances
5 Feb 1992	Tselina 2	Zenit	Baikonour	Explosion at lift-off
27 May 1993	Gorizont	Proton	Baikonour	Ran out of fuel/contaminated fuel
25 May 1994	US-P EORSAT	Tsyklon M	Plesetsk	Second and third stages failed to separate, crashed over eastern Siberia
28 Mar 1995	Gurwin UNAMSAT EKA 2	Start	Plesetsk	Premature fourth-stage shutdown Fifth stage failed to ignite
14 May 1996	Yantar Kometa	Soyuz U	Baikonour	Shroud broke at 49 seconds
20 Jun 1996	Yantar Kobalt	Soyuz U	Baikonour	Shroud broke at 50 seconds
20 May 1997	Tselina 2	Zenit	Baikonour	Failed at 48 seconds
9 Sept 1998	Globalstar 12	Zenit	Baikonour	Second stage failed at 280 seconds
5 Jul 1999	Raduga	Proton	Baikonour	Second stage failure (first Briz M)
27 Oct 1999	Ekspress A1	Proton	Baikonour	Second stage failed at 222 seconds
12 Mar 2000	ICO 1	Sea Launch	Kiribati	Second stage failure

Two of these failures could have been avoided, had committees of investigation been more effective. The two Soyuz U failures arose from the introduction of a new glue on the launcher shroud and the problem should have been identified. Similarly, two Proton failures were due to poor manufacturing procedures in a particular batch of engines and this should likewise have been identified the first time. The Tsyklon failure was an unusual one in an otherwise tested and proven series. Enough has already been said about Zenit, and the following table lists the overall reliability rates for current Russian launchers.

Reliability rates of operational Soviet/Russian launchers

Launcher	Introduction	Successes	Failures	Reliability (%)
Proton	1965	277	37	86.7
Soyuz and variants	1963	1,088	8	99.3
Cosmos 1, 3, 3M	1964	416	17	96
Tsyklon 2/M	1969	124	4	97
Tsyklon 3	1977	114	5	96
Zenit 1,2,3	1985	36	7	80
Rockot	1994	2	0	
Pioner	1993	3	0	
Topol	1995	0	1	
Shtil	1998	1	0	
Dnepr	1999	1	0	

Data for 14 August 2000. Series no longer in service not included.

The Proton reliability figures are in fact higher than these data suggest, since most of the Proton failures took place during its troublesome early phase. Of the 37 Proton failures, only three were during the 1990s. Again, the Zenit's poor performance is evident.

A more helpful indicator of current reliability is to look at performance over 1992–2000. Of 300 launches during the period of the Russian space programme after 1991, there were 11 failures or 3.6%. This gives an overall reliability rate of 96.4%.

8

The rocket engine frontier
Rocket engines in Russia from 1992

At the heart of a good space programme is a good rocket engine, once remarked Sergei Korolev. Engines are the single most important element in a rocket's design and are crucial for high performance and reliability. Was Russia able to maintain and develop the achievements of the Soviet Union in rocket engine design?

Rocket engines were very much the starting point of the entire Soviet space effort, dating back to the foundation of the Gas Dynamics Laboratory. Engine design was made a priority from the start and the engine design bureaux attracted the most talented of the Soviet Union's engineers. Leadership of the space programme often came from the engine design bureaux, most evident when the greatest engine designer of them all, Valentin Glushko, became the chief designer of the entire programme from 1974.

Russia's rocket engines have been designed in four main bureaux.[74] The most important one is the Gas Dynamics Laboratory (GDL), now NPO Energomash; the others are the Kosberg bureau (Khim Automatiki); the Isayev bureau (KhimMash); and the Kuznetsov bureau. Here, their work and role in the new Russian space programme are reviewed.

GDL: THE MOST POWERFUL ROCKETS IN THE WORLD

The most famous rocket engine design bureau was GDL, whose history has been intimately associated with Valentin Glushko. As a design bureau, it went through many evolutions, becoming merged with OKB-1 when Glushko became chief designer, but going its own way again as NPO Energomash, its current name, or, to be more precise, *Energomash imemi Valentin Glushko* (Energomash, dedicated to the memory of Valentin Glushko). It is probably still the biggest engine design bureau and 9,000 people work in its Khimki headquarters and subsidiaries. One could say that it is the world's foremost rocket engine design bureau, having produced a total of 52 operational engines during the twentieth century. Its engines are considered the best in the world—not just by the Russians, as one might expect, but by the Americans who are now using them to power their new generation of Atlas rockets.

These engines may be divided into three groups: the RD-100 series, which use liquid oxygen and non-storable propellants; the RD-200 series, which use nitrogen and storable propellants; and the RD-300 series, which use exotic propellants. RD stands for rocket motor (*raketny dvigatel*).

GDL has been making rocket engines since 1921 and made the engines that powered the Soviet version of the German A-4 rocket and subsequently the RD-107 and RD-108 engines used for the R-7 rocket, used initially to launch Sputnik and since then developed in the Vostok, Molniya and Soyuz versions. The RD-108 was used as the core stage on the R-7 (block A), the RD-107 on the four strap-on stages (blocks B, V, G, D). Each RD-107 weighed 1,155 kg dry and had a thrust of 1,000 kN.

The most recent operational engine using kerosene is the RD-171, used on the Zenit and as strap-on to the Energiya. The RD-171 weighs 8,755 kg and has a thrust of 7905 kN. The engine has a chamber pressure of 3,560 lb/in^2 and can be throttled from 40% to 105%. Its weight is 13 tonnes, height 4.3 metres and diameter 4.1 metres. An indication of the progress of rocketry may be gauged from the fact that chamber pressure rose from 16 atmospheres on the RD-100 to 250 atmospheres in the RD-171, the highest thrust of any Soviet engine. Versions of the RD-171 will be used to power the new Angara rocket.

GDL developed a second range of engines which use storable propellants. These were the RD-214 and RD-216, used on the R-12 and R-14 Cosmos rockets; the RD-219 used on the R-36 Tsyklon; and the RD-253 used on the UR-500 Proton (now made by the Motorostroitel company). Few have been built in the 300 series. An example was the RD-301, an experimental engine developed in 1969–76 using fluorine and ammonia generating a thrust of 98 kN for 750 seconds. It has not been used operationally.

POWERING THE ATLAS

American aerospace companies were quick to realize the potential of Soviet-designed rocket engines and establish commercial ventures. In 1992, NPO Energomash signed an agreement with America's prime engine manufacturer, Pratt & Whitney, to market its engines in the USA. In early 1996, a version of the RD-171 engine, called the RD-180, won the competition to power the new American Atlas III rocket, designed to be the main conventional booster rocket for the United States from 2000.

The RD-180 export turned into a great success story. Lockheed Martin, engaged in competition with Boeing for the American launcher market, bought 101 RD-180 engines from Energomash for $1bn. The programme for fitting the RD-180 to the

The Proton's RD-253 motor

new Atlas rocket was overseen by a joint company called Amross. Despite the cooperative nature of the venture, the Americans applied national security restrictions rigorously. The 25 Russian engineers overseeing the first launch of the new Atlas were not allowed in the Cape Canaveral control centre but made to follow the launch in an adjacent hangar under the supervision of the US Defence Threat Reduction Agency.

The Atlas III was designed to replace the Atlas II, which had been the workhorse of medium-size American satellites for many years. Ironically, it was based on the original Atlas which had been targeted on the Soviet Union since the late 1950s.

Now the Atlas was to be re-equipped with the engines of its adversary, both for the Atlas III and a later model, the Atlas V. Lockheed Martin reckoned that the RD-180 would give the company a 25% cost advantage on Boeing and other competitors.

The RD-180 was so powerful that its one motor replaced the two engines of the Atlas II—and provided 30% more thrust. Not only that, but it was the first American launcher, apart from the shuttle, that could throttle its engines, giving it another performance advantage. The throttle range was from 37 to 100%. The RD-180 had 15,000 fewer parts than previous Atlas engines, making a failure much less likely. Its use of oxygen enrichment in the early stages of the burn enhanced performance and kept running temperatures down.

After a string of first-launch failures, such as Boeing's new Delta and the European Ariane 5, Lockheed Martin engineers were naturally apprehensive as the first flight of the RD-180-powered Atlas III neared. Their nerves were not steadied by four aborted countdowns, called off due to weather, radar failures and, most irritatingly of all, pleasure boats straying up the Canaveral coast to the launch site.

By the time the first Atlas III had counted down to zero on 24 May 2000, the engine had already reached twice the engine pressure of any previous American launch vehicle, $3,700 \, lb/in^2$. At lift-off, the RD-180 was on 74% thrust, generating $288,716 \, kg$ thrust, kept deliberately low so as not to damage the launch complex. Only 5 seconds later, the RD-180 had accelerated to 92% thrust, burning an oxygen-rich mixture at the rate of a tonne of oxygen every second. Already, it was generating more thrust than the space shuttle with its huge solid rocket boosters and main engines combined.

Next came the crucial stage of maximum dynamic pressure, or max Q as the engineers call it. At this stage, the vehicle goes supersonic as it pushes through the densest layers of the atmosphere. Pressures on the launcher are so intense that there is a real danger of the vehicle breaking up—indeed, it was at this very point that the Challenger exploded in 1986. Now, at 33 seconds into its mission, the twin-nozzled RD-180 throttled back to 64% of thrust until 63 seconds as the vehicle went through this difficult phase. A minute after take-off, through the low atmosphere, the RD-180 was up to 87% and now rapidly accelerating skywards. In 3 minutes it had reached the same speed and altitude as the Atlas II had in 5 minutes. It was soon performing well over its 362,880-kg rated thrust. The Centaur upper stage then took over to bring the payload to geosynchronous orbit. The first-time success of the RD-180 left Lockheed and Energomash ecstatic.

KOSBERG BUREAU/KHIMAUTOMATIKI, BUILDER OF BRIZ

Formally known as the Design Bureau for Chemical Automatics or KB KhimAutomatiki, the Kosberg bureau worked initially on aircraft engines and then, from around 1956, rocket engines. The engines designed by Semion Kosberg (1903–65) are used for the upper stages of Russian rockets. The Kosberg design bureau was set up in Moscow in 1941, evacuated to Berdsk and then settled in the city of Voronezh after the war. The current anodyne name of 'chemical automatics' is not one loved by its current workers (it was probably another Soviet terminological deception, designed to distract Western analysts from its true purpose) and it is more commonly referred to as the Kosberg bureau.

The Kosberg KB developed the engines for the upper stages of the R-7, namely RD-0105 engine for the Luna probes, the RD-0109 engine for the Vostok upper stage, the RD-110 engine for the block I (Soyuz) and the RD-0107 engine (block L , Molniya), all of which used liquid oxygen and kerosene; and the upper stages of the Proton, namely the block B (RD-0210 and 0211), block V (RD-0213) and block D (RD-0225)), using nitrogen tetroxide and UDMH, which gave so much trouble in the late 1960s. More recently, Kosberg built the RD-0120 engine for the Energiya second stage.

The bureau remains in the forefront of rocket engine design and is currently developing a new generation of upper stages to replace the old block D of the 1960s—Briz. In particular, the Kosberg bureau has designed and built the new Briz M for the Proton rocket and the Briz K for the Rockot. The Briz, the bureau hopes, will bring new performance, versatility and reliability to the upper stages of Russian rockets into the new century. Other projects under development include a new oxygen–kerosene engine (RD-0124), a methane–oxygen engine (RD-0129) and tripropellant engines (RD-0120TD and RD-0700).

ISAYEV BUREAU/KHIMMASH: SECRET, HIGH-POWERED UPPER STAGE

The Isayev bureau was set up in 1943. The bureau started life as plant 293 in Podlipki in 1943, directed by Alexei M. Isayev and was renamed OKB-2 in the 1950s, being given its current name, KB KhimMash, in 1966. His engines were used in the rocket programme as small propulsion systems, manoeuvring and orientation engines. Besides spacecraft, its work has concentrated on long-range naval, cruise and surface ballistic missiles and nuclear rockets and by the early 1990s had built over 100 rocket engines, mainly small ones for upper stages, mid-course corrections and attitude control. Examples were the KDU-414 (used on the early planetary probes), the KTDU-1 (employed on Vostok), the KTDU-5A (used for the Luna soft-landings), the KTDU-53 (for Zond), the KRD-61 (for the Luna 16 ascent stage) and the KTDU-35 (used for Soyuz). The Isayev bureau made the systems of the Soyuz T series and the thrusters used on Salyut and Mir.

One of the most remarkable episodes of the Russian space programme was the emergence, out of the hangar, of a secret engine developed by the bureau in the 1960s. In the late 1980s, India was looking for a third-stage hydrogen-powered rocket motor able to get its satellites into 24-hour orbit. India especially wanted to orbit *Gramsat*, a multipurpose telecommunications satellite, to bring educational television to the villages of rural India. Knowing it would take at least 10 to 15 years to develop the technology itself, India sought help from abroad.

Originally, India made enquiries with Japan about purchasing the LE5, but nothing seems to have come of this. Hearing of these overtures, the Indians were approached by the General Dynamics Corporation in the United States offering an American engine. The cost was prohibitive as was an offer shortly thereafter from Europe's Arianespace. Just then, a third approach came, this time from the Soviet Union, offering two engines and technology transfer for the more reasonable price of Rs 2,350m or $200m.

The engine in question was the KVD-1, built by the Isayev design bureau. No one had heard of the engine, and the West was certain that the USSR had never been able to develop cryogenic engine technology. The KVD-1 engine has a thrust of 7,100 kg, a burn time of 800 seconds and a combustion chamber pressure of 54.6 atmospheres. It weighs 282 kg, is 2.1 metres tall and has a diameter of 1.6 metres. The KVD-1 had a turbopump-operated engine with a single fixed thrust chamber, two gimballed thrust engines, could operate for up to 7.5 hours and be restarted five times. It weighed 3.4 tonnes empty and 19 tonnes fuelled. The KVD-1 had unsurpassed thrust and capabilities that made it unmatched for years. Its specific impulse of 461 seconds was still the highest in the world by the end of century.[75] The KVD-1 was first test-fired in June 1967 and was tested for 24,000 seconds in six starts. Five live engines were tested over the years 1974–7 (some were made even after the

KVD-1

manned lunar programme had been cancelled). This small object was to spark off international controversy.

The KVD-1 was not a new motor—it was originally developed as part of the Soviet manned moon landing programme as far back as 1964. Under Vasili Mishin's plan, the KVD-1 would have been used to land a Soviet lunar module able to sustain a team of cosmonauts for a month. However, the KVD-1 never flew and was placed in storage.

Development of the Indian upper stage with Russian help had been underway for four years when the arrangements were denounced by American President George Bush as a violation of the Missile Technology Control Regime. In May 1992, the Bush administration announced that it was applying American sanctions on both the Indian Space Research Organization (ISRO) and Glavcosmos for two years. The sanctions involved American non-cooperation with both agencies and the introduction of sales embargoes on India. India objected strongly to the American actions, pointing out that high-powered hydrogen-fuelled upper stages which took a long time to prepare were of little military value in attacking a neighbouring country with which they already had a land border. India also pointed out that the Americans had offered them the very same technology and had made no objections throughout the years 1988–92 when the arrangements had begun.

In 1993, with the accession of Bill Clinton as president, the American attitude relented. He approved a re-opening of cooperation with ISRO and Glavcosmos if the Russians transferred individual engines, but not the production technology that would enable India to design its own cryogenic engines. In July 1993, after negotiations in Washington DC, Russia backed off its proposals to transfer technology to India and suspended its agreement, invoking *force majeure* (circumstances beyond their control), to the fury of the Indian government.

The KVD-1 had now become caught up in a much bigger game—the negotiations between America and Russia for the construction of the International Space Station. Russia suggested compensation for loss of the Indian contract and the $400m paid by the United States for seven American flights to Mir may have become part of the equation.

In a revised agreement with India in January 1994, Russia agreed to transfer three, later renegotiated by the Indians to seven, KVD-1s intact, without the associated technology, for a price of $9m. The negotiations were later described as very tough, but the Indians managed to negotiate a lower price and extra engines in exchange for the loss of technology transfer. In addition, two models would be supplied to test how they would best fit the launcher shroud. India was required by the United States to agree to use the equipment purely for peaceful purposes, not to re-export it or modernize it without Russia's consent.

According to some American sources, the Russians transferred the production technology in any case.[76] The appropriate documents, instruments and equipment were allegedly transferred in four shipments from Moscow to Delhi on covert flights by Ural Airlines. As a cover, they used 'legitimate' transhipments of Indian aircraft technology travelling the other way to Moscow for testing in Russian wind-tunnels. The plot thickened when at the same time in October 1994, two senior scientists at the Indian Liquid Propulsion Systems Centre, S. Nambi Narayanan and P. Sasikumaran, were arrested for 'spying for foreign countries'. Eventually, the Central Bureau of Investigation admitted that the charges against S. Nambi Narayanan and P. Sasikumaran were false and baseless and they were freed. Later, the United States was accused of setting them up as part of a dirty tricks campaign against the sale of the KVD-1.

Eventually, two sample models were delivered between 1997 and 2000, along with up to seven ready-to-fly stages. The first of six KVD-1s arrived on 23 September 1998 in Madras, whence it was taken to Sriharikota. The compliance deal specified that production technology *not* be transferred with the engines—in other words, no blueprints.

The first KVD-1 was due to launch in early 2001 on the new Indian Geosynchronous Launch Vehicle. Later, another version of the engine, the KVD-1M, would be used to power the Proton M upper stage, if it is built, and the Angara upper stage.

KUZNETSOV/NPO TRUD: MORE SECRETS FROM THE SIXTIES

Kuznetsov is an aviation design bureau headed by Nikolai Kuznetsov (1910–95) who became involved with the space programme when Valentin Glushko refused to cooperate with Korolev in building the engines of the Soviet Moon rocket, the N-1. Accordingly, Korolev asked the Kuznetsov bureau to make them, even though it had no experience of building rocket engines. Undaunted, Nikolai Kuznetsov developed the engines of the lower and upper stages of the N-1, called the NK-33 and NK-43 respectively (NK personalized for Nikolai Kuznetsov). Kuznetsov was essentially an aircraft engine designer and the construction of the NK rocket engines was Nikolai Kuznetsov's first and only venture into the space business. The NK engines built for the N-1 never concluded a complete flight, but their merits came to be recognized many years later.

When Glushko cancelled the N-1 programme, he ordered that the NK engines be destroyed. The engineers in Samara could not bring themselves to carry out what they regarded as technological vandalism, so they acknowledged the order and hid the engines away in a hangar—hundreds of them (at least 450). It was as well they did. In the 1990s they were discovered by visiting American engineers who could not believe what they saw—hundreds of unused, high-performance rocket engines, gathering dust.

Kuznetsov continued work on his NK-33 engine at his own expense. He decided on a duration test of 20,360 seconds on a test stand. It ran perfectly; and 14 engines logged up to 14,000 seconds in other tests. Chief designer Mishin considered it the best rocket engine ever made.

The American Aeroject company at once bought 90 of them for $450m and in 1995 sent them off to its Sacramento, California, plant for testing and evaluation. They worried if there would be any problems in relighting motors that had been in storage since 1974. They ran two tests—of 40 and 200 seconds—and there were not. Aerojet's evaluation of the engine found that it could deliver over 10% more performance than any other American engine and enthused over its simplicity, lightness and low production costs. Aerojet entered it in the competition for a new American launcher, but in the end sold it to the Kistler company for its new K-1 American–Taiwanese private launcher. Three NK-33s will provide 234 tonnes of thrust for the first stage and one NK-43 will provide 178 tonnes for the second. Launch sites were planned for Nevada and Woomera, Australia. The K-1 rocket was equipped with airbags and parachutes that made it reusable. However, progress with the launcher was slow. With a surfeit of competitive launchers in the world, investors were slow to support a new and unproven system, even if the old NK-33 was one of its selling points.

In the event, the NK-33's future may prove to be with the Japanese. In 1996, the Japanese introduced a new light booster, the J-1. Although technically successful, the J-1 rocket proved to be extremely expensive, due to the Japanese zeal for quality control and using indigenous components. As part of a comprehensive review of the programme in 1999, a redesign was ordered using less expensive, already proved, bought-in components. TsSKB and Aeroject made a successful bid for the NK-33, and the first flight was set for 2003. Kuznetsov has also promoted the NK-31 and NK-43 as the engine for a redesigned Soyuz launcher, the Yamal.

EXOTIC ROCKET TECHNOLOGIES

Angara was the most substantial innovation in Russian rocketry in the 1990s. Financial shortages mean that resources available for research and development during the period were extremely tight. The few resources that were available were devoted to the introduction of improved upper stages which gave worthwhile, if undramatic, improvements in performance.

For the United States, relatively generous resources were available to fund new rocket development. Throughout the period, the United States gave attention to what should happen in the post-shuttle period. The holy grail for 1990s rocketeers was single-stage-to-orbit, a concept pioneered in the late 1960s by Philip Bono and Kenneth Gatland.[77] The notion was that the entire spacecraft on the pad should carry enough fuel and power to reach orbit, not wastefully dropping off boosters or stages en route. Such a spacecraft would necessarily combine the features of rocket, shuttle and spaceplane. Originally, the space shuttle was to have been single-stage-to-orbit but this was not within the reach of 1970s technology.

The concept of single-stage-to-orbit continued to lurk in the background in the 1980s, especially as the operation of the shuttle gobbled up so much of NASA's budget. President Reagan pledged the United States to develop an aerospaceplane called the Orient Express, in effect a suborbital version able to fly from one hemisphere to another at 10 times the speed of sound or better. In the early 1990s, some limited but impressive take-off and landing trials were made by a small-scale SSTO demonstrator, the DC-X.

NASA eventually began to fund a demonstrator, the X-33—a blunter, smaller, shorter, fatter all-in-one version of the SSTO. If this proved successful, an operational scaled-up version called the Reusable Launch Vehicle or Venturestar would take its place by 2010, flying weekly missions out of Cape Canaveral and other launch sites to the space station and deploy regular cargo in Earth orbit. Development of the X-33 required the mastery of ultrastrong lightweight materials, aerospike engines and new thermal shields, but these proved much more difficult than all but the most hardened critics anticipated. Achieving these goals proved to be elusive: launch dates of the X-34 slipped from 1999 to 2003 and NASA had to explore new ways to upgrade the shuttle and keep it in service much longer than it had ever intended.

When the Soviet Union learned of plans for the Orient Express, it responded in the only way it knew how: by building one. On 27 January 1986 the Soviet government ordered construction of a Soviet aerospaceplane. Three design bureaux proposed to build it—one rocket company and two plane companies: Energiya, Yakovlev and Tupolev. In the event, Tupolev, which had built the Soviet supersonic transport, the Tu 144, got the contract. The Soviet aerospaceplane became known as the Tupolev 2000, but the project had been cancelled for lack of progress and funding by 1992. The only progress was three scramjet launches from Baikonour organized by the Central Institute for Aviation Motors, TsIAM. In 1989, 1990 and 1991 TsIAM flew a 9-metre-long missile 180 km downrange and 30 km high. Two short burns totalling 30 seconds were made on the hydrogen-fuelled scramjet, apparently successfully.

In the event, this was not the end of the story, for the aerospaceplane project was revived several years later as the MiG AKS project, also known as the MiG 2000. Small scramjets for the new MiG were tested in the military facilities at Shary Sagan. MiG managed to attract some NASA interest in a test flight of what it is hoped will be the powerplant for the MiG, to be flown on a Cosmos rocket out of Kapustin Yar and to touch down near Vladivostok.[78] In late 1999, St Petersburg State University presented proposals for a hypersonic aerospaceplane called Ajax which would run on a magnetic plasma chemical engine.

During the 1990s, Russia continued to carry out tests, albeit somewhat sporadically, of scramjets and others means of hypersonic transport that could pave the way for an aerospaceplane. Several such programmes continued on the books (Kholod, Oriol) but without financing they made little progress. NPO Mashinostroeniye, for example, built a 2-tonne pointed demonstrator appropriately called Igla (Russian word for 'needle') for launch on a suborbital mission on a SS-19 to an altitude of 45 km and a speed of mach 6. Igla is 2.2 tonnes and 7.9 metres long; it uses liquid hydrogen and is to be fired 50 km high on a suborbital mission at up to 10 times the speed of sound on the nose of a Rockot from Plesetsk in 2004, with recovery in the Pacific.

NUCLEAR, ELECTRIC POWER SOURCES AND ENGINES

The Soviet Union made a substantial investment in space-based nuclear power sources and engines. Small nuclear power sources were carried on the Lunokhod

RD-301: tested but never used

lunar rovers. The RORSAT US-A programme (1974–89) used nuclear power sources called Topaz, but after poor programme performance and two highly publicized accidents, the programme was ended.[79] In the late 1980s, an improved Topaz, 1,000 kg in weight, was built by the Arsenal design bureau and Zvezda, with 5–6 kW (some quote 10 kW) of power in a thermionic energy converter. It was even launched on two experimental missions—Cosmos 1818 and 1867—providing electrical power for station-keeping over a year. Also during the 1980s, NPO Energiya developed a lithium-cooled 150-kW reactor and although some ground tests were conducted, no full-up test was ever run. Officially, it was developed for a manned flight to Mars, but American intelligence linked it instead to a Soviet star wars programme for particle-beam weapons.

In addition to the development of nuclear reactors to provide electrical power for spacecraft, the Soviet Union began preliminary work on a nuclear rocket engine. The Kosberg design bureau in Voronezh began work in 1965 on the RD-0410 liquid hydrogen nuclear rocket engine. Tested between 1978 and 1991, the RD-0410 weighed 2 tonnes, was 3.5 metres high and had a thrust of 3.6 tonnes. A successor, the RD-0411, was also built. These tests were designed to pave the way ultimately for a manned Mars mission.

Russia did not continue with the development of nuclear power sources or nuclear rocket engines. Instead, Russia sold them—to the United States! The original nuclear laboratory, the Kurchatov Institute, developed a single-cell uranium crystal reactor with 6 kW output, called Enisy. Twenty-eight Enisy units were built between 1970 and 1988, one reactor being successfully tested for 14,000 hours. In 1992, the United States Ballistic Missile Defence Organization bought two Enisy reactors for $13m and shipped them to Albuquerque in New Mexico for testing, followed by a further four Enisy reactors the following year. In effect, the Kurchatov Institute traded the secrets of the Enisy programme for cash to keep the institute open. The Americans, for their part, were anxious to examine this advanced technology. Their payments would have the intended effect, they hoped, of keeping Russian nuclear experts working in Moscow, rather than seeking employment abroad with hostile powers such as Iran or Iraq.

RUSSIAN ROCKET ENGINES IN PERSPECTIVE

Of all the space-faring nations, Russia has the longest history in building rocket engines—one of theory and practice dating back to 1921. The priority given to engine design, and the pre-eminence given to design engineers like Valentin Glushko, and the success of a number of individual engines, such as the R-7's RD-107 and the Proton's RD-253, were at the heart of the Soviet Union's success in space.

The 1990s marked the rediscovery of the performance and abilities of Soviet rocket engines from the 1960s. Quite literally, they were rediscovered: the NK-33s and NK-43s in the hangar in Samara, the KVD-1 in the Isayev bureau. They were pressed into service for the Kistler and J-1 launchers and the Indian GSLV respectively. The RD-180, a design which dated to the late 1970s, made its maiden flight on the American Atlas III in 2000. Russian rocket engines from the 1960s and 1970s were so sturdy and powerful that they were better than anything the Americans could offer 30 years later.

Within their own programme, the Russians began the process of replacing the block D with the Briz upper stage. However, important though this was, it represented the limits of Russian innovation in the 1990s. It is doubtful if, due to the adverse financial situation, Russia was able to make any new investment in rocket engine design. Progress on nuclear electric, nuclear propulsion and scramjet technologies was very limited. The Kurchatov Institute sold its nuclear power sources outright to the United States. While it is true that the engine design bureaux announced a series of upgrades for existing engines, the degree to which this work entered new territory is doubtful. The financial situation created a contrast in which Russian innovation and new engine research was at a standstill while the United States put resources into the new X-technologies. The X-33 was the first project in the world to use the new aerospike engine and would become the basis for the new fully reusable space shuttles from about 2010 onwards. Even if American progress on the X-33 and its associated projects (e.g. the X-34) was slower than hoped, Russia was carrying out no parallel work at all. Although the advanced nature of Soviet engines developed long ago gave Russia a substantial cushion, Russia would, over the course of time, lose its lead in this frontier area of space development.

9

The design frontier
*The design bureaux of the Russian space programme
from 1992*

The organizational core of the Russian space programme is the design bureau. This could be classified by an experimental design bureau (*Opitnoye Konstruktorskoye Biuro*, OKB), a design bureau (KB) or a scientific research institute (NII). During the Soviet period, the great leaders of the programme built up large design offices, factories and plants. They bid for contracts, put forward their own projects and embarked on rival enterprises. Designers (*konstruktor*) such as Sergei Korolev, Vladimir Chelomei, Mikhail Yangel, Semion Kosberg, Valentin Glushko and Dmitri Kozlov exerted considerable influence over Soviet planning and politics for a long period, their gigantic role unparalleled by industrial leaders in the United States, most of whom remained relatively anonymous. The design bureaux exercised power in their own right—indeed, it was the inability of the Soviet political machine to control them and their rivalries that was a major contributor to the Soviet Union losing the Moon race. Nowadays, these organizations are called NPOs (scientific production associations), but the term 'bureau' is still widely used.[80]

What happened to the design bureaux under the Russian space programme? How was the Russian programme organized to cope with its changing circumstances?

ENERGIYA: PREMIER DESIGN BUREAU AT THE COALFACE OF THE ECONOMIC CRISIS

The premier design bureau throughout the Soviet period and into the Russian period was the original OKB-1 in Podlipki, directed by Sergei Korolev, who designed the R-7, Vostok, Voskhod, Soyuz, the N-1, the Zenit spy satellites and the first generation of lunar and interplanetary probes. When he died, OKB-1 became NPO Energiya under his successors Vasili Mishin (1966-74), Glushko (1974-88) and then Yuri Semeonov. In 1994, it was renamed *RKK Energiya imemi Sergei Korolev* (Rocket

Cosmic Corporation, dedicated to the memory of Sergei Korolev). The company was part-privatized in 1994, the government keeping 51%, but a further 13% was sold off three years later to raise funds to launch the Zvezda service module. It is still by far the largest design bureau and employs between 22,000 and 30,000 people.

Energiya, now located in the town of Korolev, remains predominant among the Russian design bureaux, is responsible for all manned-related operations and owns the Mir space station. In 1996, it published a coffee-table book history of its first 50 years and its key role in Soviet and Russian space exploration was readily apparent. The bureau is the lead organization for the development of the Soyuz, Progress and space stations. Most of those working in mission control in Korolev belong to the Energiya bureau and it is likely to maintain its high visibility for some time. It was the Energiya engineering staff who sorted out the problems on Mir in 1997 and its managers who worked hard and largely successfully to attract foreign investment for the Mir station throughout the decade. Almost half of Russia's cosmonauts are recruited from Energiya engineers.

Energiya is the lead agency for the Russian end of the space station project. The company bore the brunt of the financial shortages which crippled the Russian space programme in the late 1990s. Its managers struggled endlessly with deficits by cajoling money from politicians, begging advances from NASA and by privatizing ever more of the company's stock. Against the odds, they managed to launch the Zvezda module to the International Space Station in July 2000 and will be the lead agency on the Russian side in the station's development.

CHELOMEI'S BUREAU: A THRIVING STATE COMPANY

The main rival to Korolev was OKB-52, located at Reutov, directed by Vladimir Chelomei, who masterminded the building of the Almaz space station, TKS modules, the Proton rocket and a range of space planes and military projects. After his death in 1984, it was renamed NPO Mashinostroeniye and directed by Yuri Semeonov (1984–8) and then G. Efremov. The bureau suffered severely from the funding cuts of the 1990s and by 1997 had only 4,500 staff.

Responsibility for the Chelomei bureau's most lasting design, the Proton rocket, passed to the biggest rocket factory in Moscow, and the one best known in the West, the MV Khrunichev State Research and Production Space Centre in Moscow, which can also claim to be a descendant. Originally an automobile plant, then a production line for German Junkers monoplanes, in the Second World War a producer of Red Air Force planes, it became the Myasishchev design bureau and then the Salyut design bureau. From 1960, it received contracts for the production of rockets and spacecraft designed by Vladimir Chelomei and, most important of all, the Proton rocket, from 1965.

Khrunichev was one of the first companies to enter an arrangement with Western companies, when it signed a deal with Lockheed Martin in 1993 establishing the International Launch Services (ILS) Joint Venture. As a result, Khrunichev was soon able to attract significant foreign contracts for the launching of commercial

Lift Test of the Zenit-3SL Launch Vehicle on the Odyssey Launch Platform

and communications satellites with the Proton rocket. Capital investment from
Lockheed enabled Khrunichev to modernize its facilities, especially those for
launch preparation in Baikonour. Ironically, it was the privatized Energiya
company which experienced the greater economic difficulties in the 1990s, while
the Khrunichev company, which continued to be a state enterprise, prospered
from the foreign venture. The FGB module for the International Space Station
was built there—on schedule—and, with American financial assistance, the first
elements of the new Angara rocket began to appear there in 2000. Khrunichev,
which developed the new Briz upper stage and the Rockot smaller launcher, has
also branched out into satellite design and manufacture, proposing new, small Earth
resources satellites called Monitor and Yacht.

LAVOTCHKIN ADAPTS TO CHANGED TIMES

As OKB-1 became overloaded with work in the mid-1960s and the urgency to beat
America became overwhelming, many elements were hived off to new or existing
design bureaux. The unmanned lunar programme was first to be devolved—going in
1965 to what had been the old Lavotchkin aircraft design bureau, OKB-301 in
Khimki, which dated to 1937. Georgi Babakin, chief designer there from 1965
to 1971, was succeeded by Sergei Kryukov (1971–8) and then by Vyacheslav
Kovturenko (1923–95). Lavotchkin had some spectacular successes, like the Luna
Moonscoopers, the Lunokhod Moon rovers, and the Venera landers on Venus. With
the end of the lunar programme and the virtual termination of planetary flights, one
might have expected the Lavotchkin bureau to sink out of sight. Lavotchkin built
Phobos 1-2 and Mars 8. The upper stages Ikar and Fregat are built there, giving
Lavotchkin much new business in the late 1990s. It also obtained the contract for a
number of military programmes such as the Oko early warning satellite, Prognoz
and Arkon, as well as the civilian programmes Kupon, Spektr and Interball.

NPO YUZHNOYE: MISSILE LINES 'LIKE SAUSAGES'

Traditionally, the third largest design bureau in the Soviet system, after those of
Korolev and Chelomei, was the NPO Yuzhnoye in Dnepropetrovsk in the Ukraine.
Originally a car and tractor factory, a government resolution established the plant as
a rocket design bureau and factory in 1954 as OKB-586 under the leadership of
Mikhail Yangel. Originally, OKB-586 specialized in military rockets, but later
branched out into other rockets and satellites. This bureau designed and produced
principally military rockets and payloads. OKB-586 built the R-12 (small Cosmos),
R-14 (Cosmos 3), R-36 (Tsyklon) and Zenit rockets, though some construction has
since been contracted out to other plants in Russia.

Once this was under way, the OKB designed and built a number of Cosmos
scientific satellites: RORSATs, EORSATs, the Tselina electronic intelligence
satellite, Okean and Sich. After Yangel's death, the OKB was led by Vladimir

Made by NPO Yuzhnoye: the Tsyklon

Utkin. In the 1970s, it was renamed NPO Yuzhnoye (the Russian word for 'southern' or 'Pivdenne' in Ukrainian). One of its most prominent activities at present is the manufacture of the Zenit rocket used for the Sea Launch programme.[81] Although the total number of programmes handled by Yuzhnoye was not high, the production runs were often very long. At one stage about 50,000 worked in the space division but that number is probably closer to 6,000 now. At one stage Yuzhnoye was running six rocket production lines simultaneously, so much so that Khrushchev once boasted to the West that 'we produce rockets in a line like the way you make sausages'

During the 1990s, Yuzhnoye continued to work with the Russian design bureaux and contractors with whom it had developed interlocking relationships for so many years, while at the same time attempting to promote an independent Ukrainian space programme. Any Ukrainian contribution to the International Space Station, such as a scientific module, is likely to be led by NPO Yuzhnoye. The bureau continues to supply Russia with the Tselina 2 electronic intelligence satellite, a key part in the military space programme and its associated Zenit rocket. Overall, though, its role in the Russian space programme is diminishing, with declining production of the Cosmos rocket transferred to NPO Polyot in Omsk and the Tsyklon rocket near the end of its career. With the building of the new Angara rocket, Russia will have less need to turn to Ukraine for the troublesome Zenit booster.

NPO-PM: BUILDER OF COMSATS

The principal builder of communications satellites is NPO-PM in Krasnoyarsk, located in forests on the banks of the Yenisei river. The founder was Mikhail Reschetnev. When the work of designing Russia's first communications satellites overwhelmed Korolev in the mid-1960s, he passed responsibility for the new Molniya programme to what was the No. 2 branch of OKB-1, then OKB-10, receiving its current designation in 1961 as the Scientific and Production Association (NPO) *Prikladnoi Mechaniki* (Applied Mechanics). Little was known of NPO-PM until the 1990s, because it operated in a closed area (called Krasnoyarsk 26) and because many of its products were military communications satellites. During the

NPO-PM in Krasnoyarsk, Siberia

Russian period, the company also began to use the civilian address of Zelenogorsk. Having started with Molniya, NPO-PM was then given the responsibility for the development of all Soviet communications satellites.

Since then NPO-PM has built no less than 27 different space systems and over 1,000 individual satellites. These include Molniya, Gorizont, Ekran and the new comsats (e.g. Ekspress), as well as the military comsats (e.g. Raduga). The NPO-PM factory in Krasnoyarsk is a true cradle-to-grave operation, bringing each satellite through from design to fabrication, ground testing, integration and orbital control until the mission is concluded.[82] NPO-PM had a virtual monopoly on communications satellites until the appearance of Yamal and Kupon, built by Energiya and Lavotchkin respectively, though the poor performance of both suggests that the complexities of building comsats should not be underestimated.

In its efforts to keep ahead, NPO-PM brought in Western expertise from the Alcatel company. Sesat, launched by Proton in April 2000, was constructed by NPO-PM in Krasnoyarsk with Alcatel's assistance for the European satellite communications organization, Eutelsat. The advanced Sesat had 18 channels, and the ability to carry video, internet, mobiles, paging and software retransmission, with eight electric motors for station-keeping and a design lifetime of 10 years.

About 10,000 people now work there. Its founder, Mikahil Reschetnev, died in January 1996. Granted the size of the expanding communications market in Russia, it should be able to secure for itself a strong future.

ARSENAL: THE OLDEST DESIGN BUREAU

Traditionally, the main builder of military satellites was the Yuzhnoye design bureau in the Ukraine. From 1980, construction of the US-P series of EORSATs was transferred to the MV Frunze Arsenal design bureau of St Petersburg. Despite the importance of St Petersburg as an industrial centre in Russia, remarkably little space work is carried out there. For sheer longevity in weapons-building, the Arsenal factory is probably unequalled, for its was set up in 1711 by Tsar Peter I as a canon foundry but became OKB-7 in 1949. As a result, it can claim to be the

Arsenal's most famous product, the US-P EORSAT

oldest design bureau connected to space research. Arsenal is now promoting several new series of satellites—Predvestnik for earthquake prediction and prevention, and Obzor for space-based radar sounding.

TSSKB SAMARA: CONTINUOUS PRODUCTION FROM 1957

Outside Moscow itself, the greatest concentration of rocketry in the Russian Federation may be found in Samara. Responsibility for building the Korolev's R-7 rocket went to OKB-1's No. 3 branch in Kyubyshev on the river Volga. The No. 3 branch eventually became independent as the TsSKB, or the *Tsentralnoye Spetsializorovannoye Konstruktorskoye Buro* or Central Specialized Design Bureau. The plant actually comprises three elements: the design bureau, the rocket factory for the R-7 (the Progress plant or the Progress works) and an affiliate design bureau, KB Foton. In following these naming complexities, one must also be alert to the location changing as well, for Kyubyshev itself was later renamed Samara.

Originally the TsSKB was the Duks bicycle factory, set up by a German businessman Jules Muller in the Baltics in 1884, which by the First World War had expanded into the production of motorcycles, cars and trolleys. Duks was nationalized in 1918 as State Aviation Plant No. 1, evacuated to Kyubyshev in 1941 (Goebbels once trumpeted its destruction by the Luftwaffe) and during the war it turned out Il-2 dive bombers. TsSKB was the creation of Dmitri Kozlov and it was under his guidance that the plant assumed responsibility for manufacturing the R-7, which has been in continuous production there since 1956. Korolev was obviously pleased with Kozlov's progress, for he then gave him responsibility for the Zenit. Over time, TsSKB became the design centre for the Yantar and subsequent spy satellites like the Orlets 2. Naturally, the TsSKB also developed the associated civilian models of Zenit and Yantar, like Bion, Foton and Resurs.

By 1999, TsSKB had turned out over 1,500 rockets and over 900 satellites in the Zenit, Yantar, Orlets, Bion, Foton and Resurs series. Up to 25,500 people work in the Progress factory. Not all work in the rocket or satellite sections—about 5,000 do, of whom 360 are on the R-7 production line at any time. In the spirit of post-Soviet diversification, the factory has also branched into machine tools, vodka and sweets! In 1995, it began a profitable joint venture with the French company Starsem.

NPO POLYOT

Moving further away from Moscow, Omsk is the location of the Polyot Production Corporation. Originally, Polyot made Tupolev bombers in Moscow, but was evacuated to Omsk, beyond the Urals, in 1941. Polyot began to manufacture rockets and spacecraft from the 1960s and became involved in the manufacture of the R-12 small Cosmos rocket and engine production for the Energiya rocket (RD-170). During the 1990s, it produced the Cosmos 3M rocket and navigation satellites (Tsikada, Nadezhda, Parus, Strela, Gonetz and GLONASS). Polyot has

Cosmos 3M production line, Omsk

arrangements for the marketing of the Cosmos rocket abroad through the German company OHB. Up to 20,000 work there now.

NPO MOLNIYA

NPO Molniya is the most recent of the design bureaux, having been formed in February 1976 as an offshoot of the MiG aircraft design bureau. It is currently involved in projects to launch a small spaceplane on the back of the Antonov 225 transport aircraft. Molniya was founded by Gleb Lozino-Lozinsky, one of the designers of the Energiya–Buran system. As yet, it has not met the commercial success to match its design expertise and promotional enthusiasm.

We have discussed the design bureaux and their evolution over the 1990s, but how were they coordinated?

ORGANIZATION OF THE SPACE PROGRAMME: INTRODUCING RKA

The organization of the Soviet space programme required a massive managerial effort—in the case of the Moon race, not a very successful one. During the Soviet period, considerable Western intelligence effort went into identifying the organs which took decisions on Soviet space policy.

Western observers made the mistake of searching for NASA-style executive agencies which would issue commands for plants to implement. In practice, the Soviet space programme was directed by the Ministry of General Machine Building (MOM, or *Ministerstvo Obshchego Machinostroyeniye*) and the Commission on Military Industrial Issues (VPK), with key decisions being taken jointly by the party and government (for example, the key decisions on the Moon race). The chain of command was confused by other powerful actors such as the Academy of Sciences and the practice of rival design institutes appealing to ministers and officials in MOM, the VPK, the government and the central committee to change, amend and cancel decisions, which they often did. The Ministry of General Machine Building was of course a misnomer (the Ministry of Medium Machine Building was the cover for the nuclear industry; only the Ministry of Heavy Machine Building had a vaguely truthful title, being responsible for cranes and excavators). Not until 1985 were the space roles of MOM and the VPK brought together when Oleg Baklanov was appointed by Mikhail Gorbachev as combined Minister of General Machine Building and the head of the commission on Military Industrial Issues—a reward he reciprocated by joining the putsch against Gorbachev in 1991. (Eventually, however, he ended up in prison.) The Soviet space programme did not have executive agencies and was poorly coordinated, making its achievements, despite this, all the more remarkable. Only chief designer (*Glavnykonstruktor*) Korolev had been able, through his bullying and force of personality, to overcome the confused and diffused organizational architecture.

Soviet disorganization was, as already noted, one of the key factors contributing to the Soviet Union losing the Moon race. In 1985, Gorbachev had tried to bring cohesion and coordination to the Soviet space programme not only by the appointment of Baklanov but by the creation of the Glavcosmos agency, directed by Alexander Dunayev. In practice, Glavcosmos was more of a marketing and coordinating body than an executive agency. The following year, Glavcosmos sent marketing teams abroad to promote the sale of the Proton rocket, eventually meeting with success in 1992.

Two other bodies were important during the Soviet period. The first was the Academy of Sciences of the USSR. It was never under direct governmental control, for the Academy is a self-perpetuating body of learned men and women of science—about 300 full members and 300 corresponding members—that long predated the Revolution of 1917. It had only a limited formal role in the organization of the Soviet space programme but was influential, especially during the presidency of Mstislav Keldysh when it was called several times to adjudge programmes and projects for manned and unmanned spaceflights. After 1991, the academy was renamed RAN (the Russian Academy of Sciences). The second was the Institute for Space Research in Moscow (IKI). Again, Western observers imagined that IKI 'directed' the Russian space effort. Although in practice it had a role in the design of projects, its real function was to be the public face of the programme for Western visitors and, later, journalists interested to learn more about the programme. But it was not the Soviet equivalent of NASA.

NEW SPACE AGENCY

Originally, it had been thought that the space programme would continue as a unified effort by the former republics of the Soviet Union. These were, effectively, the 15 republics bar the Baltics. The agreement setting up the successor body to the Soviet Union, namely the Commonwealth of Independent States (CIS), specified space research as a defined area of cooperation between the member states. Indeed, on the second last day of the Soviet Union, 30 December 1991, the prime ministers of the former Soviet Union, excepting the Baltic states, Moldova and the Ukraine, signed an agreement for their civilian space activities to be governed by a new interstate council, for the joint strategic forces to run the military programme and that Baikonour be used as a common spaceport.

After the fall of the Soviet Union, both the Ministry of General Machine Building and Glavcosmos disappeared. In their place, Boris Yeltsin quickly created the Russian Space Agency. Now, Westerners could at last identify the NASA-type body for which they had long been searching. The CIS council never got going as an effective body and in a short time departing cosmonauts sported Russian flags and little pretence was made that the programme was a united effort of the CIS.

The Russian Space Agency, RKA, was formed by the decree of President Yeltsin on 25 February 1992. Yuri Koptev, formerly involved in managing the space programme through the now-abolished Ministry of General Machine Building, became its first director. This grey and burly administrator took over MOM's old building, but was allocated a fraction of its staff, between 200 and 300 (interestingly, a week later, on 2 March 1992, Ukrainian President Leonid Kravchuk decreed the formation of the National Space Agency of Ukraine (NSAU), with centres in Kharkiv and Dnepropetrovsk). In 1997, the entire Russian space industry was brought into the Ministry for Space and Telecommunications under 56-year-old Vladimir Bulgat. He was one six vice-premiers appointed by President Yeltsin after he won re-election as President of the Russian Federation and was responsible for a super-ministry of space, telecommunications, information, transport, the nuclear industry, natural resources and the environment, including the Russian Academy of Sciences (RAN). The Russian Space Agency, which soon developed an attractive patch and logo to rival NASA's, has nine divisions: state programmes, manned projects, launch facilities, science and commercial, international, ground, external and legal, resources and business.

Nice logo or not, the RKA had none of the authority, never mind the budget of NASA. Yuri Koptev, appointed at around the same time as NASA's new administrator Dan Goldin, worked hard to maintain and develop the space programme. However, it was difficult for him to make his mark when budgets were shrinking; even agreed financial allocations did not arrive and long-term planning was necessarily replaced by the month-by-month effort to survive. The powerful and far-flung design bureaux each fought their battles to protect their own spheres of operation and to survive—an art which they had practised for many years. The survival of the manned space programme depended on the actions and decisions of the RKK Energiya, owner of Mir, much more than anything the Russian Space Agency might decide. The lack of RKA authority was apparent when the Buran project was cancelled: it was not a decision of the agency, as one might have expected, but rather a decision of the council of designers. Later, the decision on whether to keep Mir in operation beyond 1999 was made by the (private) board of Energiya, not by Koptev, who was equivocal about the matter. Although the RKA issued plans and project lists, their relative weighting and priority was unclear and in no way matched the hard choices which the reforming Dan Goldin was imposing on his NASA colleagues at the same time. The development of the Angara rocket was, again, a function of Khrunichev's ability to attract resources rather than a conscious long-term planning decision by the RKA. This is not to be unfair to Koptev, for he was probably the right man for an impossible job at a terrible time. It was Yuri Koptev who saw the opportunity presented in the summer of 1993 to merge the Mir 2 project with the American space station and it was Koptev who steered that merger through and who kept the International Space Station on track, despite many difficulties, ever since. His peer, Dan Goldin, had much to thank him for, something he often did.

DESIGN BUREAUX IN PERSPECTIVE

The Soviet design bureaux, the organizational bedrock of the Soviet space programme, survived the transition to the Russian space programme. The largest,

Energiya, experienced the most financial difficulties, but remained undisputedly pre-eminent. Effectively, it owned the manned space programme, Mir, the cosmonauts, mission control and the new international space station project. Its rival of old, the Chelomei design bureau, spawned one of two descendants, Khrunichev, which, continuing in state hands, became a profitable international space corporation selling Proton rockets on the world market and marketing new models, like Rockot. Accelerated by the break-up with Ukraine, the NPO Yuzhnoye became a less important element, while still supplying key rockets and satellites, especially for the military programme. Other design bureaux, like TsSKB, Lavotchkin and NPO-PM, managed to adapt by developing their existing products and diversifying into new areas.

The table below summarizes the leading Russian design bureaux, few of whom still kept their Soviet-style names (indeed, some of their locations had changed name too). They had proved remarkably impervious to the arrival of the new Russian Space Agency, which promised, later if not sooner, a greater sense of coordination and coherence to the national space effort. The old post-war *glavnykonstruktori* had passed on, but apart from that, little else had changed during the transition to the Russian space programme, except for one crucial thing, the money.

Russia's design bureaux

Original name	Founded	Current name	Location	Chief designers
OKB-1	1946	RKK Energiya	Podlipki, now Korolev	Sergei Korolev Vasili Mishin Valentin Glushko
OKB-52	1955	NPO Machinostroyeniye/ Khrunichev	Reutov/Podlipki	Vladimir Chelomei
OKB-1 branch No. 3	1959	TsSKB/Progress	Samara	Dmitri Kozlov
OKB-301	1937	OKB Lavotchkin	Khimki	SA Lavotchkin
OKB-155 branch No. 4	1976	NPO Molniya	Moscow	Gleb Lozino-Lozinsky
OKB-1 branch No. 2	1959	NPO-PM	Krasnoyarsk	Mikhail Reschetnev
KB Arsenal	1719	KB Arsenal	St Petersburg	PA Tyurin
OKB-586	1954	NPO Yuzhnoye	Dnepropetrovsk	Mikhail Yangel
GDL	1921	NPO Energomash	Leningrad, now Khimki	Valentin Glushko Boris Katorgin
OKB-2	1943	KB KhimMash	Podlipki	Alexei Isayev
OKB-154	1941	KB KhimAutomatiki	Voronezh	Semion Kosberg
OKB-276	1946	NPO Trud	Samara	Nikolai Kuznetsov

10

The commercial frontier
Financing the Russian space programme

Many allusions have already been made to the financial difficulties which beset the Russian space programme. They were evident in the problems in sustaining the Mir programme, in the delays to the International Space Station, in the contraction of military and civilian unmanned programmes, in the deterioration of the cosmodromes, in the scrapping of the comships, and in the virtual abandonment of pure research. But how serious were these difficulties? How did the programme shrink? How were decisions taken? How did Russia find new means of funding its space programme? Were these temporary expedients—and will they last?

FUNDING THE SPACE PROGRAMME

The contraction of the space programme took place in three phases. The space programme had actually begun to retreat at the end of the Soviet period, in the course of 1989–90. In 1989, Soviet space spending reached a peak of R6.9bn.* At the time, it was estimated to account for 1.5% of gross national product. Of this amount, R3.9bn went to the military, R1.3bn went to Buran and the rest, R1.7bn, went to civilian programmes.

In 1990, the budget was cut for the first time. This did not have an immediate effect on operational capacity, but was most evident when the second flight of Buran began to recede and unmanned probes began to be long-fingered (Mars 94, Spektr). The budget cut was not applied across the board, but fell on programmes in the

---

* Currencies present a difficult problem from this period onward. At the end of the Soviet period, the ruble was set at parity with the £ sterling but later subsequently slid to £1 = 8,000 and $1 = 5,000. Several rates operated over 1992–2000 period and these are converted at prevailing rates. As a result, readers should treat all figures given here with appropriate caution and context. Where possible, Western currencies will be given to enable comparisons to be made ($ or euros). The euro is quoted here at $0.90 = €1.

pipeline. Indeed, it was the Buran project which most severely felt the chill and the rapid deterioration of Baikonour was largely attributed to the exodus from the Buran sheds, which had hitherto been a hive of activity. Buran allocations fell from R300m in 1989 to R220m in 1990, at a time when they should have been increasing as the first manned flight was being prepared. Finance and resources were increasingly explained as the reason for delays, rather than operational considerations, which would always have been the case in the past. Launchings that had been free previously were now charged, and where customers had been charged before, the charges rose dramatically. Filling the third seat on Soyuz with a paying visitor became a commercial imperative, not an international nicety. By late 1992, the new director of the Russian Space Agency, Yuri Koptev, noted how, that year, only half the country's planned civilian launchings had been carried out, and less than a third of the planned military launchings. Employment had begun to fall from the high point of 400,000 in the late 1980s.[83]

The second contraction took place soon after the demise of the Soviet Union. Interestingly, this coincided with the time space budgets in Russia began to be quantified and published for the first time. From 1992 onwards came the first reports of serious shortages, delays, postponements and cancellations and these had risen to a chorus by 1993–4. As early as February 1992, the tracking fleet was recalled permanently, the first of many retrenchments. It was too costly to operate and hard currency must be paid whenever it put in at a foreign port. In summer 1993 Yuri Koptev convened a press conference to express his fears that the programme was on the verge of collapse. The programme had lost 30% of its personnel in the past three years, it was said. That summer marked the cancellation of the Energiya/Buran programme. Both the Russian and Western press featured a series of articles from spring 1993, questioning whether the once-great space programme could survive.

The following year, 1994, government allocations to the space programme were sharply reduced and arrived late, but payments did arrive by the year's end (hence a number of launch surges in late December). The sharper Western analysts pointed out that the high continued rate of Soviet space launchings in 1993–4 was deceptive, for the programme had been running on equipment built in earlier and more stable times: a radical operational decline was imminent. Employment was now down to less than 300,000. Russian economists calculated that space spending was already down to 0.23% of the national budget (compared to 0.97% spent on space research in the United States).[84]

A sharp downturn in the performance of the space programme was indeed seen in 1995, with a radical reduction in the rate of military and unmanned space activities. The next year, there were gaps between the recovery of one reconnaissance spacecraft and the launch of the next. Soyuz missions were stretched out, rockets were not available in time, nose cones were not available and production lines had slowed. Staff in many enterprises had not been paid for months. Several space enterprises were, on paper, bankrupt. Rocket launchings were held up because subcontractors, still unpaid, would not deliver components and fuel until accounts had been settled. R-7 launchers were borrowed from the military stock in order to keep Mir aloft.

The R-7: 1950 design, still flying in 2000

There were warnings that many comsats had exceeded their lifetimes and that, without new launches, many communications services, especially across Siberia, would simply break down. The main space magazine in Russia, *Novosti Kosmonau-tiki*, commented in October 1996 that hardly any satellites were available to launch, and even where there were, there were no rockets to launch them. The magazine predicted that Russian space activities would probably end in a year and a half.

The third contraction took place with the collapse of the ruble in autumn 1998. By then, it had been hoped that the worst of the contraction of the programme had bottomed out and that it might begin to recover, but these hopes were quickly dispersed. Now, this time, less than half the agreed but already shrunken space budget failed to arrive, creating financial chaos and further demoralization. Programmes and projects had been contracted for, which meant that clients were unpaid and debts began to mount. Contractors in turn could not pay their suppliers and many activities slowed down or even grinded to a halt. This time, there was a resigned acceptance that if government money did not arrive, it would not be coming late: it just would not be coming at all. Employment was down to about 100,000. For those who remained, wages were at rock bottom: the makers of the Energomash world-beating RD-180 engine had an average wage of R3,000 a month, or $104. The military were no better off: wages in the Golitsyno control centre were only R2,000 a month, or just over $60.

RUSSIAN SPENDING IN GLOBAL PERSPECTIVE

There is broad agreement among Western observers that there has been a staggering fall in the level of investment in the space programme in Russia, in the order of 80% over 1989–99. The Russian space programme is now one of the least well funded in the world. At least 200,000 technicians have left the programme, and the workforce has fallen by between half and three-quarters, some even say seven-eighths.[85,86] How sharp was the decline? What pattern did it follow? The diagram attempts to provide an estimate,[87] showing Russian space budgets falling from a high point of 7 trillion rubles in 1989 to about 3.4 trillion at present. These figures understate the fall, for they only partially take into account the fall in the ruble value on international exchanges, so the rate of decline should be much steeper.

The picture is complicated by the growing distinction between budgets agreed and money actually delivered and the form in which it is delivered. An endemic problem in the 1990s was that the proportion of the budget which actually arrived fell far short of the programming agreed. In 1995, only 77% of the agreed budget arrived, falling to 54% in 1997 and 49% in 1998. Things improved a little in 1999, when 63% actually arrived. As a result, the Russian place in the world table was actually even worse, since the budgets of the other, more economically stable countries, generally did arrive. Even when payment was made, this was sometimes in the form of loans,

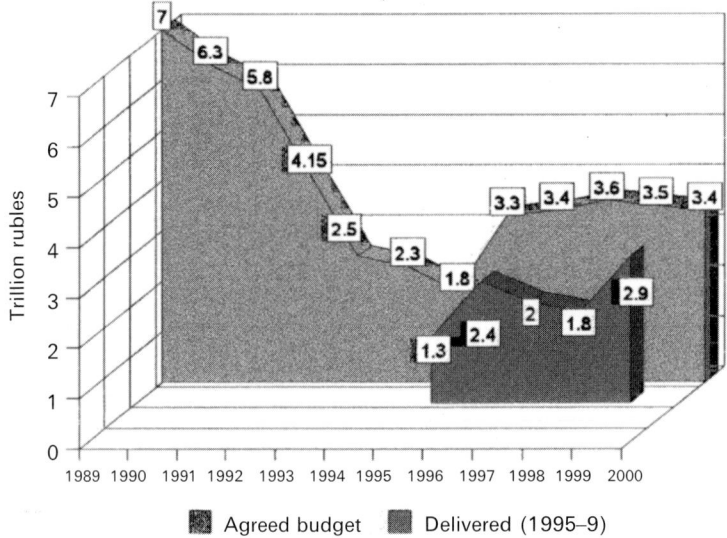

Russian civilian space budget, in trillion rubles, 1989–2000

notes and undertakings rather than 'real' money. The government once paid NPO-PM promissory notes on a bankrupt bank. This was not appreciated.

The ruble collapsed in 1998, which seems to have been the worst year for non-arrival of funding. There was a modest improvement in 1999 and there are early indications that 2000 may be better.

RUSSIA'S CURRENT SPACE BUDGET

Calculating space budgets between diverse countries is a notoriously difficult exercise. By 2000, space spending in Russia had fallen to such an extent that Russia was occupying the lowest place in the world league. The diagram shows the figures for government allocations to the civilian space budget.

Several comments should be made on these figures. At one level, they do not make sense, for the level of Russian space activity is clearly at a level of magnitudes greater than that of Japan, Germany or France. However, if one makes adjustment for the exchange rate, depressed economic conditions, wages in Russia and the absence of new government or capital investment, the figures become more understandable. To these figures should be added military spending, which would almost double the amounts shown for Russia and the United States.

The Russian figures are further complicated by the level of foreign investment, which is not factored in. The level of investment from Western companies through joint ventures may be considerable at this stage, even though its impact on the programme is uneven, with some parts benefiting significantly and others not at all. The level of foreign investment is estimated to be in the order of $600m annually. Energiya is estimated to receive half its earnings from abroad and some

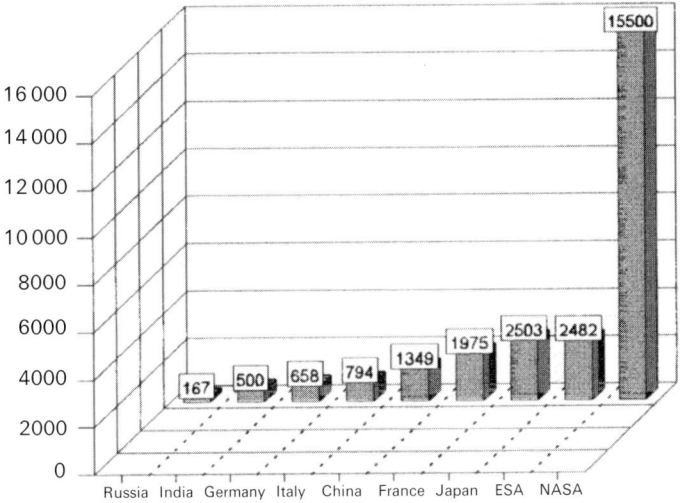

World space budgets, 2001 (projected), in millions of euros

estimates have even been given that 74% of earnings come from abroad. While not large by global standards, it is impressive if one considers that there was no such investment at all 10 years ago.

Granted these disconcerting figures, it is extraordinary that any programme was maintained. The launch rate is one of the main means of measuring space activity. How did the financial crisis impact on the launch rate?

LEVEL OF RUSSIAN SPACE ACTIVITY

At first sight, it is easy to measure the contraction of Soviet/Russian space activity by comparing the number of launches at the height of the Soviet period (102 in 1982) with a rate a quarter of that at the end of the following decade (26 in 1999). However, to attribute the rapid fall in launch rate to the change in government in 1992 alone is too simplistic: the rate had already begun to fall in the Soviet period, and the fall continued in the Russian period because fewer launches were required as satellites had much longer operating lives.

By the end of the Soviet period, launches had fallen by a quarter, from the peak of 102 to 75 launchings in 1990. There was a sharp fall to 59 launches in the last year of Soviet government, 1991. The decline continued until it hit bottom point in 1996, with 23 launches, a level from which it has recovered.

Equally significant is that Russia's relationship with the United States on the table shifted. From launching twice as many satellites as the United States in 1992, the two countries had drawn even by 1995. The following year, 1996, the Americans moved ahead for the first time since the early days of the space race. By the end of the decade, the two space superpowers still made 80% of launchings. Although Europe had a steady, but modest rate of launchings, and although several other countries had satellite programmes, the launch rate for the rest was low.

Even though its space budgets were much depleted and, by all accounts, down to the level of the new spacefaring nations, the table below shows that Russia still accounted in 1999 for 37% of world launches, not far behind the United States' 41%.

So, how does one explain the catastrophic state of Russian space finances with its

World-wide rate of rocket launchings, 1992–2000

	1992	1993	1994	1995	1996	1997	1998	1999	2000
Russia	54	47	46	32	23	26	24	26	20
USA	28	24	26	27	32	37	34	29	
Europe	7	7	6	11	10	12	11	10	
China	4	1	5	2	2	6	6	4	
Japan	1	1	2	2	1	2	2	–	
India			2		1	1		1	
Israel				1					
Total	94	80	87	75	69	84	77	70	

ability to maintain a space programme? The explanation lies in commercialization. The first socialist space programme was forced to become, in the shortest possible time, one of the most competitive, capitalist and global in the world.

COMMERCIALIZATION

Even during the Soviet period, space planners had begun to open their space programme to the West and search for commercial opportunities. Mikhail Gorbachev's Glavcosmos had attempted to sell the Proton rocket to Western manufacturers and produced glossy promotional material, and even issued a Proton manual, to assure investors. Friendly countries that expected launches from the Soviet Union now had to pay more. In 1988, the USSR had put India's first Landsat satellite into orbit for $2m, but by the time of the second launch, just before the dissolution of the USSR, the price had risen to $14m. Even still, old habits died hard. When the financial package to put Briton Helen Sharman up to Mir collapsed in 1991, they went ahead and flew her anyway.

As the economic crisis intensified in 1992, the design bureaux individually and collectively began an all-out effort to draw in external sources of funding. Some of their efforts were crude, clumsy and even bizarre. The cosmonaut training centre, TsPK, offered cosmonaut training to tourists, several days costing up to $10,000, including several minutes of weightlessness in an Il-76 trainer. The following year, 1993, Sotheby's auctioneers sold off 200 rare items of Russian space history, including space suits (Leonov's, Gagarin's and the Moon suit); space cabins (e.g. Cosmos 1443); the dummy which flew on Spaceship 4; Vasili Mishin's diaries; and

Training in weightlessness for cosmonauts—sold as an adventure activity

Sergei Korolev's slide rule. To the surprise of the auctioneer trade, the hardware fetched the least money and the personal mementoes of cosmonauts and designers the most (many were bought up by entrepreneur and politician Ross Perot, who public-spiritedly put them on public display at the National Air & Space Museum on the Mall in Washington DC). Cosmonauts filmed inflated Pepsi cans in orbit. Talgat Musabayev and Nikolai Budarin advertised BMW cars for Spanish television. The Cosmos pavilion at the Exhibition for Economic and Scientific Achievements in northern Moscow was turned into a car salesroom.

Ultimately, the survival of the space programme would depend not on orbital advertisements or the swinging of the auctioneer's hammer but on successful hard-headed commercial ventures.[88] In April 1994, the Russian government approved a *Space Programme for the Period 1994–2000*, drawn up in consultation with the Russian Academy of Sciences (RAN) and the Russian Space Agency, RKA. Unlike most space plans, it said little about actual or aspirational space projects. Instead, it concentrated on the conversion of defence production for economic benefit (in Russian, *konversiya*); and commercialization of the space industry. The authors might have argued that, without either, any such projects would remain an aspiration.

Privatization was an early feature of this process. The largest design bureau in the Russian space industry, Energiya, was ordered to be part-privatized by presidential decree, and 49% of its R1bn equity (then valued at $650m at prevailing exchange rates) was offered to its management, employees, citizens, institutional investors and foreign interests. In July 1998, in an effort to maintain Russia's commitments to the International Space Station and finish off building the Zvezda service module, a further 13% of Energiya was privatized, bringing in $120m, and reducing the government shareholding to 25%. Overall, privatization was not the success that was hoped for: the companies were grossly undervalued, there was little money to invest and there was a flood of other, bigger privatizations underway at the time. Some state companies, notably Khrunichev, managed much better without privatization.

The core of the commercialization process was the establishment by the design bureaux of joint ventures and other forms of commercial arrangements with European and American companies. By 2000, 87 NPOs in the Russian space industry had commercial partnerships or agreements with the West.

THE AMERICAN DEAL

The first, and to this day the most outstanding, Russian–American commercial enterprise was initiated by the American aerospace giant Lockheed, which, following a hastily arranged visit by Lockheed officials to the Khrunichev factory in December 1992, quickly formed a joint company with the Khrunichev plant, the manufacturers of the Proton rocket, called Lockheed Khrunichev Enterprises (LKE). When Lockheed merged with Martin Marietta and Energiya was formally tied into the company, it became Lockheed Khrunichev Energiya International

(LKEI). LKEI was renamed yet again in 1995 as International Launch Services (ILS).

Proton now became part of a giant Russian–American company able to offer Atlas and Titan rockets as part of a world launcher family for prospective customers, probably communications satellite companies trying to find the most economic way of getting their satellites to 24-hour orbit. LKEI offered Proton on the world market at $70m (Europe's Ariane rates were around $70–120m). LKEI won its first foreign contract in 1992, when the international maritime satellite organization, INMARSAT, paid $75m for it to launch an INMARSAT navigation satellite on the Proton. The company quickly attracted contracts for the European Astra 1, Loral's Tempo and PanAmsat 6. In late 1993, the Loral company signed a $250m deal with Lockheed Khrunichev to launch its small new communications satellites, Iridium, and by 1995 Proton had 20 confirmed orders, representing 15% of the world market.

Proton was undoubtedly one of the big success stories of commercialization. By 2000, the annual production rate of Protons was increased from 12 a year to 18 a year. For Khrunichev, the deal with Lockheed brought a steady flow of orders, relative economic certainty, access to the world market and new investment. In 1995, LKEI announced plans to invest $23m in Baikonour. These involved upgrading the airport, dressing the roads between the airport and the Proton pads and improvements in the clean rooms—making the assembly and integration area one of the best in the world.

In a separate programme, Khrunichev set up a joint venture for the Rockot cold war missile, offering it as a civilian programme for small satellite launches. In 1995, Khrunichev formed a company with German Daimler-Benz Aerospace to market Rockot, and it acquired a new company name, Eurockot. In 1999, it won its first large-scale orders and, in 2000, the Eurockot company was part-bought in turn by the German company Astrium.

The Lockheed Khrunichev deal set a marker for Russia's other rockets. Lockheed's great American rival, Boeing, not to be outdone, entered a rival joint venture with NPO Yuzhnoye for the marketing in the West of the Zenit booster, paving the way for the Yuzhnoye/Energiya Boeing Sea Launch project (see Chapter 6). Rockwell established a joint venture for marketing the Tsyklon while Assured Access made a similar arrangement with NPO Polyot for the Cosmos 3M. Rocket-maker Energomash made the deal of the century, earning over $1bn from the sale of its RD-180 rocket engine to the American company Pratt & Whitney for the new Atlas III and V, with Kuznetsov not far on its heels, selling the NK-31 and NK-43 to Aerojet. Western rocket manufacturers were thus able to offer a range of rockets to meet clients' needs, including former rival rockets.

THE EUROPEAN DEAL

The main European deal was between the TsSKB Progress plant in Samara and the French Arianespace company in 1996. A joint company, Starsem, was set up in July

Soyuz in Starsem colours

1996 to develop and market the Soyuz rocket to launch commercial payloads. Starsem comprises TsSKB (25%), the Russian Space Agency (25%), Arianespace (15%) and Aerospatiale (35%). Starsem was able to sell launcher space on the R-7 at a cost of 40% less than that of the American Delta 2. Within months, it had found orders to loft American Globalstar communication satellites and, later, the European scientific satellite series Cluster.

Originally, the idea was that Starsem would market the existing Soyuz rocket while developing a successor called the Soyuz 2 or Rus. In the event, the Rus project receded, Starsem instead fitting ever more powerful upper stages to Soyuz—Ikar (for Globalstar) and Fregat (for Cluster). Fregat was to be replaced in due course by the ST with new Ariane-based fairings, uprated engines and digital controls. In mid-2000, the Soyuz ST won its first contract—for 32 SkyBridge comsats to be launched at monthly intervals from 2002 to 2004.

Starsem provided capital to improve facilities at Samara but especially at Baikonour. The old and increasingly rundown assembly hall for the R-7, which dated to Sputnik, was replaced by new facilities in the Energiya assembly building. Here, Starsem invested €35m in three white rooms of world standard—a payload preparation facility, hazardous processing facility, and integration facility—and a hotel for staff on site.[89] Later, additional resources were invested in improving the two Soyuz launch pads, 1 and 31. By 2000, TsSKB could look forward to ramping up its production line.

GLOBALIZATION

In the course of time, these ventures became part of the ever-more interlocked global economy, spanning Russian, European and American corporations. Rockot was part of Eurockot, which in turn was part of Khrunichev-Daimler Chrysler; Soyuz was produced by Starsem which was part-owned by Aerospatiale. Thus Ariane was able to offer its own Ariane 5 large rocket, Soyuz and Eurockot for large, medium and small-sized satellite customers respectively. International Launch Services was likewise able to offer Titan, Proton and Atlas as a launch family, with the prospect of Angara later.

Russia's increasing integration into the world economy was further evident when its socialist-bloc economic associations were disbanded. Intercosmos was disbanded in 1994 and Russia became the 41st member of the European telecommunications satellite organization, Eutelsat.

These joint ventures and partnerships have not necessarily met with total approval in Russia. Some resent the way in which the rocket programme's family silver has been sold off at bargain basement prices to rivals who stand to gain huge profits from their lifetime's investment. The joint ventures have drawn criticism that they will lead to a brain, patent and knowledge drain to the United States and that the once-great Russian rocket industry will lose its ingenuity and ability to innovate. Their colleagues counter and say that they have no choice if they are to survive. Most Russian engineers are sanguine about the prospects, one in Samara being recently overheard saying that they had coped with two revolutions already this century (1917 and 1991) and 'we're still here'. The following table summarizes the principal joint ventures.

By the new century, links between Western companies and Russia were as extensive as they were complicated. A study by the Centre for the Analysis of European Security compiled an inventory in 1999 of European–Russian coopera-tion, and itemized 87 joint projects and enterprises. These ranged from rocket

Cluster in final assembly: a Russian–European scientific success

Joint ventures

The main launcher joint ventures

Launcher	Joint venture name	Russian partner	Western partner
Proton	International Launch Services	Khrunichev	Lockheed
Soyuz	Starsem	TsSKB Samara	Aerospatiale
Cosmos	Cosmos USA	KB Polyot	Assured Access Inc.
Zenit 3SL	Sea Launch	Energiya	Boeing
Rockot	Eurockot	Khrunichev	Daimler Benz, Astrium

The main rocket engine joint ventures

Engine	Joint venture name/project	Russian partner	Western partner
RD-170	Atlas III	Energomash	Pratt & Whitney
RD-170	Sea Launch	Energiya/Yuzhnoye	Boeing
Ikar, Fregat, ST	Starsem	TsSKB Samara	Aerospatiale
NK-31,43	Kistler K-1	NPO Trud	Aerojet

The main satellite/module joint ventures

Satellite	Joint venture name	Russian partner	Western partner
Foton	Foton programme	Foton/TsSKB	Kaiser Threde
Ekspress AM, K	Comsats	NPO-PM	Alcatel, Nippon Electric

Sesat—a project between NPO-PM and Alcatel

services (Starsem/TsSKB, Eurockot) to scientific projects (Integral) and communications satellites (Alcatel with NPO-PM).

These business achievement would have been difficult to achieve in normal circumstances, still less the problematic environment of the Russian space programme in the 1990s. They were even more remarkable, considering the ongoing problem which dogged them: quotas.

QUOTAS PROBLEM

Throughout this period, the number of commercial Western satellites which could be launched by Russia was limited, restricted, regulated and supervised by the United States.

During the cold war, the United States had effectively prohibited American companies from flying their satellites on Soviet rockets. The principal reason given was national security, rather than commerce, the United States fearing that the USSR would pick apart a satellite in transit to the launchpad in its hangar late at night and learn its secrets (something the Americans themselves were adept at doing: they once kidnapped Luna 3's back-up model en route to an exhibition in Mexico, disassembled it and put it back together overnight and the Russians never found out). Satellites were classified as munitions. Any company wishing to export a satellite, or satellite parts, was required to obtain an export licence and this would, as a matter of course, be refused. During the cold war, this was an academic issue, for the Soviet Union was not trying to launch Western satellites and Western communications companies in turn saw no reason to engage a Soviet launching organization.

By 1992, the political situation had of course changed. The Proton won its first contract, to fly INMARSAT, but in order to actually carry it out, had to obtain an export licence, a step which required presidential approval from George Bush, which was forthcoming. With Russia pressing Proton sales aggressively at low prices, and with an explosion in demand for new communication satellites, new arrangements were required.

The United States decided that it would agree on principle to American satellites being launched on Russian rockets, but would limit the number and control the prices. In fact, these arrangements were effectively a world-wide regulation, for the term 'American satellite' included any satellite with any American components, however small, since these components required an export licence too. Apart from Russia's own satellites, hardly any other commercial communications satellites in the world did not use at least some American components.

Accordingly, American export requirements constituted a form of regulation of the world launcher market. In effect the commerce department (later, this responsibility was transferred to the state department) would issue so many export licences, providing that other agreements over technology transfer and nuclear proliferation were also agreed. The old paranoias persisted and the Russians had to give guarantees that satellites en route to the launch pad were under American supervision at all times. All business dealings between American defence and space companies and Russian ones were subject to controls designed to prevent the transfer of advanced technology to the Russians (as the Americans were buying Russian know-how, any such transfer was more likely to be in the other direction!). In the late 1990s these controls grew ever tighter, because of US fears of advanced technology falling into the hands of rogue states. Enforcing the regulations presented major problems for Russian–American companies, for they found that deals were held up for months as a result of checking and bureaucratic delay by short-staffed American control offices. Energomash in particular complained that it had never transferred as much as a bolt or a rivet to a rogue state, yet a licensing agreement with the Americans was held up for 400 days while bureaucrats picked the company apart.[90] In one incident, jumpy technology transfer officers objected to Russia supplying Iran with a fire-fighting pump on the basis that the Iranians could re-engineer the technology into a rocket engine.

The first quota negotiations in Moscow led to an intergovernmental agreement on 2 September 1993. Under its terms, the United States permitted Russia nine commercial launches between 1996 and 2001, with the added conditions that the launch prices offered must be not less than 7% below the world average and that there must not be more than two launches in any given year. Granted that the world satellite market during this period could be numbered in hundreds, this was hardly a generous arrangement.

In an amendment to the intergovernmental agreement on 30 January 1996, following a meeting between Al Gore and Viktor Chernomyrdin, this was relaxed to 20 commercial launchings, with the prices not more than 15% below the world average. Eventually, Russia was restricted to not more than 23 commercial launchings by 2001, although the Russians argued that without such restrictions it could win at least 35 American launch orders. Inevitably, negotiations on quotas became caught up in wider issues of non-proliferation and the Russian price for launchers, which invariably undercut American launchers. In June 2000, Russia made another appeal to the United States to lift the quotas when they were due to expire at the end of the year, but the Americans inevitably linked the quotas to a range of other international concerns. However, if the Russians had a difficult time with quotas, for the Chinese it was even worse. Following a highly publicized spy scandal, the Americans effectively grounded the Chinese fleet of Long March launchers and put them out of the world commercial launcher business.

Although the main battleground with quotas was over national security and technology transfer, these issues were largely a proxy for a domestic American economic struggle between satellite-makers and rocket-makers. It was very much in the interest of American rocket-makers to limit the penetration of Russian rockets on the world market and prevent them from undercutting their own prices and higher labour costs. Throughout the 1990s, the satellite-makers fought against the quotas, for they wanted to get their product to 24-hour orbit for their customer at the lowest possible price, be that on an American, Russian or Chinese rocket. But the satellite-makers had a weak lobby on Capitol Hill and were, for example, quite unable to prevent the ban on Chinese launchings. The quotas had numerous perverse effects. NASA complained that Russia lacked funds to build the International Space Station, at the same time as Russia was prevented from earning the resources to build it—by the United States.

Quotas or not, Proton's first commercial Western launch took place in April 1996 when it lofted a Luxembourg communications satellite called Astra 1F into orbit. Over the ensuing years, Proton was to win a significant number of commercial contracts, keeping its production line open, attracting new investment, and earning profits for Lockheed Khrunichev.

Thus by summer 2000, Proton had made 17 successful commercial satellite launches, Sea Launch one and Soyuz/Ikar six. Most of the commercial launches have been telephone and television communications satellites to 24-hour orbits, though Sirius marked an innovation, being the first of a new generation of three

Commercial launches

8 Apr 1996	Astra 1F	Proton	Baikonour	35,772–35,790	1,436	0.04	
6 Sep 1996	Inmarsat 3-2	Proton	Baikonour	35,958–36,136	1,449	2.6	
28 Aug 1997	PAS-5	Proton	Baikonour	35,745–35,830	1,436	0.06	
14 Sep 1997	Iridium 27-33	Proton	Baikonour	776–779	100.41	86.4	
2 Dec 1997	Astra 1G	Proton	Baikonour	35,775–35,819	1,436	0.03	
24 Dec 1997	Early Bird 1	Start	Svobodny	480–488	94.28	97.3	
24 Dec 1997	Asiasat 1	Proton	Baikonour	361–35,999	638	51.1 (fail)	
7 Apr 1998	Iridium 62-68	Proton	Baikonour	776–779	100.39	86.3	
7 May 1998	Echostar 4	Proton	Baikonour	35,778–35,795	1,436	0.04	
30 Aug 1998	Astra 2A	Proton	Baikonour	7,932–35,991	791	15	
4 Nov 1998	PAS-8	Proton	Baikonour	35,600–36,075	1,438	0.34	
9 Feb 1999	Globalstar (4)	Soyuz/ Ikar	Baikonour	1,337–1,356	112.61	52	
15 Feb 1999	Telstar 6	Proton	Baikonour	35,776–35,798	1,436	0.06	
15 Mar 1999	Globalstar (4)	Soyuz/ Ikar	Baikonour	1,409–1,417	114.07	52	
21 Mar 1999	Asiasat 3S	Proton	Baikonour	35,762–35,823	1,436	0.07	
15 Apr 1999	Globalstar (4)	Soyuz/ Ikar	Baikonour	902–944	103.48	51.94	
28 Apr 1999	Abrixas (+ 1)	Cosmos 3M	Kapustin Yar	554–603	96.25	48.44	
20 May 1999	Nimiq 1	Proton	Baikonour	35,777–35,794	1,436	0.06	
18 Jun 1999	Astra 1H	Proton	Baikonour	33,593–35,804	1,380	0.52	
22 Sep 1999	Globalstar (4)	Soyuz/ Ikar	Baikonour	901–957	103.6	51.98	
26 Sep 1999	LM1	Proton	Baikonour	35,782–35,794	1,436	0.03	
10 Oct 1999	Direct TV1R	Zenit 3SL	Kiribati	35,783–35,786	1,435	0.06	
18 Oct 1999	Globalstar (4)	Soyuz/ Ikar	Baikonour	1.394–1,419	113.93	51.96	
22 Nov 1999	Globalstar (4)	Soyuz/ Ikar	Baikonour	897–942	103.41	51.97	
18 Feb 2000	Garuda	Proton	Baikonour	35,776–35,801	1,436	0.08	
18 Apr 2000	Sesat	Proton	Baikonour	36,038–36,090	1,450	0.05	
30 Jun 2000	Sirius 1	Proton	Baikonour				
15 Jul 2000	Champ	Cosmos 3M	Plesetsk	460		87.3	
	Mita Rubin						
16 Jul 2000	Cluster 1	Soyuz/ Fregat	Baikonour	240–18,000			
28 Jul 2000	PAS-9	Zenit 3SL	Kirbati	1,900–35,814 (transfer orbit) 1.2			
9 Aug 2000	Cluster 2	Soyuz/ Fregat	Baikonour				

Piggyback launches for which commercial fees were charged are not included.

radio satellites, designed to provide global digital quality radio. They have been for a mixture of American, Canadian (Nimiq) and Asian customers (Garuda was Indonesian). Lockheed's faith in the reliability of the Proton was vindicated. There was only one failure (Asiasat). In addition to communications satellites, Russia was increasingly contracted to orbit scientific payloads, the best example being the Cluster satellites for the European Space Agency. Another project was Champ, a €40m programme standing for Challenging Minisatellite Payload, developed by the German Aerospace Agency DLR and the Geophysical Research Centre in Potsdam, Berlin. The aim of 522-kg Champ is to measure the Earth's magnetic field most precisely through the use of GPS, laser reflectors, ion-drift meter, star sensor, accelerometer and magnetometer—and an American instrument was carried to measure the choppiness of the seas.

Commercialization was a two-way channel and could also work in reverse. When the independent Russian media company MediaMost wanted to launch its communications satellite Bonum in 1998, it did not go domestic for either its satellite or the launcher, opting for Hughes to build the satellite and a Delta 2 for launcher (22 November 1998). MediaMost came under sharp criticism from the government for its lack of patriotism for not choosing either Russian satellite builders or launchers. MediaMost, which has been considered critical of the government, retorted by pointing out that American satellites were better and that to launch them on a Russian rocket would have involved long bureaucratic delays and heavy import taxes.

TECHNOLOGY TRANSFER TO ROGUE STATES

Were American fears about technology transfer justified? With its economy under pressure, the temptation for Russia to sell military space technology to other countries must have been irresistible, despite international agreements against the proliferation of missiles and nuclear technology.

During the mid-1990s, there were sporadic reports that Russia was selling rocket parts, even entire missiles, to Iran, Iraq and North Korea. These appeared to be confirmed when in August 1998, North Korea astonished the world by announcing that it had placed its first Earth satellite in orbit, though the claim was challenged in the West and few except the North Koreans ever saw the satellite or picked up its signals. During heated congressional hearings on Mir in autumn 1997, congressman Weldron dramatically produced an SS-18 accelerometer and gyroscope retrieved from a river in Iraq where they had been dumped by smugglers: 'Why are we using American dollars to fund technology programmes that are leaking technology to America's enemies?' he asked.[91] In June 2000, the American National Security Agency again accused Russia of selling parts and providing technical advice to rogue states. A late 1999 Central Intelligence Agency report leaked in autumn 2000 alleged that Russian companies were continuing to supply ballistic missile technology to Iran.

It is very difficult to establish whether Russia has actively aided developing countries in their attempts to build rockets which can launch intercontinental ballistic missiles or put satellites into orbit. Four countries—North Korea, Iraq, Pakistan and Syria—had developed long-range versions of the Scud missile, which is the old R-11 rocket developed by the Korolev and Makeev bureaux from 1953 to 1959 and subsequently exported long before anti-proliferation restrictions were agreed. It is just possible, using upper stages, to modify the Scud in such a way that it can put a very small satellite into orbit, as the Iraqis may have attempted to do (the *Al Abbas* launcher) and which the North Koreans undoubtedly tried to do in 1998 (*Taepo Dong*).[92] However, there is a considerable difference between exporting missiles in the 1960s and the government actively aiding these countries now to build up an intercontinental ballistic missile capacity. A nuclear warhead would, by definition, be a heavy object and far beyond the lifting capacities of a Scud's intercontinental range.

TECHNOLOGY TRANSFER TO CHINA

By contrast, one country where technological cooperation is proven and well documented is China. This has, without question, earned additional resources for the Russians.

There was close cooperation between the Soviet Union and China during the 1950s. The USSR gave the Chinese two German A-4s and invited Chinese students in groups of 50 to study Soviet rockets. They were sent packing when Khrushchev and Mao Zedong split in 1960, and not until early 1992 were working relations restored and the Chinese returned to Moscow. This set in train a series of regular exchanges and visits, and Chinese specialists participated in the Bion 10 mission later that year.

Chinese engineers and scientists came to Moscow to study the Soyuz spacecraft, Russian ground and tracking facilities and the environmental control systems for manned spacecraft. In spring 1995, Russia and China reached an intergovernmental agreement on space cooperation, specifying Russian assistance to China in the area of manned spaceflight. The Chinese bought a number of hardware items, principally a docking system and environmental control systems, both for hard currency. Two cosmonaut instructors, Wu Tse and Li Tsinlung, underwent a two-year-long training course in the Yuri Gagarin cosmonaut training centre. It was unclear if they would pilot the first Chinese spaceship or whether they would pass on their skills and become the trainers of the Chinese cosmonaut squad. At the conclusion of their training, they left Star Town on 19 November 1998 and two years later it was reported that a Chinese cosmonaut squad comprising fighter pilots was in formation in Beijing. Two Russian cosmonauts, Anatoli Berezovoi and Anatoli Filipchenko, went to Beijing to provide technical consultancy for the Chinese.

An unmanned version of the first manned Chinese spaceship, called the Shenzhou ('heavenly vessel'), made its maiden flight a year later. The Chinese later announced that, after a second rehearsal, they would be ready to fly a manned mission in 2001.

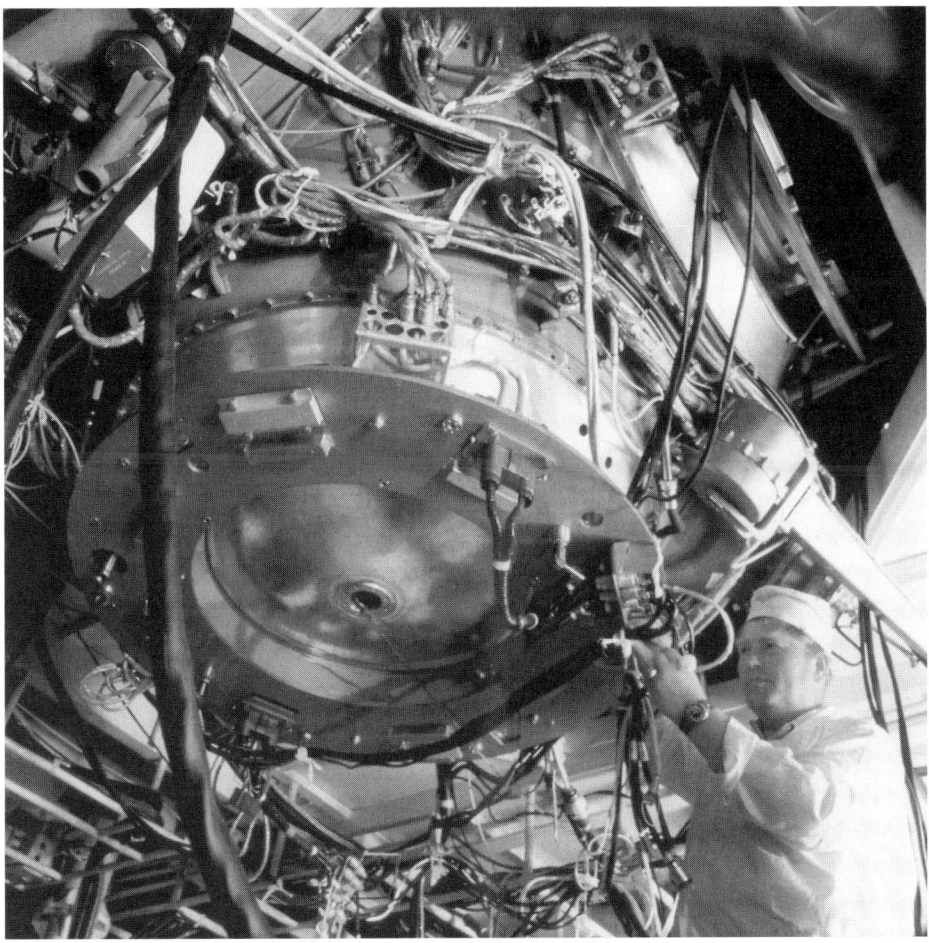

Docking systems under test

Shenzhou resembled Soyuz, as did the upper part of its Long March 2F launcher which had a shroud and escape system quite similar to Soyuz. In the West, it was alleged that China had simply copied the Soyuz. Such criticism ignored the manner in which the Chinese had always worked hard to develop their own indigenous technology. Yuri Koptev described the 1995 agreement as one in which Russia helped China 'fill in some of the gaps'. The orbital module of the Shenzhou was twice the size of the Soyuz orbital module and had its own solar panels, indicating that it would later be used as a small space station module in its own right.

Further talks took place between the Russian and Chinese space agencies in Beijing in May 2000. Soon afterwards, China spoke of its plans to develop first a Salyut-class orbital space station and then a Mir-class station. During spring 2000, there were reports that Russia might even sell the Mir space station to the Chinese. The Chinese seem to have baulked at this and confined themselves to purchasing defined and limited items of hardware and technical know-how for aspects of manned spacecraft and space station design. The commercial value of these links to the Russian space programme is not known, but after 10 years of international trading, the Russians probably knew at this stage how to strike a deal.

COMMERCIALIZATION IN PERSPECTIVE

The Russian space programme was the least commercialized in the world during the Soviet period and quickly became one of the most commercialized. Although the process of transition had begun in the final two or three years of the rule of Mikhail Gorbachev, commercialization proceeded with a vengeance from early 1992 onwards. Rather than allow the programme to sink without trace in financial collapse and bankruptcy, the leadership of the Russian space programme and the design bureaux quickly reoriented themselves around the new economic realities. While some aspects of commercialization were clumsy, by the end of the decade most Russian space companies had managed to arrange joint ventures or other forms of partnership with Western companies. The outcomes have been uneven, with modest results for some companies and substantial successes for others (e.g. Khrunichev, Energomash). Several design bureaux and companies like Energiya now had significant export earnings. The immediate outcome of the transition was the sale of the RD-180 and NK engines, 17 commercial Proton launches, six Globalstars and a number of minor missions, with many more promised. These missions were a growing proportion of the Russian space programme, as the table below shows.

Commercial launches as proportion of all Russian launches

Year	Commercial launches	Proportion of all launches
1996	2	7%
1997	5	18%
1998	5	21%
1999	12	46%
2000	7	33%
	31	24%

By 2000, Russia had what the Soviet Union never had: a commercial space programme. This had been achieved in the face of economic turmoil and an American regulatory regime which restricted the global launcher market and ensured that it was not, at least for the Russians, a level playing field. Launches alone understate the level and importance of joint ventures, for they have also been important in other, less visible, areas of space technology, like communications satellites (NPO-PM), engines (Energomash, Kuznetsov) and materials processing (Foton).

11

The failed frontier?
Concluding comments

At the end of the Soviet period of the space programme, the official press agency published a glossy brochure of the years that lay ahead: a new orbital complex, regular resupply missions by the Buran space shuttle, Mars landers and rovers, orbiting observatories like Spektr. Soviet scientists projected a return to the Moon to map the chemical composition of its surface, aerospaceplanes, the Tsiolkovsky probe to loop around Jupiter and fly to the Sun and eventually a manned mission to the planet Mars to plant there the red flag.

None of this happened. From 1992 onwards, under the Russian space programme, Buran was cancelled, the Mars 96 probe crashed ignominiously, the unmanned scientific programme was almost wound up, Russia's military could barely keep watch on its enemies from above and there was no longer any talk about going to Mars. Tsiolkovsky's visions of exploring the solar system were relegated to the archive. The historical artefacts of the golden years of Soviet rocketry were pawned off, and ageing, retired designers mused nostalgically about the good old days. When cosmonauts did fly into space, now a less frequent event, the two or three helmeted and spacesuited adventurers were accompanied to the launch pad by a man in the black flowing gowns of a priest of the orthodox church to bless them on their way. They needed all the luck they could get: quality control was no longer what it was. Rockets failed for sheer carelessness. The cosmodromes rotted before their very eyes and the always reliable rocket troops were now mutinous. The staff and technicians had left for greener pastures abroad and more enterprising companies at home. The beautiful ships of the tracking fleet had been cut up for scrap. The exhibition hall in Moscow, which had proudly displayed Soviet space achievements and where millions had gasped in wonderment at what had been achieved, was now a car salesroom. Nobody knew who the cosmonauts were, and nobody cared. The space shuttle was a children's amusement in Gorky Park. If ever there was a nemesis of a great and noble project, the new Russian space programme surely was. The cosmic frontier had turned out to be as illusory as the communist ideal which had been its host and mentor.

Or was it?

From 1992 onwards the Russian space programme went through a traumatic adjustment to radically different economic circumstances, almost all of which were entirely beyond its control. Its budget, we now know, fell substantially, probably by at least three-quarters in real terms. The programme lost between a half and seven-eighths of its staff—probably about three-quarters. By 2000, the budget of the civil Russian space programme was running at about the same level as that in India. Russia lost all its overseas space facilities and had to spend $115m a year to rent its own cosmodrome from the Kazakhs, money it did not really have. However, despite this, Russia

- managed to keep the Mir space station in continuous operation until 1999, recommissioning it briefly the following year;
- built, launched and docked the two core modules of the International Space Station, the Zarya Functional Control Block and the Zvezda service module;
- maintained a military space programme, running an ongoing programme of Kobalt and Neman spy satellites and introducing new generations of photo-reconnaissance satellites, Orlets 2 and Arkon;
- maintained a vigil of electronic intelligence satellites, Tselina 2 and US-P and sustained its systems of navigation satellites (Parus) and early warning spies-in-the sky (Oko); and
- maintained a civilian space programme, continuing to fly missions for Earth resources (Resurs), materials processing (Foton), ocean surveillance (Okean) and science (Interball and Koronas).

The Soviet generation of communications satellites was phased out and replaced by new generations of much improved communications satellites with longer lifetimes (Ekspress).

Despite financial collapse, Russia was able to maintain a reasonable launch rate. Although, from 1996, Russia slipped from being the world's largest launcher of satellites to second place, it was a close second place. In 1999, Russia was launching 37% of all the satellites in the world, not far behind the United States at 41%, still making it a space superpower, and not falling to the level of the new spacefaring nations.

The Russian space programme demonstrated a high level of adaptability to the new, difficult and uncertain economic conditions. This was most clearly shown when:

- Russia appointed a new space agency within months of the fall of communism;
- the Energiya corporation attracted in a regular set of foreign funds from Europe and the United States to maintain missions to Mir;
- 87 space-based companies entered joint ventures with American and European companies to sustain and develop their projects;
- Russia developed a commercial space programme within four years, a feature absent from the Soviet space programme;

Mir in 1992: the last achievement of the Soviet space programme

- Plesetsk cosmodrome was adapted for new launchers, Volgograd station was reopened and an entirely new cosmodrome, Svobodny, was commissioned and fired its first satellites into orbit;
- four new types of rocket were introduced: Rockot, Start, Dnepr and Shtil, the last involving submarine-based launches; and a new launch platform, Sea Launch, was opened in the Pacific Ocean; and
- four new upper stages were introduced—Ikar, Fregat, Briz K and Briz M—with a new stage forthcoming, the Soyuz ST.

The most dramatic evidence of the ability of the programme to survive was the Mir space station. Although the station had been designed for five years' operation, it was still orbiting 14 years later. Despite a succession of problems—nosecone shortages, delays in rockets or their engines not being available, the loss of sintine fuel, difficulty of access to the landing sites, the fire, the collision, the lack of communication relays through comships and Luch relay satellites—the programme managed to carry on. Mir was not just the last great achievement of the Soviet space programme, but the first great achievement of the Russian space programme. During the Russian period, Mir hosted 16 long-stay crews, docked with two modules, hosted eight shuttle visits, welcomed a host of international visitors, received 32 Progress supply ships and was the home for 52 spacewalks. Mir pushed back the frontiers of medical science, making six-month flights routine, one doctor

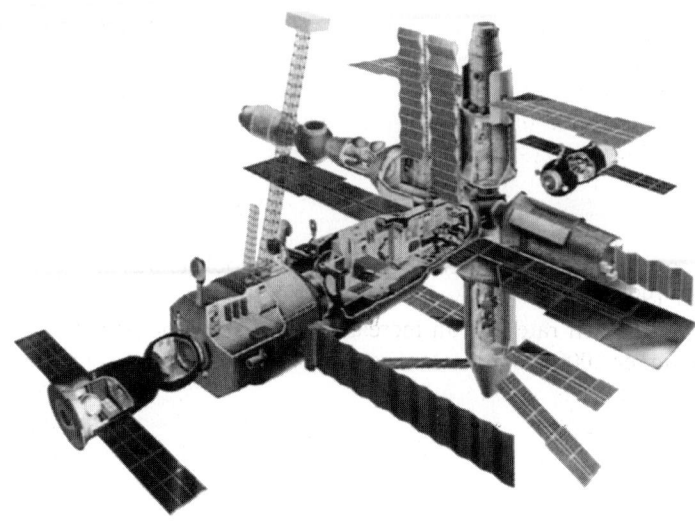

The completed MIR station

staying on Mir for a year and a quarter. Despite the catastrophic events of the summer of 1997, the station continued to operate as the only permanent scientific station above our planet, and by the time Sergei Zalotin and Alexander Kaleri left Mir in the summer of 2000, it was in good working order.

The International Space Station provided a lifeline equally to both American and Russian space programmes. Thankfully, both had leaders, Goldin and Koptev, and presidents, Clinton and Yeltsin, who were able to grasp the potential for the joint venture and quickly translate it into a viable project. For Russia, just as for the United States, the International Space Station involved a substantial level of commitment in terms of hardware, facilities and station crews. The considerable difficulty which Russia experienced in trying to find the resources for the completion of the Zvezda module obscured the fact that Russia remained the second largest financial contributor to the project, the provider of two of the four core modules (Zarya, Zvezda, to join with Unity and Destiny) and would be supplying a stream of Soyuz manned spacecraft, Soyuz taxis, Progress supply ships and, later, the Universal Docking Module.

THE FINAL COLLAPSE?

During 1992–3, several leaders of the space programme, along with independent commentators, predicted for certain the collapse of the Russian space programme in 12 to 18 months. Many renewed their predictions as Russia's launch rate fell dramatically in 1995, when Zvezda remained obdurately on the ground and when the ruble collapsed in 1998. Cosmonauts would no longer fly and the cosmodromes would fall silent, except possibly for a few contracted-out Western launches on some old, bargain-basement rockets. So far, these Cassandras have not yet been proved right. For an enterprise supposedly on its final legs, the Russian space programme has been showing unusual signs of life:

- The first new group of cosmonauts for many years completed their training in 2000.
- The production line of the Soyuz and Proton rockets was increased for the first time in 1999.
- The 1999 launch rate was an increase over that of the previous year.
- The summer 2000 launch rate was the fastest for several years.
- The first hardware of the new Angara launcher appeared in 2000 and what had seemed to be a paper project made progress towards its first flight.
- Yet another version of the Soyuz is being introduced, the improved TMA model.

It is possible that 1998 marked the low point of the extreme financial and organizational pressure inflicted on the Russian space programme. The most defining moment in the decline of the Russian space programme probably took place during that fateful, frighteningly fast and terrifying moment of 25 June 1997 when the Progress M-34 space freighter crashed into Mir, the hull depressurized, the sirens rang out and station commander Vasili Tsibliev ordered Michael Foale into the Soyuz for an emergency escape and the irrevocable abandonment of the space station. It was while he waited and waited there that he wondered why his colleagues were not coming to join him for their plunge to Earth—until it occurred to him that they had no intention of leaving and that you did not just abandon the pride of the Russian space fleet without a fight. Quickly back in action, the three men went on to save the station—and probably the Russian space programme with it. The rest we know. Had Mir been abandoned there and then, as indeed the safety rules said it should have been, it would have been a very steep mountain indeed for the demoralized Russia ever to put cosmonauts back into space. If anyone saved the Russian space programme, it was Vasili Tsibliev, Sasha Lazutkin and their good friend Michael Foale who did so that fateful day.

The commitment of the Russians to their space programme in general, and the Mir station in particular, was something which many outside observers found hard to understand. For many Americans, involved in the International Space Station project and exasperated by the endless delays, the attachment to Mir was obsessive and irrational. You junked it and you moved on. Not so. By contrast, an enduring feature of the Soviet and the Russian space programme is its sense of history. It is not one universally shared in a country which has endured much hardship and where many people have more immediate and pressing concerns on their mind, but it is one held by enough people to matter. Soviet and Russian achievements in space had been built up painstakingly, painfully, over many years. Their founders had learned in the hard school of the camps, the wartime frontline, the early

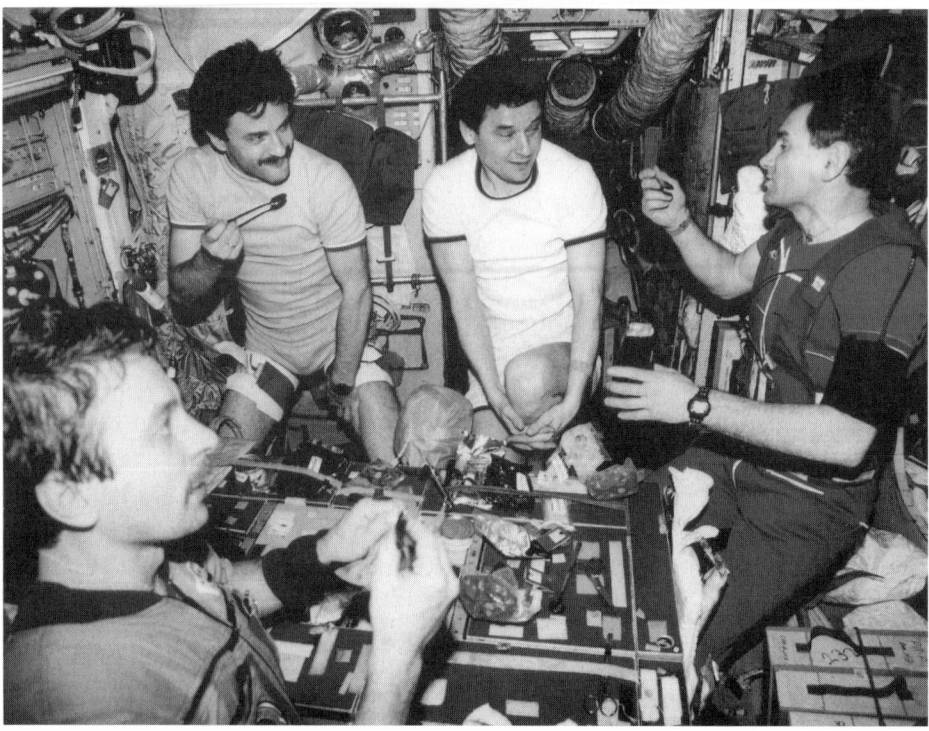

Cosmonauts on board Mir, including Vasili Tsibliev (right). Is he the man who saved the Russian space programme?

rockets that always exploded and the Brezhnevite bureaucracy. They had known the heartbreaking failures, the loss of two Soyuz crews, the satellites that went silent, the upper stages that would not be tamed, the Moon race they could not win, and the Mars probes that disappeared. But they also remembered the night the Sputnik was launched, the day they hit the Moon, the glory of Gagarin's flight, Terreskhova, the spacewalk, the soft-landing on the Moon, the pictures from Venus, the first space station and now Mir. These things had enabled the Soviet Union and Russia to walk tall in the world, to mark out space exploration as a unique arena of accomplishment.

It was a space programme in which its participants and admirers could justifiably take immense pride—a programme built on a potent mixture of courage, endurance, daring, engineering genius, quality and imagination. It was a programme which had deep historical roots—going back to Tsiolkovsky in the 1890s, Kondratyuk's writings during the First World War, Tsander's plans to go to Mars, and Glushko's first experiments in the Gas Dynamics Laboratory in 1921. It was a programme which predated the communist experiment and not one which should be bound by its time frame. In keeping the Russian space programme alive, its

'Man will ascend into the expanse of the heavens and found a settlement there' (Tsiolkovsky)

engineers and scientists, now joined by its managers and accountants, were keeping alive a dream that went back two centuries. Tsiolkovsky said it best, in 1897:

Mankind will not remain forever on the Earth. In pursuit of light and space he will timidly at first probe the limits of the atmosphere and later extend his control throughout the solar system. Man will ascend into the expanse of the heavens and found a settlement there. The impossible of today will become the possible of tomorrow.

Appendix

Russian space launchings, 1992–2000

1992

Date	Name	Vehicle	Site	Orbit	Period	Incl.	Type
21 Jan 1992	Cosmos 2175	Soyuz U	Plesetsk	184–337	89.74	67.14	Kobalt
24 Jan 1992	Cosmos 2176	Molniya M	Plesetsk	604–39,747	717.68	62.8	Oko
25 Jan 1992	Progress M-11	Soyuz U	Baikonour	264–304	90.13	51.6	Freighter
27 Jan 1992	Cosmos 2177–2178	Proton	Baikonour	19,111–19,149	675.74	64.81	GLONASS
17 Feb 1992	Cosmos 2180	Cosmos 3M	Plesetsk	962–1,016	104.94	82.93	Parus
4 Mar 1992	Molniya 1-83	Molniya M	Plesetsk	620–39,731	717.69	62.81	Comsat
9 Mar 1992	Cosmos 2181	Cosmos 3M	Plesetsk	973–1,013	105.03	82.95	Nadezhda
17 Mar 1992	Soyuz TM-14	Soyuz U2	Baikonour	263–303	90.11	51.6	Manned
1 Apr 1992	Cosmos 2182	Soyuz U	Plesetsk	166–315	89.33	67.15	Kobalt
2 Apr 1992	Gorizont 25	Proton	Baikonour	35,769–35,796	1,435	1.44	Comsat
8 Apr 1992	Cosmos 2183	Soyuz U2	Baikonour	240–294	89.85	64.87	Neman
15 Apr 1992	Cosmos 2184	Cosmos 3M	Plesetsk	967–1,014	104.95	82.94	Parus
19 Apr 1992	Progress M-12	Soyuz U	Baikonour	279–321	90.46	51.6	Freighter
29 Apr 1992	Resurs F-14	Soyuz U	Plesetsk	231–235	89.22	82.09	Resources
29 Apr 1992	Cosmos 2185	Soyuz U	Baikonour	211–279	89.43	69.97	Kometa
28 May 1992	Cosmos 2186	Soyuz U	Plesetsk	182–352	89.94	62.85	Kobalt
3 Jun 1992	Cosmos 2187–2194	Cosmos 3M	Plesetsk	1,402–1,480	114.93	74.01	Strela 1
23 Jun 1992	Resurs F-15	Soyuz U	Plesetsk	226–233	89.15	82.32	Resources
30 Jun 1992	Progress M-13	Soyuz	Baikonour	391–414	92.51	51.6	Freighter
1 Jul 1992	Cosmos 2195	Cosmos 3M	Plesetsk	958–1,011	104.85	82.93	Parus
8 Jul 1992	Cosmos 2196	Molniya M	Plesetsk	590–39,733	717.14	62.95	Oko
13 Jul 1992	Cosmos 2197–2202	Tsyklon 3	Plesetsk	1,397–1,416	113.99	82.59	Strela 3
14 Jul 1882	Gorizont 26	Proton	Baikonour	35,792–35,799	1,435	1.48	Comsat
24 Jul 1992	Cosmos 2203	Soyuz U	Plesetsk	190–312	89.53	62.81	Kobalt
27 Jul 1992	Soyuz TM-15	Soyuz U2	Baikonour	223–344	90.12	51.6	Manned
30 Jul 1992	Cosmos 2204	Proton	Baikonour	19,121–19,140	675.76	64.83	GLONASS
30 Jul 1992	Cosmos 2207	Soyuz U	Plesetsk	237–355	90.49	82.33	Oblik
6 Aug 1992	Molniya 1-84	Molniya M	Plesetsk	632–39,721	717.73	62.85	Oko
12 Aug 1992	Cosmos 2208	Cosmos 3M	Plesetsk	788–807	100.86	74.04	Strela 2
15 Aug 1992	Progress M-14	Soyuz U	Baikonour	288–352	90.87	51.63	Freighter

19 Aug 1992	Resurs F-16	Soyuz U	Plesetsk	221–238	89.15	82.57	Resources
	Pion 5, 6						
10 Sep 1992	Cosmos 2209	Proton	Baikonour	35,764–35,806	1,435	1.32	Prognoz
22 Sep 1992	Cosmos 2210	Soyuz U	Plesetsk	150–349	89.76	67.26	Kobalt
8 Oct 1992	Foton 8	Soyuz U	Plesetsk	220–359	90.31	62.81	Materials
14 Oct 1992	Molniya 3-42	Molniya M	Plesetsk	477–39,877	717.75	62.84	Comsat
20 Oct 1992	Cosmos 2211–2216	Tsyklon 3	Plesetsk	1,400–1,415	114	82.59	Strela 3
21 Oct 1992	Cosmos 2217	Molniya M	Plesetsk	599–39,757	717.79	62.95	Oko
27 Oct 1992	Progress M-17	Soyuz U	Baikonour	269–323	90.38	51.6	Freighter
29 Oct 1992	Cosmos 2218	Cosmos 3M	Plesetsk	968–1,015	105	89.92	Parus
30 Oct 1992	Ekran 20	Proton	Baikonour	35,575–35,692	1,428	1.58	Comsat
15 Nov 1992	Resurs 500	Soyuz U	Plesetsk	187–303	89.58	89.46	Commem.
17 Nov 1992	Cosmos 2219	Zenit 2	Baikonour	849–855	102	71.01	Tselina 2
20 Nov 1992	Cosmos 2220	Soyuz U	Plesetsk	167-342	89.62	67.14	Kobalt
24 Nov 1992	Cosmos 2221	Tsyklon 3	Plesetsk	636–665	97.8	82.51	Tselina D
25 Nov 1992	Cosmos 2222	Molniya M	Plesetsk	591–32,288	708.15	62.88	Oko
27 Nov 1992	Gorizont 27	Proton	Baikonour	35,814–36,528	1,435	0.1	Comsat
2 Dec 1993	Molniya 3-43	Molniya M	Plesetsk	416–39.939	717.77	62.83	Comsat
9 Dec 1992	Cosmos 2223	Soyuz U2	Baikonour	241–293	89.85	64..66	Neman
17 Dec 1992	Cosmos 2224	Proton	Baikonour	35,877–36,179	1,448	2.3	Prognoz
22 Dec 1992	Cosmos 2225	Soyuz U	Baikonour	214–309	89.73	64.91	Orlets 1
22 Dec 1992	Cosmos 2226	Tsyklon 3	Plesetsk	1,479–1,526	116.03	73.63	Geodetic
25 Dec 1992	Cosmos 2227	Zenit 2	Baikonour	849–854	101.96	71.02	Tselina 2
25 Dec 1992	Cosmos 2228	Tsyklon 3	Plesetsk	633–669	97.75	82.53	Tselina D
29 Dec 1992	Cosmos 2229	Soyuz U	Plesetsk	218–376	90.45	62.81	Bion 10

1993

12 Jan 1993	Cosmos 2230	Cosmos 3M	Plesetsk	973–1,007	104.91	82.94	Nadezhda
13 Jan 1993	Molniya 1-85	Molniya M	Plesetsk	606–39,746	717.72	62.85	Comsat
19 Jan 1993	Cosmos 2231	Soyuz U	Plesetsk	179–376	90.06	67.13	Kobalt
24 Jan 1993	Soyuz TM-16	Soyuz U2	Baikonour	391–394	92.41	51.6	Manned
26 Jan 1993	Cosmos 2232	Molniya M	Plesetsk	591–39,363	709.66	62.78	Oko
9 Feb 1993	Cosmos 2233	Cosmos 3M	Plesetsk	954–1,009	104.72	82.94	Parus
17 Feb 1993	Cosmos 2234–2236	Proton	Baikonour	19,115–19,138	675	64.85	GLONASS
21 Feb 1993	Progress M-16	Soyuz U2	Baikonour	388–392	92.36	51.6	Freighter
25 Mar 1993	Raduga 29	Proton	Baikonour	35,981–36,97	1,454	1.42	Comsat
25 Mar 1993	Start 1	Start 1	Plesetsk	684–970	101.44	75.76	Demo
26 Mar 1993	Cosmos 2237	Zenit 2	Baikonour	849–853	101.95	71.02	Tselina 2
30 Mar 1993	Cosmos 2238	Tsyklon M	Baikonour	404–418	92.78	65	US-P
31 Mar 1993	Progress M-17	Soyuz U	Baikonour	255–310	90.16	51.6	Freighter
1 Apr 1993	Cosmos 2239	Cosmos 3M	Plesetsk	967–999	104.75	82.93	Parus
2 Apr 1993	Cosmos 2240	Soyuz U	Plesetsk	190–322	89.62	62.85	Kobalt
6 Apr 1993	Cosmos 2241	Molniya M	Plesetsk	641–39,772	718.95	62.88	Oko
16 Apr 1993	Cosmos 2242	Tsyklon 3	Plesetsk	634–668	97.74	82.53	Tselina D
21 Apr 1993	Molniya 3-44	Molniya M	Plesetsk	614–40,618	735.63	62.83	Comsat
27 Apr 1993	Cosmos 2243	Soyuz U2	Baikonour	192–236	88.77	70.35	Neman
28 Apr 1993	Cosmos 2244	Tsyklon M	Baikonour	404–418	92.78	65.03	US-P
11 May 1993	Cosmos 2245–2250	Tsyklon	Plesetsk	1,398–1,417	113.95	82.59	Strela 3

21 May 1993	Resurs F-17	Soyuz U	Plesetsk	230–237	89.17	82.57	Resources
22 May 1993	Progress M-18	Soyuz U	Baikonour	390–391	92.36	51.6	Freighter
26 May 1993	Molniya 1-86T	Molniya M	Plesetsk	401–40,884	736.72	62.88	Comsat
16 Jun 1993	Cosmos 2251	Cosmos 3M	Plesetsk	781–806	100.74	74.04	Strela 2
24 Jun 1993	Cosmos 2252–2257	Tsyklon 3	Plesetsk	1,409–1,481	114.77	82.59	Strela 3
25 Jun 1993	Resurs F-18	Soyuz U	Plesetsk	223–241	89.14	82.58	Resources
1 Jul 1993	Soyuz TM-17	Soyuz U	Baikonour	388–395	92.39	51.6	Manned
7 Jul 1993	Cosmos 2258	Tsyklon M	Baikonour	404–418	92.78	65.05	US-P
14 Jul 1993	Cosmos 2259	Soyuz U	Plesetsk	170–366	89.99	67.12	Kobalt
22 Jul 1993	Cosmos 2260	Soyuz U	Baikonour	241–297	89.89	82.29	Resurs T
4 Aug 1993	Molniya 3-45	Molniya M	Plesetsk	412–39,931	717.53	62.8	Comsat
10 Aug 1993	Cosmos 2261	Molniya M	Plesetsk	582–39,760	717.52	62.89	Oko
10 Aug 1993	Progress M-19	Soyuz U	Baikonour	387–393	92.36	51.6	Freighter
24 Aug 1993	Resurs F-19	Soyuz U	Plesetsk	224–234	89.08	82.59	Resources
31 Aug 1993	Meteor 2-21	Tsyklon 3	Plesetsk	938–969	104.12	82.55	Weather
7 Sep 1993	Cosmos 2262	Soyuz U	Baikonour	207–325	89.83	64.89	Orlets 1
16 Sep 1993	Cosmos 2263	Zenit 2	Baikonour	849–855	101.96	71	Tselina 2
17 Sep 1993	Cosmos 2264	Tsyklon M	Baikonour	403–418	92.77	65.03	US-P
30 Sep 1993	Raduga 30	Proton	Baikonour	35,849–35,920	1,441	1.52	Comsat
11 Oct 1993	Progress M-20	Soyuz U	Baikonour	388–397	92.41	51.6	Manned
26 Oct 1993	Cosmos 2265	Cosmos 3M	Plesetsk	291–1,474	103.68	82.94	At density
28 Oct 1993	Gorizont 28	Proton	Baikonour	35,752–35,789	1,435	1.51	Comsat
2 Nov 1993	Cosmos 2266	Cosmos 3M	Plesetsk	950–1,019	104.79	82.95	Parus
5 Nov 1993	Cosmos 2267	Soyuz U2	Baikonour	240–304	89.95	70.39	Neman
18 Nov 1993	Gorizont 29	Proton	Baikonour	35,039–35,100	1,399	1.48	Comsat
22 Dec 1993	Molniya 1–87	Molniya M	Plesetsk	440–39,182	702.98	62.82	Comsat

1994

8 Jan 1994	Soyuz TM-18	Soyuz U2	Baikonour	385–392	92.24	51.6	Manned
20 Jan 1994	Gals 1	Proton	Baikonour	35,888–35,959	1,443	0.21	Comsat
25 Jan 1994	Meteor 3-6 Tubsat	Tsyklon 3	Plesetsk	1,186–1,208	109.36	82.56	Weather
28 Jan 1994	Progress M-21	Soyuz U	Baikonour	309–334	90.96	51.6	Freighter
5 Feb 1994	Raduga 1-3	Proton	Baikonour	24hr			Comsat
12 Feb 1994	Cosmos 2268–2273	Tsyklon 3	Plesetsk	1,412–1,426	114.2	82.58	Strela 3
18 Feb 1994	Raduga 31	Proton	Baikonour	24hr orbit			Comsat
2 Mar 1994	Koronas 1	Tsyklon 3	Plesetsk	487–528	94.78	82.49	Solar obs.
17 Mar 1994	Cosmos 2274	Soyuz U	Plesetsk	163–350	89.64	67.13	Kobalt
22 Mar 1994	Progress M-22	Soyuz U	Baikonour	255–318	90.24	51.6	Freighter
11 Apr 1994	Cosmos 2275–2277	Proton	Baikonour	19,114–19,143	675	64.82	GLONASS
23 Apr 1994	Cosmos 2278	Zenit 2	Baikonour	849–855	101.97	71.02	Tselina 2
26 Apr 1994	Cosmos 2279	Cosmos 3M	Plesetsk	957–1,007	104.73	82.95	Parus
28 Apr 1994	Cosmos 2280	Soyuz U2	Baikonour	241–306	89.98	70.38	Neman
20 May 1994	Gorizont 30	Proton	Baikonour	35,773–37,787	1,435	1.33	Comsat
22 May 1994	Progress M-33	Soyuz U	Baikonour	398–400	92.53	51.6	Freighter
7 Jun 1994	Cosmos 2281	Soyuz U	Plesetsk	237–296	89.84	82.58	Oblik
14 Jun 1994	Foton 9	Soyuz U	Plesetsk	221–364	90.36	62.81	Materials
1 Jul 1994	Soyuz TM-19	Soyuz U2	Baikonour	396–400	92.52	51.6	Manned

7 Jul 1994	Cosmos 2282	Proton	Baikonour	35,775–35,813	1,435	2.29	Prognoz
14 Jul 1994	Nadezhda 4	Cosmos 3M	Plesetsk	954–1,005	104.68	82.95	Navigation
20 Jul 1994	Cosmos 2283	Soyuz U	Plesetsk	169–330	89.5	67.11	Kobalt
29 Jul 1994	Cosmos 2284	Soyuz U	Baikonour	214–277	89.41	70.38	Kometa
2 Aug 1994	Cosmos 2285	Cosmos 3M	Plesetsk	974–1,013	104.98	74.03	Obzor
5 Aug 1994	Cosmos 2286	Molniya M	Plesetsk	569–39,767	717	62.89	Oko
11 Aug 1994	Cosmos 2287–2289	Proton	Baikonour	19,112–19,134	675	64.88	GLONASS
23 Aug 1994	Molniya 3-46	Molniya M	Plesetsk	605–39,756	717	62.79	Comsat
25 Aug 1994	Progress M-25	Soyuz U	Baikonour	394–398	92.40	51.6	Freighter
26 Aug 1994	Cosmos 2290	Zenit 2	Baikonour	220–315	89.55	64.81	Orlets 2
21 Sep 1994	Cosmos 2291	Proton	Baikonour	35,758–35,817	1,436	1.53	Potok
27 Sep 1994	Cosmos 2292	Cosmos 3M	Plesetsk	400–1,954	108.93	82.99	Density
3 Oct 1994	Soyuz TM-20	Soyuz U2	Baikonour	393–397	92.46	51.6	Manned
11 Oct 1994	Okean 4	Tsyklon 3	Plesetsk	632–666	97.7	82.55	Resources
13 Oct 1995	Ekspress 1	Proton	Baikonour	35,777–35,808	1,4236	0.21	Comsat
31 Oct 1994	Elektro 1	Proton	Baikonour	35,851–35,992	1,441	1.3	Weather
2 Nov 1994	Cosmos 2293	Tsyklon M	Baikonour	404–417	92.78	65.03	US-P
4 Nov 1994	Resurs O1	Zenit 2	Baikonour	661–663	97.98	98.05	Resources
11 Nov 1994	Progress M-25	Soyuz U	Baikonour	393–395	92.43	51.6	Freighter
20 Nov 1994	Cosmos 2294–2296	Proton	Baikonour	18,782–19,135	669	64.89	GLONASS
24 Nov 1994	Cosmos 2297	Zenit 2	Baikonour	849–854	101.97	71	Tselina 2
29 Nov 1994	Geo-1K	Tsyklon 3	Plesetsk	1,480–1,527	116	73.61	Geodetic
14 Dec 1994	Molniya 1-88T	Molniya M	Plesetsk	441–39,910	717	62.78	Comsat
16 Dec 1994	Luch 1	Proton	Baikonour	35,775–35,815	1,436	2.58	Relay
20 Dec 1994	Cosmos 2298	Cosmos 3M	Plesetsk	786–810	100.83	74.03	Strela 2
26 Dec 1994	Rosto	Rockot	Baikonour	1,875–2,253	128.66	64.8	Demo
28 Dec 1994	Cosmos 2299-2304	Tsyklon 3	Plesetsk	1,402–1,416	113.97	82.57	Strela 3
28 Dec 1994	Raduga 32	Proton	Baikonour	35,796–35,900	1,439	1.5	Comsat
29 Dec 1994	Cosmos 2305	Soyuz U2	Baikonour	240–298	89.89	64.91	Neman

1995

24 Jan 1995	Tsikada Astrid Fasat	Cosmos 3M	Plesetsk	965–1,021	104.97	82.97	Tsikada
15 Feb 1995	Progress M-26	Soyuz U	Baikonour	391–397	92.45	51.6	Freighter
16 Feb 1995	Foton 10	Soyuz U	Plesetsk	220–369	90.4	62.81	Materials
2 Mar 1995	Cosmos 2306	Cosmos 3M	Plesetsk	469–517	94.47	65.85	Romb
7 Mar 1995	Cosmos 2307–2309	Proton	Baikonour	19,113–19,149	675	64.8	GLONASS
14 Mar 1995	Soyuz TM-21	Soyuz U2	Baikonour	391–396	92.42	51.6	Manned
22 Mar 1995	Cosmos 2310	Cosmos 3M	Plesetsk	980–1,011	105.02	82.94	Parus
22 Mar 1995	Cosmos 2311	Soyuz U	Plesetsk	168–336	89.54	67.19	Kobalt
9 Apr 1995	Progress M-27	Soyuz U	Baikonour	389–396	92.41	51.6	Freighter
20 May 1995	Spektr	Proton	Baikonour	393–400	92.49	51.6	Mir module
24 May 1995	Cosmos 2312	Molniya M	Plesetsk	604–39,545	717	62.89	Oko
8 Jun 1995	Cosmos 2313	Tsyklon M	Baikonour	403–418	92.77	65.04	US-P
28 Jun 1995	Cosmos 2314	Soyuz U	Plesetsk	166–340	89.56	67.13	Kobalt
5 Jul 1995	Cosmos 2315	Cosmos 3M	Plesetsk	970–1,014	104.94	82.91	Tsikada
20 Jul 1995	Progress M-28	Soyuz U	Baikonour	394–398	92.48	51.6	Freighter

24 Jul 1995	Cosmos 2316–2318	Proton	Baikonour	19,104–19,131	675	64.85	GLONASS
3 Aug 1995	Interball 1	Molniya M	Plesetsk	870–191,752	5,458	62.9	Solar obs
	Magion 4						
9 Aug 1995	Molniya 3-48	Molniya M	Plesetsk	427–39,969	718	62.8	Comsat
30 Aug 1995	Cosmos 2319	Proton	Baikonour	35,806–35,960	1,441	1.48	Potok
31 Aug 1995	Sich	Tsyklon 3	Plesetsk	632–669	97.73	82.53	Resources
	FAISAT						
3 Sep 1995	Soyuz TM-22	Soyuz U2	Baikonour	393–398	92.46	51.6	Manned
26 Sep 1995	Resurs F-20	Soyuz U	Plesetsk	231–235	89.16	82.32	Resources
29 Sep 1995	Cosmos 2320	Soyuz U2	Plesetsk	242–302	89.95	64.92	Neman
6 Oct 1995	Cosmos 2321	Cosmos 3M	Plesetsk	258–793	95.14	82.94	Parus fail
8 Oct 1995	Progress M-29	Soyuz U	Baikonour	393–396	92.45	51.6	Freighter
11 Oct 1995	Luch 1-1	Proton	Baikonour	35,767–35,810	1,436	3.07	Relay
31 Oct 1995	Cosmos 2322	Zenit 2	Baikonour	849–852	101.94	71.02	Tselina 2
17 Nov 1995	Gals 2	Proton	Baikonour	35,787–35,949	1,440	0.17	Comsat
14 Dec 1995	Cosmos 2323–2325	Proton	Baikonour	18,679–19,133	666	64.83	GLONASS
18 Dec 1995	Progress M-30	Soyuz U	Baikonour	391–399	92.46	51.6	Freighter
20 Dec 1995	Cosmos 2326	Tsyklon M	Baikonour	407–415	92.78	65.02	US-P
28 Dec 1995	IRS-1C	Molniya M	Baikonour	816–818	101.24	98.59	Resources
	Skipper						

1996

16 Jan 1996	Cosmos 2327	Cosmos 3M	Plesetsk	952–1,021	104.82	82.98	Parus
25 Jan 1996	Gorizont 31	Proton	Baikonour	36,763–35,882	1,436	1.52	Comsat
19 Feb 1996	Gonetz D1,2,3	Tsyklon 3	Plesetsk	1,400–1,414	113.94	82.58	Strela 3
	Cosmos 2328–2330						
19 Feb 1996	Raduga 33	Proton	Baikonour	242–36,502	645	48.6	Comsat fail
21 Feb 1996	Soyuz TM-23	Soyuz U	Baikonour	391–398	90.22	51.6	Manned
14 Mar 1996	Cosmos 2331	Soyuz U	Plesetsk	164–358	89.73	67.14	Kobalt
8 Apr 1996	Astra 1F	Proton	Baikonour	35,772–35,790	1,436	0.04	Commercial
23 Apr 1996	Priroda	Proton	Baikonour	391–396	92.43	51.6	Mir module
24 Apr 1996	Cosmos 2332	Cosmos 3M	Plesetsk	295–1,565	103.62	82.96	Density
5 May 1996	Progress M-31	Soyuz U	Baikonour	391–396	92.43	51.6	Freighter
5 May 1996	Gorizont 32	Proton	Baikonour	35,725–35,852	1,436	1.47	Comsat
31 Jul 1995/6	Progress M-32	Soyuz U	Baikonour	293–333	90.78	51.6	Freighter
14 Aug 1996	Molniya 1-89T	Molniya M	Plesetsk	464–39,887	717	62.84	Comsat
17 Aug 1996	Soyuz TM-24	Soyuz U	Baikonour	375–390	92.2	51.6	Manned
29 Aug 1996	Interball 2	Molniya M	Plesetsk	769–19,211	347	62.77	Solar obs
	Magion 5						
	Musat						
4 Sep 1996	Cosmos 2333	Zenit 2	Baikonour	849–852	101.94	71.01	Tselina 2
5 Sep 1996	Cosmos 2334	Cosmos 3M	Plesetsk	970–1,000	104.8	82.94	Parus
6 Sep 1996	Inmarsat 3-2	Proton	Baikonour	35,958–36,136	1,449	2.65	Commercial
26 Sep 1996	Ekspress A2	Proton	Baikonour	35,838–35,908	1,440	0.21	Comsat
24 Oct 1996	Molniya 3-48	Molniya M	Plesetsk	610–39,768	718	62.83	Comsat
16 Nov 1996	Mars 8	Proton	Baikonour	139–155	87.43	51.5	Mars (fail)
19 Nov 1996	Progress M-33	Soyuz U	Baikonour	371–390	92.16	51.65	Freighter
11 Dec 1996	Cosmos 2335	Tsyklon M	Baikonour	403–419	92.79	65.05	US-P

20 Dec 1996	Cosmos 2336	Cosmos 3M	Plesetsk	979–1,012	105.03	97.83	Parus
24 Dec 1996	Bion 11	Soyuz U	Plesetsk	217–379	90.48	62.8	Biology

1997

10 Feb 1997	Soyuz TM-25	Soyuz U	Baikonour	378–394	92.28	51.6	Manned
14 Feb 1997	Gonetz D4,5,6 Cosmos 2337–2339	Tsyklon 3	Plesetsk	1,413–1,423	96.88	82.6	Strela 3
4 Mar 1997	Zeya	Start 1	Svobodny	467–480	94.06	97.28	Demo
6 Apr 1997	Progress M-34	Soyuz U	Baikonour	384–397	92.36	51.6	Freighter
9 Apr 1997	Cosmos 2340	Molniya M	Plesetsk	541–39,815	717	62.94	Oko
17 Apr 1997	Cosmos 2341	Cosmos 3M	Plesetsk	978–1,014	105.03	82.92	Parus
14 May 1997	Cosmos 2342	Molniya M	Plesetsk	513–39,381	708	62.83	Oko
15 May 1997	Cosmos 2343	Soyuz U	Baikonour	197–292	89.39	64.86	Orlets 1
7 Jun 1997	Cosmos 2344	Proton	Baikonour	1,509–2,747	130	63.42	Arkon
5 Jul 1997	Progress M-35	Soyuz U	Baikonour	386–392	92.34	51.6	Freighter
5 Aug 1997	Soyuz TM-26	Soyuz U	Baikonour	386–392	92.33	51.6	Manned
14 Aug 1997	Cosmos 2345	Proton	Baikonour	34,295–37,274	1,435	1.3	Prognoz
28 Aug 1997	PAS 5	Proton	Baikonour	35,745–35,830	1,436	0.06	Commercial
14 Sep 1997	Iridium 27-33	Proton	Baikonour	776–779	100.41	86.4	Commercial
23 Sep 1997	Cosmos 2346 Fasat	Cosmos 3M	Plesetsk	939–996	104.92	82.92	Parus
24 Sep 1997	Molniya 1-90	Molniya M	Plesetsk	449–39,906	717	62.84	Comsat
4 Oct 1997	Progress M-36	Soyuz U	Baikonour	382–391	92.29	51.	Freighter
9 Oct 1997	Foton 11	Soyuz U	Plesetsk	218–375	90.45	62.81	Biology
12 Nov 1997	Kupon 1	Proton	Baikonour	33,849–38,086	1,445	0.06	Comsat
18 Nov 1997	Resurs F21 (F1M)	Soyuz U	Plesetsk	180–236	88.66	82.33	Resources
2 Dec 1997	Astra 1G	Proton	Baikonour	35,775–35,819	1,436	0.03	Commercial
9 Dec 1997	Cosmos 2347	Tsyklon M	Baikonour	403–419	92.79	65.04	US- P
15 Dec 1997	Cosmos 2348	Soyuz U	Plesetsk	165–345	89.61	67.15	Kobalt
20 Dec 1997	Progress M-37	Soyuz U	Baikonour	378–389	92.63	51.6	Freighter
24 Dec 1997	Early Bird 1	Start 1	Svobodny	480–488	94.28	97.3	Commercial
24 Dec 1997	Asiasat 1	Proton	Baikonour	361–35,999	638	51.1	Commer. fail

1998

28 Jan 1998	Soyuz TM-27	Soyuz U	Baikonour	379–385	92.19	51.6	Manned
17 Feb 1998	Cosmos 2349	Soyuz U	Baikonour	212–278	89.41	70.37	Kometa
14 Mar 1998	Progress M-38	Soyuz U	Baikonour	376–383	92.14	51.6	Freighter
7 Apr 1998	Iridium 62-68	Proton	Baikonour	776–779	100.39	86.39	Commercial
29 Apr 1998	Cosmos 2350	Proton	Baikonour	35,958–36,089	1,448	2.3	Prognoz
7 May 1998	Cosmos 2351	Molniya M	Plesetsk	525–39,829	717	62.96	Oko
7 May 1998	Echostar 4	Proton	Baikonour	35,778–35,795	1,436	0.04	Commercial
14 May 1998	Progress M-39	Soyuz U	Baikonour	371–379	92.05	51.6	Freighter
15 Jun 1998	Cosmos 2352–2357	Tsyklon 3	Plesetsk	1,310–1,875	118.03	82.59	Strela 3
24 Jun 1998	Cosmos 2358	Soyuz U	Plesetsk	167–334	89.52	67.13	Kobalt
25 Jun 1998	Cosmos 2359	Soyuz U	Baikonour	240–303	89.94	64.71	Neman
1 Jul 1998	Molniya 3-49	Molniya M	Plesetsk	432–39,943	718	62.78	Comsat
7 Jul 1998	Tubsat N, N1	Shtil 1	Barents Sea	401–777	96.45	78.92	Demo

10 Jul 1998	Resurs O1–4	Zenit 2	Baikonour	817–818	101.24	98.79	Resources
	Safir 2						
	Techsat						
	MYSat						
	Fasat Bravo						
	Reflector						
28 Jul 1998	Cosmos 2360	Zenit 2	Baikonour	848–854	101.95	71.02	Tselina 2
13 Aug 1998	Soyuz TM-28	Soyuz U	Baikonour	364–374	91.92	51.6	Manned
30 Aug 1998	Astra 2A	Proton	Baikonour	7,932–35,991	791	15	Commercial
29 Sep 1998	Molniya 1-91	Molniya M	Plesetsk	420–40,657	732	62.82	Comsat
25 Oct 1998	Progress M-40	Soyuz U	Baikonour	355–365	91.74	51.6	Freighter
4 Nov 1998	PAS-8	Proton	Baikonour	35,600–36,075	1,438	0.34	Commercial
20 Nov 1998	Zarya	Proton	Baikonour	388–401	92.44	51.6	Space station
10 Dec 1998	Nadezhda 5	Cosmos 3M	Plesetsk	977–1,013	105.01	82.95	Navigation
	Astrid 2						
24 Dec 1998	Cosmos 2361	Cosmos 3M	Plesetsk	969–1,013	104.91	82.94	Parus
30 Dec 1998	Cosmos 2362–2364	Proton	Baikonour	19,125–19,129	675	64.8	GLONASS

1999

9 Feb 1999	Globalstar (4)	Soyuz/Ikar	Baikonour	1,337–1,356	112.61	52	Commercial
15 Feb 1999	Telstar 6	Proton	Baikonour	35,776–35,798	1,436	0.06	Commercial
20 Feb 1999	Soyuz TM-29	Soyuz U	Baikonour	346–364	91.65 5	1.6	Manned
28 Feb 1999	Raduga 1-4	Proton	Baikonour	36,430–36,563	1,472	1.48	Comsat
15 Mar 1999	Globalstar (4)	Soyuz/Ikar	Baikonour	1,409–1,417	114.07	52	Commercial
21 Mar 1999	Asiasat 3S	Proton	Baikonour	35,762–35,823	1,436	0.07	Commercial
28 Mar 1999	Sea Launch	Zenit 3SL	Kiribati	639–36,064	645	1.21	Demo
2 Apr 1999	Progress M-41	Soyuz U	Baikonour	339–355	91.48	51.6	Freighter
15 Apr 1999	Globalstar (4)	Soyuz/Ikar	Baikonour	902–944	103.48	51.94	Commercial
21 Apr 1999	UOSAT 12	Dnepr	Baikonour	649–652	97.75	64.56	Demo
28 Apr 1999	Abrixas (+1)	Cosmos 3M	Kapustin Yar	554–603	96.25	48.44	Commercial
20 May 1999	Nimiq 1	Proton	Baikonour	35,777–35,794	1,436	0.06	Commercial
18 Jun 1999	Astra 1H	Proton	Baikonour	33,593–35,804	1,380	0.52	Commercial
8 Jul 1999	Molniya 3-50	Molniya M	Plesetsk	468–40,811	736	62.82	Comsat
16 Jul 1999	Progress M-42	Soyuz U	Baikonour	346–353	91.53	51.66	Freighter
17 Jul 1999	Okean 5	Zenit 2	Baikonour	661–663	97.97	98.06	Resources
18 Aug 1999	Cosmos 2365	Soyuz U	Plesetsk	167–343	89.6	67.14	Kobalt
26 Aug 1999	Cosmos 2366	Cosmos 3M	Plesetsk	963–1,008	104.81	82.93	Parus
6 Sep 1999	Yamal 101, 102	Proton	Baikonour	35,512–36,295	1,442	0.01	Comsat
7 Sep 1999	Foton 12	Soyuz U	Plesetsk	217–384	90.53	62.8	Materials
22 Sep 1999	Globalstar (4)	Soyuz U	Baikonour	901–957	103.6	51.98	Commercial
26 Sep 1999	LM1	Proton	Baikonour	35,782–35,794	1,436	0.03	Commercial
28 Sep 1999	Resurs F-22	Soyuz U	Plesetsk	222–230	89.02	82.32	Resources
10 Oct 1999	Direct TV1R	Zenit 3SL	Kiribati	35,783–35,786	1,435	0.06	Commercial
18 Oct 1999	Globalstar (4)	Soyuz/Ikar	Baikonour	1.394–1,419	113.93	51.96	Commercial
22 Nov 1999	Globalstar (4)	Soyuz/Ikar	Baikonour	897–942	103.41	51.97	Commercial
26 Dec 1999	Cosmos 2367	Tsyklon M	Baikonour	404–418	92.78	65.04	US-P
27 Dec 1999	Cosmos 2368	Molniya M	Plesetsk	554–39,720	716	62.83	Oko

2000

Date	Payload	Vehicle	Site	Orbit	Period	Incl.	Type
1 Feb 2000	Progress M1-1	Soyuz U	Baikonour	302–320	90.75	51.6	Freighter
3 Feb 2000	Cosmos 2369	Zenit 2	Baikonour	848–854	101.95	71.01	Tselina 2
8 Feb 2000	Dumsat/Fregat	Soyuz/Fregat	Baikonour	581–607	96.56	64.86	Demo
18 Feb 2000	Garuda	Proton	Baikonour	35,776–35,801	1,436	0.08	Commercial
12 Mar 2000	Ekspress A3	Proton	Baikonour	35,779–35,794	1,436	0.19	Comsat
20 Mar 2000	Dumsat/Fregat	Soyuz/Fregat	Baikonour	243–18,021	320	64.64	Demo
6 Apr 2000	Soyuz TM-30	Soyuz U	Baikonour	327–330	91.1	51.6	Manned
18 Apr 2000	Sesat	Proton	Baikonour	36,038–36,090	1,450	0.02	Commercial
26 Apr 2000	Progress M1-2	Soyuz U	Baikonour	328–330	91.13	51.6	Freighter
3 May 2000	Cosmos 2370	Soyuz U	Baikonour	241–303	89.95	64.76	Neman
16 May 2000	Simsat	Rockot	Plesetsk	544–558	95.67	86.39	Demo
6 Jun 2000	Gorizont 33	Proton	Baikonour				Comsat
24 Jun 2000	Ekspress A6	Proton	Baikonour				Comsat
28 Jun 2000	Nadezhda M-6 Tsinghua 1 SNAP-1	Cosmos 3M	Plesetsk	684–708		83	Navigation
30 Jun 2000	Sirius	Proton	Baikonour				Commercial
5 Jul 2000	Cosmos 2371	Proton	Baikonour				Potok
12 Jul 2000	Zvezda	Proton	Baikonour	185–355	89.6	51.62	Space station
15 Jul 2000	Champ Mita Rubin	Cosmos 3M	Plesetsk	450–460		87.3	Commercial
16 Jul 2000	Cluster 1	Soyuz/Fregat	Baikonour	240–18,000			Commercial
28 Jul 2000	PAS-9	Zenit 3SL	Kirbati				Commercial
6 Aug 2000	Progress M1-3	Soyuz U	Baikonour				Freighter
9 Aug 2000	Cluster 2	Soyuz/Fregat	Baikonour				Commercial

This listing is for payloads reaching orbit, including an incorrect orbit. For a list of launch failures from 1992 to 2000, see page 250.

In the case of multiple launches, the data given are for the first satellite.

Notes and references

1. Thomas Y. Canby: A generation after Sputnik—Are the Soviets ahead in space? *National Geographic Magazine*, vol. 170, no. 4, October 1986.
2. For a history of Mir, see Rex Hall (Ed.): *The history of Mir, 1986–2000*. London, British Interplanetary Society, 2000.
3. Bart Hendricks: The origins and evolution of Mir and its modules. *Journal of the British Interplanetary Society*, vol. 51, 203–222, 1988.
4. See Bert Vis: Crewing history for the Buran programme. *Journal of the British Interplanetary Society*, vol. 50, 3–7, 1997.
5. Thomas Marold: Space shuttle and Buran orbiters – an overview of tests and flights of all space shuttles and Buran orbiters to the end of 1996. *Spaceflight*, vol. 39, December 1997.
6. See Florence David: Le défi du docteur Poliakov. *Ciel & Espace*, janvier 1995.
7. Florence David: Femmes cosmonautes—dûr métier—les dessous de la conquête. *Ciel & Espace*, février 1994.
8. Yuri Usachov: Cosmonaut diary—taking risks without fear. *Aerospace Journal*, March/April 1997.
9. There are several accounts of the epic events of the next hours, days and weeks. Michael Foale's story is told by his father in Colin Foale: *Waystation to the stars*. Headline, London, 1999. Jerry Linenger's story is told in his own book *Off the planet—surviving five perilous months aboard the station Mir* (New York, McGraw-Hill, 2000). A comprehensive account is given in Bryan Burrough: *Dragonfly—NASA and the crisis aboard Mir*. Fourth Estate, London, 1999. A clear and useful summary is given in David J. Shayler: *Disasters and accidents in manned spaceflight*. Springer/Praxis, 2000.
10. Dwayne Allen Day: Mir and the ISS hearings on the hill. Internet posting, 19 September 1997.
11. For a more detailed account of this landing, see Craig Mellow: Aiming for Arkalyk. *Air & Space*, August/September 1998.
12. See Andy Salmon: Science on board the Mir space station. *Journal of the British Interplanetary Society*, vol. 50, no. 8, 283–295, August 1997; David M.

Harland: *The Mir space station – precursor to space colonization*. Wiley/Praxis, 1997.

13. Anders Hansson: *Detecting earthquakes from Mir*. Paper presented to the British Interplanetary Society, 6 June 2000.

14. Anders Hansson: *Wheat in space*. Paper presented to the British Interplanetary Society, 6 June 1998.

15. For EVAs until April 1997, see D. J. Shayler, Outside Mir: Ten years of EVA operations. *Journal of the British Interplanetary Society*, vol. 51, 29–38, 1998.

16. See G. I. Severin, I. P. Abramov, M. N. Doudnik and V. I. Svertshek: *History of creation of Russian space suits, escape and life support means for space vehicle and space station crews*. Paper presented to the International Astronautical Federation Congress, Amsterdam, 4–8 October 1999.

17. See David S. F. Portree and Robert Trevion: *Walking to Olympus—on EVA chronology*. NASA, 1997.

18. D. J. Shayler: The proposed USSR Salyut and US shuttle docking mission c. 1981. *Journal of the British Interplanetary Society*, vol. 44, no. 11, 1991.

19. Bart Hendrickx: The origins and evolution of Mir and its modules. *Journal of the British Interplanetary Society*, vol. 51, 203–222, 1998.

20. For a detailed background and technical description of the ISS, see David M. Harland: *The Mir space station – precursor to space colonization*. Wiley/Praxis, 1997.

21. See David Shayler: *Disasters and accidents in manned spaceflight*. Springer/Praxis, 2000.

22. See Beth Dickey: The captain, the pro and the fighter pilot. *Air & Space*, February/March 2000.

23. See Sven Grahn: *Soyuz emergency landing zones – the Ugol Pasadki story*. Sven's place, 2000 (http://www.users.wineasy.se/svengrahn/histind/ugol/ugol.htm)

24. Jennifer Green: Russian participation in the International Space Station and issues arising. *Journal of the British Interplanetary Society*, vol. 50, 296–302, 1997.

25. See Rex Hall: Civilians in the cosmonaut team. *Journal of the British Interplanetary Society*, vol. 46, 405–408, 1993.

26. For a history of the Zenit programme, see Peter Gorin: Zenit—the first Soviet photoreconnaissance satellite. *Journal of the British Interplanetary Society*, vol. 50, 441–448, 1997.

27. Sven Grahn: *Soviet/Russian reconnaissance satellites*. Sven's place, 1999. http://www.users.wineasy.se/svengrahn/histind/reccess/reccess.htm.

28. For a comprehensive description of this programme, see Peter Gorin: Black amber—Russian Yantar-class optical reconnaissance satellites. *Journal of the British Interplanetary Society*, vol. 51, 309–320, 1998.

29. See Desmond Ball: *The intelligence war in the Gulf*. Australian National University, Canberra, 1991.

30. For an analysis of this programme, see P. S. Clark: Russian fifth generation photoreconnaissance satellites. *Journal of the British Interplanetary Society*, vol. 52, 133–150, 1999.

31. See P. S. Clark: Space débris incidents involving Soviet/Russian launches. *Journal of the British Interplanetary Society*, vol. 47, 379–391, 1994.

32. Phillip S. Clark: The flight of Cosmos 2290. *Journal of the British Interplanetary Society*, vol. 49, 259–266, 1996.

33. Geoff Perry, Stuart Eves and Grant Thomson: And whoever heard about projects Zerkalo, Nord and Arkon-1? *News bulletin of the Astronautical Society of Western Australia*, vol. 23, no. 3, 27–30 December 1998.

34. Asif Siddiqi: Staring at the sea—the Soviet RORSAT and EORSAT programme. *Journal of the British Interplanetary Society*, vol. 52, 397–416, 1999.

35. Maxim Ermakov: *Russian non-geostationary communication satellite systems*. Paper presented to the 50th International Astronautical Federation, 4–8 October 1999.

36. Asif Siddiqi: Staring at the sea—the Soviet RORSAT and EORSAT programme. *Journal of the British Interplanetary Society*, vol. 52, 397–416, 1999.

37. See P. Daly: An introduction to the USSR GLONASS navigation satellites. *Journal of the British Interplanetary Society*, vol. 46, 385–390, 1993.

38. See Christian Lardier: La constellation GLONASS n'est plus complete. *Air & Cosmos*, 1674, 16 octobre 1998.

39. Christian Lardier: La Russie lance un satellite OKO. *Air & Cosmos*, 2 juin 1995.

40. Phillip S. Clark: Russian minor military satellites. *Spaceflight*, vol. 39, July 1997.

41. A most useful description of the series may be found in Phillip S. Clark: Resurs F and related recoverable polar orbit satellites. *Zenit*, 53, July 1991.

42. See Christian Lardier: Les satellites civiles du TsSKB Progress. *Air & Cosmos*, 1696, 26 mars 1999.

43. The best summary of the series may be found in Phillip Clark: Russian and Ukranian satellites to observe the oceans. *Space Policy*, vol. 15, 13–17, 1999.

44. P. S. Clark: Recoverable satellites flown at 81.3–82.6 deg orbital inclinations. *Journal of the British Interplanatary Society*, vol. 50, no. 1, January 1997.

45. Jonathan's space report, no. 385, 1999 Jan. 13, at http://www.harvard.edu/~jcm/space/jar/back/news.385

46. See Joel Powell: German reentry vehicle flies on Foton 11. *Spaceflight*, vol. 40, June 1998.

47. See The Kettering Group: History of the Molniya orbit. *Journal of the British Interplanetary Society*, vol. 52, 1999.

48. Phillip S. Clark: Soviet geosynchronous satellite activity October 1991–May 1992. *Journal of the British Interplanetary Society*, vol. 46. no. 10, 1993.

49. Phil Clark: *Russian geosynchronous satellites from 1990*. Paper presented to the British Interplanetary Society, 6 June 2000.

50. Phillip S. Clark: A review of the Soviet geodetic satellite programme. *Zenit*, vol. 44, October 1990.

51. See La lettre d'information du CNES, no. 27, dec. 1996.

52. See Craig Covault: Moon mission to lead Russian space reforms. *Aviation Week & Space Technology*, 19 May 1997.

53. Roald Z. Sagdeev: *The making of a Soviet scientist*. John Wiley, New York, pp. 232–243, 1994.

54. For a description of the project, see Andy Salmon: *Mars '96*. Paper presented to the British Interplanetary Society, 1 June 1996; and Darren Burnham and Andy Salmon: Mars '96—the long and winding road to Mars. *Spaceflight*, vol. 38, November 1996.

55. Christian Lardier: Retour en force de la propulsion electrique. *Air & Cosmos*, 5 mai 2000.

56. P. Mills: Energiya and Buran at Baikonour. *Spaceflight*, November 1989.

57. See Nigel Evans: *Baikonour—a personal view*. Paper presented to the British Interplanetary Society, 4 June 1994.

58. Neil da Costa: *Star City and the cosmonaut training centre*. Paper presented to the British Interplanetary Society, 5 June 1999.

59. Jaap Tervej: *Kaliningrad*. Paper presented to the British Interplanetary Society, 6 June 1998.

60. See Christian Lardier: Ménaces sur la défence spatiale russe. *Air & Cosmos*, 1705, 28 mai 1999.

61. Russian tracking systems are reviewed by Phillip S. Clark: Space débris incidents involving Soviet/Russian launchers. *Journal of the British Interplanetary Society*, vol. 47, no. 9, 1994; and by Henk H. F. Snid: Soviet command and control. *Journal of the British Interplanetary Society*, vol. 44, no. 11, 1991.

62. For a history of the R-7, see Timothy Varfolomeyev: Soviet rocketry that conquered space. *Spaceflight*, in ten parts: Part 1, vol. 37, August 95; Part 2, vol. 37, February 1996; Part 3, vol. 37, June 1996; Addendum, vol. 38, September 1996; vol. 39, October 1997; Part 4, vol. 40, January 1998; Part 5, vol. 40, March 1998; Part 6, vol. 40, May 1998; Part 7, vol. 40, September 1998; Part 8, vol. 40, December 1998; Part 9, vol. 41, May 1999; Part 10, vol. 42, April 2000.

63. See Christian Lardier: Fiabilité record pour le lanceur semiorka. *Air & Cosmos*, 1997, 2 avril 1999.

64. Stefan Wotzlaw, Ferdinand Kasman and Michael Nagel: Proton—development of a Russian launch vehicle. *Journal of the British Interplanetary Society*, vol. 51, 3–18, 1998.

65. The introduction of the Zenit may be traced in Phillip S. Clark: The first missions of the Soviet medium-lift launch vehicle. *Journal of the British Interplanetary Society*, vol. 41, no. 3, 1988. See also Phillip S. Clark: New SL-16 and Energiya information released. *Zenit*, vol. 28, June 1989.

66. See Per Olav Sanner: 1999—a sea odyssey. *Spaceflight*, vol. 41, January 1999; Daren Burnham and Andy Salmon: On the waterfront – Scottish built ship sets sail for space. *Spaceflight*, vol. 39, August 1997; A. R. Thompson: Sea Launch—the ship and platform. *Spaceflight*, vol. 39, June 1997.

67. Japp Tervej: *New Russian rockets*. Paper presented to the British Interplanetary Society, 6 June 1999; Phillip Clark: Russian proposals for launching satellites from the oceans. *Space Policy*, vol. 15, 9–12, 1999.

68. See Andy Salmon: Volna—a new facility for micro-g research. *Spaceflight*, vol. 38, August 1996.

69. Hormuz P. Mama: Khrunichev—world's largest satellite launch vehicle producer. *Spaceflight*, vol. 39, February 1997.

70. See Anatoli Kisilev, Alexander Medvedev, Yuri Trufanov and Vasili Yuriev: *Angara launch vehicle family – state of development and prospects for introduction of reusable elements*. Paper presented to 50th International Astronautical Federation, 4–8 October 1999, Amsterdam. For a description of the current state of the Russian launcher industry, see also Craig Covault and Michael Taverna: Russians advance with Angara, Proton & Soyuz. 21st century launch vehicles, *Aviation Week & Space Technology*, 13 December 1999.

71. See Michael A. Taverna: Soyuz to test revolutionary inflatable reentry system. *Aviation Week & Space Technology*, 31 January 2000.

72. See P. S. Clark: *Launch profiles ussed by the Block D*. Paper presented to the British Interplanetary Society, 5 June 1999.

73. William Scott: Ukraine OKs sales of Tu-160s as space launch platforms. *Aviation Week & Space Technology*, 11 January 1999; William Scott: Tu-160 launch programme revamped to cut costs. *Aviation Week & Space Technology*, 12 June 2000.

74. Phillip S. Clark: Soviet rocket engines. *Zenit* special, Astro Info Service, Halesowen, 1989; Phillip S. Clark: Soviet rocket engine overview. *Spaceflight*, July 1990. See Christian Lardier: *Liquid propellant engines in the Soviet Union*. Paper presented to the 50th International Astronautical Congress, Amsterdam, 4–8 October 1999.

75. Christian Lardier: La propulsion future sera cryogénique. *Air & Cosmos*, no. 1509, 10 mars 1995.

76. David S. Cloud: Warheadache. *The New Republic*, 20 April 1998.

77. Philip Bono and Kenneth Gatland: *Frontiers of space*. Blandford, London, 1969.

78. Phil Mills: *Russian aerospaceplane projects*. Paper presented to the British Interplanetary Society, 6 June 2000.

79. Gary Bennett: *A look at the Soviet nuclear power programme*. NASA, Washington DC, 1989, reprinted in *Zenit*, vol. 40, June 1990.

80. For a recent description of the bureaux, see Asif Siddiqi: Soviet design bureaux. *Spaceflight*, vol. 39, August 1999.

81. Phillip S. Clark: The history and projects of the Yuzhnoye design bureau. *Journal of the British Interplanetary Society*, vol. 49, 267–276, 1996.

82. See Christian Lardier: La fabrication de satellites à Krasnoyarsk. *Air & Cosmos*, 1649, 13 mars 1998.

83. For an assessment of Soviet space capacity in its last years, see Nicholas L. Johnson: *The Soviet year in space, 1990*. Teledyne Brown Engineering, Colorado Springs, 1991.

84. Valentin Khalin: Entrepreneurship—a commercial wheel for the space wagon. *Delovoi Mir*, 30 April 1994.

85. Christian Lardier: Ménaces sur le programme spatial russe. *Air & Cosmos*, 17 mars 1995.

86. Craig Covault: 95,000 Russian layoffs, launch breakdown feared. *Aviation Week & Space Technology*, 15 November 1993.

87. This table is based on compositing a range of Western sources, which are generally in broad agreement with each other. The interpretation was done by

the author. Among the sources consulted were *Air & Cosmos*, especially Christian Lardier: La NASA en hause, la RKA en berne. *Air & Cosmos*, vol. 11 février. 2000, no. 1736; *Zenit*, June 1992; *Ciel & Espace, Aviation Week & Space Technology, Spaceflight*; and Dennis Newkirk: *Russian space review 1996* in Russian aerospace guide. Roselle, Illinois.

88. Alan Postlethwaite: Opening doors. *Flight International*, 18–24 April 1990; Tom Harpole: Can Russia's space program survive? *Air & Space*, February/March 1993; Time Furniss: Red star wars. *Flight International*, 3–9 March 1993; Tim Furniss: Culture shock. *Flight International*, 4–10 November 1992; Michael Lemonick: Space programme for sale. *Time*, 16 March 1992; Craig Covault: Russian Proton challenges Ariane. *Aviation Week & Space Technology*, 24 April 1995.

89. See Christian Lardier and Pierre Langereux: Première visite des usines de Samara avec Starsem. *Air & Cosmos*, 1691, 19 février 1999.

90. See Craig Covault: Russian ventures face tech-transfer gauntlet. *Aviation Week & Space Technology*, 20 March 2000.

91. Dwayne Allen Day: Mir and the ISS hearings on the hill. Internet posting, 19 September 1997.

92. See Berry Saunders: *Scud-based launchers*. Paper presented to the British Interplanetary Society, 5 June 1999.

Index